Decompositions
of Manifolds

This is a volume in
PURE AND APPLIED MATHEMATICS

A Series of Monographs and Textbooks

Editors: Samuel Eilenberg and Hyman Bass

The complete listing of books in this series is available from the Publisher upon request.

Decompositions of Manifolds

ROBERT J. DAVERMAN

Department of Mathematics
The University of Tennessee
Knoxville, Tennessee

 1986

ACADEMIC PRESS, INC.
Harcourt Brace Jovanovich, Publishers

Orlando   San Diego   New York   Austin
Boston   London   Sydney   Tokyo   Toronto

ACADEMIC PRESS, INC.
Orlando, Florida 32887

United Kingdom Edition published by
ACADEMIC PRESS INC. (LONDON) LTD.
24–28 Oval Road, London NW1 7DX

Library of Congress Cataloging in Publication Data

Daverman, Robert J.
  Decompositions of manifolds.

  (Pure and applied mathematics series)
  Bibliography: p.
  Includes index.
  1. Manifolds (Mathematics)   2. Decomposition
(Mathematics)   I. Title.   II. Series: Pure and
applied mathematics (Academic Press)
QA3.P8   [QA613]   510 s   [514'.3]   86-3485
ISBN 0–12–204220–4 (alk. paper)

PRINTED IN THE UNITED STATES OF AMERICA

86 87 88 89   9 8 7 6 5 4 3 2 1

CONTENTS

v

IV   The Cell-Like Approximation Theorem

V   Shrinkable Decompositions

VI   Nonshrinkable Decompositions

VII   Applications to Manifolds

PREFACE

This book is about decompositions, or partitions, of manifolds, usually into cell-like sets. (These are the compact sets, similar to the contractable ones, that behave homotopically much like points.) Equivalently, it is about cell-like mappings defined on manifolds. Originating with work of R. L. Moore in the 1920s, this topic was renewed by results of R. H. Bing in the 1950s. As an unmistakable sign of its importance, the subject has proved indispensable to the recent characterization of higher-dimensional manifolds in terms of elementary topological properties, based upon the work of R. D. Edwards and F. Quinn.

Decomposition theory is one component of geometric topology, a heading that encompasses many topics, such as PL or differential topology, manifold structure theory, embedding theory, knot theory, shape theory, even parts of dimension theory. While most of the others have been studied systematically, decomposition theory has not. Filling that gap is the overriding goal. The need is startlingly acute because a detailed proof of its fundamental result, the cell-like approximation theorem, has not been published heretofore.

Placing the subject in proper context within geometric topology is a secondary goal. Its interrelationships with the other portions of the discipline nourish its vitality. Demonstrating those interrelationships is a significant factor among the intentions. On one hand, material from other topics occasionally will be developed for use here when it enhances the central purpose; on the other hand, applications of decomposition theory to the others will be developed as frequently as possible. Nevertheless, this book does not attempt to organize all of geometric topology, just the decomposition-theoretic aspects, in coherent, linear fashion.

Uppermost in my thinking, from the earliest stages of the book's conception, has been the belief it should be put together as a text, with as few prerequisites as possible, and so it has evolved. Not intended for experts, it aims to help students interested in geometric topology bridge the gap between entry-level graduate courses and research at the frontier. Along the way it touches on many issues embraced by decomposition theory but makes no attempt to be encyclopedic. It depicts foundational material in fine detail, and as more of the canvas is unveiled, it employs a coarser brush. In particular, after the proof of the climactic result, the cell-like approximation theorem, it tends to present merely the cruder features of later topics, to expose items deserving further individual pursuit. All in all, it should equip mature readers with a broad, substantial background for successfully doing research in this area.

ACKNOWLEDGMENTS

There are many people to whom I am indebted for help with this manuscript. Phil Bowers, David Wright and, especially, Jim Henderson read large pieces of it, noticing misprints, correcting errors, and offering a multitude of valuable suggestions. So also did Charles Bass, Dennis Garity, and Sukhjit Singh.

Students in classes taught during the academic years 1978-1979 and 1981-1982 made substantial contributions, partly by reacting to material presented, but often by voicing their own insights; included among them are Jean Campbell, Jerome Eastham, Terry Lay, Steve Pax, Kriss Preston, Phil Bowers, Jung-In Choi, Rick Dickson, and Nem-Yie Hwang. Preston, Lay, Pax, and Bowers, who all wrote dissertations in geometric topology, deserve specific recognition. In addition, Mladen Bestvina and Zoran Nevajdić are two others who more recently contributed pertinent comments.

Cindi Blair typed much of the manuscript. Craig Guilbault and David Snyder scrutinized page proofs and spotted countless mistakes.

Friends and colleagues provided a great deal of encouragement. Particularly significant was the prodding by two reigning department heads at the University of Tennessee, Lida Barrett and Spud Bradley.

The National Science Foundation bestowed research support throughout the time this project was underway.

To all of the above, and to those whom I have neglected, many thanks. Finally and most importantly, a note of appreciation to my wife, Lana, who aids in untold ways.

INTRODUCTION

What is an n-manifold? A popular answer is a recitation of the definition: an n-manifold is a metric space covered by open sets homeomorphic to Euclidean n-space E^n. From a foundational perspective, that answer merely suggests another question: what is the topological nature of E^n? Such questions form the central theme around which this book is organized. The real line is characterized by a short list of simple properties; the plane, by properties almost as simple. What about the other Euclidean spaces? The goal here is to explore the topological structure of E^n in case $n \geq 3$, almost invariably paying attention only to the cases $n \geq 5$.

The book is about decompositions, or partitions, of manifolds. The typical object of study will be the decomposition space, or quotient space, associated with a decomposition of some manifold. Technically speaking, the decompositions are all upper semicontinuous ones, meaning that the decomposition elements fit together in a fashion nice enough to ensure metrizability of the decomposition spaces. The latter can be fairly pathological; nevertheless, they are not totally removed from the more familiar world of Euclidean topology, for in this context they arrive on the scene equipped with an explicit connection, via the decomposition mapping, to manifolds. The decompositions themselves will be restricted somewhat, for it happens to be well known that every Peano continuum is the continuous image of a manifold, and the subject here will be more restricted than the study of Peano continua. Cell-like decompositions, in which the partition elements behave homotopically like points, will be the predominant topic. In n-manifolds such decompositions form the class for which it is reasonable to expect the product of E^1 with the decomposition space to be an $(n + 1)$-manifold.

There have been three distinct periods during which decomposition theory has flourished. The first of them, the early period, occurred in the first half of this century, in the 1920s and 1930s, led by R. L. Moore and G. T. Whyburn. Next, the classical period started in the early 1950s, headed by R. H. Bing; supported by S. Armentrout, it continued on through the 1960s and beyond. The current era, the modern period, began in 1977 with the work of R. D. Edwards, spurred by earlier results of J. W. Cannon.

The three periods are distinguished by their characteristic emphases on certain dimensions as the realm of study. During the first one the central results pertained to decompositions of the plane and of 2-manifolds; next, of 3-manifolds; and, last, of higher-dimensional manifolds. ("Higher-dimensional" usually means of dimension $n \geq 5$, because the $n = 4$ case so often demands its own, separate treatment.)

In addition to such chronological and numerical differences, the three are clearly distinguished by significant methodological differences. The explanation demands a bit of the standard notation: consider a nice decomposition (or partition) G of an n-manifold M and also the natural map $\pi: M \to M/G$ of M to the quotient space, called M/G. Usually one hopes to prove that, under sufficient conditions about the sets comprising G, M/G is topologically equivalent to M. The strategy of the early period was to use topological characterizations of the objects involved, the familiar plane, 2-sphere, or other 2-manifold, to deduce the desired equivalence. Customarily that plan did not work at all for 3-manifolds, due in part to the comparative difficulty of distinguishing one 3-manifold from another, but also due to the more complicated problem, which still remains unsatisfactorily resolved, of understanding when a space is a 3-manifold. The classical period got under way when Bing invented a new, workable strategy embodied in his shrinkability criterion, introduced here in Section 5. It posits the existence of a homeomorphism from M to itself sending the elements of G to sets of small size, the key feature, while simultaneously submitting to certain mild cover controls. When this shrinkability criterion holds, the quotient space M/G turns out to be homeomorphic to M, under a naturally arising function obtained as the limit of a sequence of such shrinking homeomorphisms. Accordingly, under Bing's strategy, one investigated the source manifold M to see whether the shrinkability criterion was valid. That strategy has proved effective for solving a multitude of decomposition problems, not just in 3-dimensional manifolds but also in higher-dimensional ones. The modern period began abruptly, not so much because of a successful attack on higher-dimensional manifolds, for powerful results about decompositions of higher-dimensional manifolds had already been discovered at the time, but because of a new strategy, a synthesis of its predecessors, developed by Edwards. He studied the decomposition map $\pi: M \to M/G$ with an eye toward approximating it by homeomorphisms. While the possibility of obtaining such approximations was a by-product of the shrinkability criterion, it was not a fundamental tenet of the prevailing philosophy. Operating with these new tactics, Edwards was able to approximate π via successive maps that were 1-1 over larger and larger subsets of M/G, culminating at the final stage in a limiting homeomorphism.

The essence of each period, for the most part, can be distilled into a single, typifying result. In the early period it was the famous theorem of R. L. Moore.

Theorem (Moore). *If G is an upper semicontinuous decomposition of the plane E^2 into continua, none of which separates E^2, then E^2/G is homeomorphic to E^2.*

From the classical period, the most difficult era to recapture in just one epitomizing theorem, a reasonable candidate is the following combination of the work of Bing [2] and Armentrout [6].

Theorem. *Let G be an upper semicontinuous decomposition of a 3-manifold M into cellular sets. Then M/G is homeomorphic to M if and only if G is shrinkable.*

Examples from Bing made it plain that not all nice decompositions of E^3 reproduce E^3, so several people, Bing among them, engaged in wholesale testing, particularly during the early portions of the period, testing of geometric conditions imposed on the decomposition elements to see which implied shrinkability and which did not. That effort generated a lot of empirical data, no one piece of which stands out as characteristic of the period, although the entire effort might. The theorem mentioned above reflects Bing's methodology characteristic of that era and also Armentrout's complementary contribution that, in order for M/G to be homeomorphic to M, G must be shrinkable. The modern period is set off by the following breakthrough result of Edwards, marking the change from the earlier era to the current one.

Theorem (cell-like approximation). *Let G denote an upper semicontinuous decomposition of an n-manifold M, $n \geq 5$, into cell-like sets. Then the decomposition map $\pi: M \to M/G$ can be approximated by homeomorphisms if and only if M/G is finite-dimensional and satisfies the following disjoint disks property: any two maps of the 2-cell B^2 to M/G can be approximated by maps having disjoint images.*

The point, of course, is that M/G is topologically equivalent to M when the latter two conditions hold; it is also the case that then G is shrinkable.

The paramount result treated in this book is Edwards's cell-like approximation theorem, established here as Theorem 24.3. Although Edwards's proof has been available in manuscript form and has been disseminated publicly in his lectures at the 1978 CBMS Regional Conference at Stillwater, Oklahoma, it has never been published. (A brief but splendid outline appears in Edwards [5].) A primary function of this book is to rectify that matter.

Structurally the book is divided into seven chapters. Chapter I, the preliminaries, introduces the basic terminology and studies some of the elementary consequences. Functioning throughout at a level of difficulty no higher than what is encountered in an elementary general topology text, it provides a hint of the methodology, though none of the major 2-dimensional results, prevalent during the early period of decomposition theory. Chapter II, which is more demanding, pushes ahead into the classical period. Delving into a wide variety of results and examples about decompositions of 3-manifolds, without attempting to be exhaustive, Chapter II presents a fairly large sample of Bing's 3-dimensional work. In particular, in Section 9 it sets forth several key examples of interesting, unusual, historically significant decompositions of 3-space, some of which yield 3-space again and others of which do not yield any manifold whatsoever; taken on the whole, these examples absolutely must be understood if one is to fully appreciate the pitfalls in this subject or to recognize potential circumvention techniques. Chapter II is not exclusively 3-dimensional in focus, however, for it also includes an elementary proof of the important result (also an immediate consequence of the cell-like approximation theorem studied later) that non-combinatorial triangulations of n-manifolds, $n \geq 5$, do exist. The unifying device is Bing's shrinkability criterion. At the beginning this part lays out several refined notions of shrinkability and at the end, based on the local contractibility of manifold homeomorphism groups, shows all of them to be equivalent in topological manifolds. Chapter III, somewhat more technical in nature, involves an investigation of properties preserved by the typical decompositions and sets the stage for the substantial effort in the sequel. Moving from the classical to the modern period, Chapter IV steadily builds up to its climax, the proof of the cell-like approximation theorem. A concluding, almost parenthetical note for this part demonstrates how the original aspects of the subject are reinvigorated by the most contemporary; it makes use of Edwards's methodology and the planar Schönflies theorem to derive the chief result of the early period, Moore's theorem. Chapter V deals with the consequences of Edwards's result, mainly for products involving such decomposition spaces, either with a line or with another such space. Positive in tone, it stresses the conditions under which a decomposition under consideration is shrinkable. By contrast, Chapter VI is more negative, setting forth techniques for constructing pathological decompositions of high-dimensional manifolds. Finally, Chapter VII treats the far-reaching, grander applications of decomposition theory to the rest of geometric topology. Not at all self-contained, nowhere close to it, Chapter VII displays the power, centrality, and diversity potentially available in what has gone on before. It does so by suddenly bringing into play several big results from other branches of geometric topology, something done as infrequently as possible in the earlier parts of the book.

This brings us to the twin issues of textual scope and exegetical style. Two limitations confine the material to manageable size. The first, of course, pertains to the subject that titles the book; the emphasis involves the part of geometric topology concerning decompositions. As stated in the Preface, this book strives to organize linearly decomposition theory, not all of geometric topology. Accordingly, in the interest of efficiency, occasionally one will encounter, with little explanation or justification, invocation of some profound result from one of these collateral topics (e.g., the Kirby–Edwards local contractibility theorem in Section 13, the Bing–Kister isotopy theorem in Section 21, and the Lickorish–Siebenmann PL regular neighborhood classification theorem in Section 38). While invocation of a *deus ex machina* can detract from the stark beauty of a rigorous, orderly mathematical development, the demands to maintain finiteness and to make progress seem to allow no other course. Construed positively, this practice can highlight for the reader the significance of certain major theorems while identifying subjects for future study.

The second limitation pertains to the residence of the decompositions to be considered. Almost invariably these decompositions live in finite-dimensional manifolds, largely in those of dimension $n \geq 5$. The classical $n = 2$ case, which presents little difficulty, receives scant attention. The $n = 3$ case, which carries a great deal of significance and which provides both satisfaction and motivation to the visual imagination, receives much more. We study the salient examples in painstaking detail. Full exposition of this case, however, entails a variety of powerful but uniquely 3-dimensional techniques, which by choice we avoid, preferring instead to move forward into the less well-charted world of higher-dimensional manifolds. That is why we give no proof, for instance, of Armentrout's homeomorphic approximation theorem, alluded to earlier, or of the Denman–Starbird theorem that upper semicontinuous decompositions of E^3 into points and countably many starlike-equivalent compacta are shrinkable. We pay even scarcer attention to the $n = 4$ case, encountering the 4-dimensional world mostly through results that hold in other dimensions as well.* At another end of the spectrum is the $n = \infty$ case, which also is totally ignored here. While some methods for studying decompositions of infinite-dimensional manifolds are similar to those developed for the $n \geq 5$ case, many others are intrinsic to that subject.

---

* For more information about this highly intriguing situation, probably the toughest finite-dimensional case of them all, keep an eye out for a forthcoming book by M. H. Freedman and F. Quinn.

T. A. Chapman's book [2] and H. Toruńczyk's characterizations [1, 2] provide a good introduction.

What background is needed to be able to read this material? At the outset, nothing beyond an introduction to point-set topology. As the text progresses, increasingly more collateral material is brought into play, primarily taken from classical dimension theory and the beginnings of algebraic topology and of PL topology. It probably is imperative that the reader know about the PL regular neighborhood theorem and something about general position, or transversality, techniques. Perhaps the single best reference is T. B. Rushing's *Topological Embeddings*—indeed, we conceived this text as a kind of companion volume to Rushing's. The PL preliminaries laid out there should suffice for this as well, and the engulfing methodology presented there in elaborate and careful detail is extremely valuable for efforts here involving higher-dimensional manifolds.

In light of the above, the ideal personal reference library would contain:

Dimension Theory, by W. Hurewicz and H. Wallman
Introduction to Piecewise-Linear Topology, by C. P. Rourke and B. J. Sanderson
Topological Embeddings, by T. B. Rushing
Algebraic Topology, by E. H. Spanier

For later use, upon completion of this text, it might also include:

Lectures on Hilbert Cube Manifolds, by T. A. Chapman

Designed as a text, this book includes many problems, ranging from simple to fairly difficult. Some preview future subjects, others require filling in steps omitted from a proof, and still others call for reapplication of techniques developed in the body of the text. None of them, except by accident, should be impossibly hard; those at the level of recent thesis problems are suggested only after substantial groundwork has been laid. The reader is urged to do the problems, as the optimal method to begin reworking this particular canvas to reflect one's own insight and vision.

Two closing remarks about typographical shortcuts. First, within any one section specification of a result from another section is given, for example, as "Lemma 30.4", referring to Lemma 4 of Section 30. Specification of any result from the same section is made without using the section number. Second, bibliographic references are given by author's name plus the number of the item in the list of References here, as in, say, "(Bing [5])". When it is totally obvious that it is Bing's work being discussed, the reference may be given omitting "Bing" and appearing solely as "[5]".

I ————————————————————————————

PRELIMINARIES

Chapter I lays out the basic concept of upper semicontinuity and explores its basic features. When attention is restricted to a class of reasonably nice spaces, like metric spaces, this part also exhibits why the study of upper semicontinuous decompositions coincides with the study of proper mappings. Finally, sampling a bit of the flavor from the early period of decomposition theory, it sets forth a quick overview of monotone decompositions (decompositions into connected subsets).

1. ELEMENTARY PROPERTIES OF UPPER SEMICONTINUOUS DECOMPOSITIONS

Throughout this text we will reserve the term *n-manifold* for a separable metric space modeled on Euclidean n-space E^n. Accordingly, manifolds have no boundary. When we wish to allow boundary, we will say so explicitly. For precision, we declare that an *n-manifold with boundary M* is a separable metric space in which each point has a closed neighborhood homeomorphic to the standard n-cell B^n, consisting of all points in E^n whose distance from the origin is no more than 1; the *interior of M*, denoted as Int M, is the subset of points having Euclidean neighborhoods, and the *boundary of M*, denoted as ∂M, is the remainder $M - $ Int M.

Given a metric space (X, ρ), a subset A of X, and a positive number ε, we shall use $N(A; \varepsilon)$ to denote the *ε-neighborhood of A*; that is,

$$N(A; \varepsilon) = \{x \in X \mid \rho(x, A) < \varepsilon\}.$$

A *decomposition G* of a topological space S is a partition of S. Explicitly, G is a subset of the power set of S, and its elements are pairwise disjoint nonempty sets that cover S.

Associated with any decomposition G of a space S is the *decomposition space* having underlying point set G but denoted as S/G (to be read as "S mod G") in order to emphasize the distinction between the generating partition and the resultant space, which often warrants attention as an entity distinct from G. Its topology is prescribed by means of the *decomposition map* $\pi: S \to S/G$ sending each $s \in S$ to the unique element of G containing s; the topology on S/G is the quotient space topology induced by π—namely, the richest topology for which π is continuous. In this topology a subset A of S/G is open (closed) if and only if $\pi^{-1}(A)$ is open (closed) in S. Expanding the vocabulary, one says that a subset X of S is *saturated* (or, to prevent ambiguity, is *G-saturated*) if $\pi^{-1}(\pi(X)) = X$. Accordingly, the image under π of a saturated open (closed) subset of S is open (closed) in S/G.

Without further conditions governing the alignment of the decomposition elements in S, there is no reason to expect that S/G will have a reasonable topological structure. In fact, there seems to be just one connection, and not a powerful one at that, between properties possessed by elements of G and separation properties satisfied by S/G: S/G is a T_1-space if and only if each element of G is closed. Our desire is to deal with topologies richer than those of arbitrary T_1-spaces, like those of metric spaces. To focus on them, we must identify the kind of decompositions worthy of consideration.

The following is a weak form of the definition eventually to be employed. A decomposition G of a space S is (provisionally) said to be *upper semicontinuous* if each $g \in G$ is closed in S and if, for each $g \in G$ and each open subset U of S containing g, there exists another open subset V of S containing g such that every $g' \in G$ intersecting V is contained in U.

Here is the basic elementary characterization of upper semicontinuity.

Proposition 1. *Let G be a decomposition of a space S into closed subsets. The following statements are equivalent*:

G is upper semicontinuous.
For each open set U in S, the set $U^* = \bigcup \{g \in G \mid g \subset U\}$ is an open subset of S.
The decomposition map $\pi: S \to S/G$ is closed.

The proof is an exercise.

In case G is a decomposition of S, we use H_G to denote the set of nondegenerate elements of G (that is,

$$H_G = \{g \in G \mid g \text{ contains more than one point}\})$$

and N_G to denote the union of the elements of H_G. We call N_G the *nondegeneracy set* of G.

Below we list some examples.

(1) S is the plane E^2 and G is the partition consisting of the origin O and the circles centered at O.

(2) S is the plane E^2 and H_G consists of a single element, the y-axis.

(3) S is the unit circle and G consists of all pairs of diametrically opposite points on S.

(4) S is the unit circle in E^2 and H_G consists of a single element, $\{\langle -1, 0\rangle, \langle 1, 0\rangle\}$.

(5) S is the real line E^1 and H_G consists of all intervals $[2n - 1, 2n]$, where n ranges over the integers.

All of the ones above are upper semicontinuous. Next, we have some examples that are not.

(6) S is the plane E^2 and G consists of all (vertical) lines parallel to the y-axis.

(7) S is the real line and H_G consists of a single element—$H_G = \{\{x \in E^1 \,|\, x > 0\}\}$.

(8) S is the subset of E^2 given by $S = \{\langle x, y\rangle \in E^2 \,|\, |x| \leq 1, |y| \leq 1\}$ and H_G consists of the vertical line segments in S having length 2 and not containing the origin.

In a sense example (8) best illustrates the essence of upper semicontinuity: a sequence of small elements may converge to a big element, but no sequence of big elements may converge to a small element, as occurs there. This loose conceptual notion can be sharpened with the aid of some special limits.

Let $\{A_n\}$ denote a sequence of subsets of a space S. The *inferior limit of* $\{A_n\}$, written as $\liminf A_n$, is

$$\{x \in S \,|\, \text{each neighborhood of } x \text{ intersects all} \\ \text{but a finite number of the sets } A_n\},$$

and the *superior limit of* $\{A_n\}$, written as $\limsup A_n$, is

$$\{x \in S \,|\, \text{each neighborhood of } x \text{ intersects infinitely many of the sets } A_n\}.$$

Clearly, $\liminf A_n$ is a subset of $\limsup A_n$. Figure 1-1 reveals the distinction.

Proposition 2. *Suppose G is an upper semicontinuous decomposition of a T_3-space S and $\{g_n\}$ is a sequence of elements of G such that $\liminf g_n$ contains a point of some $g \in G$. Then $\limsup g_n \subset g$.*

Proof. Suppose to the contrary that $s \in g \cap \liminf g_n$ and that $s' \in g' \cap \limsup g_n$, where $g' \in G$ and $g' \neq g$. Since S is T_3, there exist disjoint open subsets U and W in S with $g' \subset U$ and $s \in W$. Invoking upper

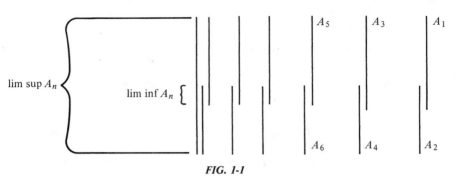

FIG. 1-1

semicontinuity, we can assume that U is saturated (by passing to the maximal saturated subset U^* as in Proposition 1). According to the definition of lim inf, there exists a positive integer N such that $i > N$ implies g_i intersects W and, thus, cannot meet U, which contradicts the supposition that $s' \in \limsup g_n$. ∎

Remark.   Let f denote a function from E^1 to the nonnegative reals. In classical analysis f is called an *upper semicontinuous function* if for each $x \in E^1$ and each $\varepsilon > 0$ there exists a neighborhood V of x such that $f(x') < f(x) + \varepsilon$, for all $x' \in V$. One way to explain the evolution of the decomposition-theoretic term is to consider the decomposition G of E^2 into singletons and the vertical line segments $\{x \times [0, f(x)] \mid x \in E^1\}$: then G is an upper semicontinuous decomposition if and only if f is an upper semicontinuous function.

Just as there are both upper and lower semicontinuous functions, so also there are both upper and lower semicontinuous decompositions. The quick definition involves the decomposition map: a decomposition G of a space S is said to be *lower semicontinuous* if $\pi: S \to S/G$ is open. In addition, G is said to be *continuous* if it is both upper and lower semicontinuous. Lower semicontinuous decompositions rarely come up in geometric topology, except in conjunction with continuous ones, which do play a role. Neither term will reappear here.

If the source S satisfies a fairly strong separation property, so also does each decomposition space S/G associated with an upper semicontinuous decomposition G.

Proposition 3.   *If G is an upper semicontinuous decomposition of a normal space S, then S/G is normal.*

Proof.   Let A_1 and A_2 denote disjoint closed subsets of S/G. Then $\pi^{-1}(A_1)$ and $\pi^{-1}(A_2)$ are disjoint closed subsets of S, and the normality of S guarantees disjoint open sets U_1 and U_2 is S containing them. The characterization of upper semicontinuity in Proposition 1 implies that U_1^* and U_2^* are saturated open sets, necessarily disjoint and necessarily containing $\pi^{-1}(A_1)$ and $\pi^{-1}(A_2)$, respectively. Consequently, $\pi(U_1^*)$ and $\pi(U_2^*)$ act as the required open sets in S/G. ■

Considered set-theoretically, the next definition is an abuse of language, but as a descriptive tool it is too efficient to pass up. We say that a decomposition G is *finite* (or *countable*) if H_G is a finite (countable) set. With this terminology we record a simple condition implying upper semicontinuity.

Proposition 4. *Every finite decomposition of a space S into closed subsets is upper semicontinuous.*

Proof. Let C denote an arbitrary closed subset of S. Then $\pi^{-1}(\pi(C))$, which is the union of C and those elements of H_G that intersect C, is also closed. Hence, $\pi(C) = \pi(\pi^{-1}(\pi(C)))$, which is the image of a saturated closed set, is closed in the decomposition space. Upper semicontinuity follows from Proposition 1. ■

Throughout what lies ahead, a fundamental concern will be the determination of the topological type of specific decomposition spaces. Toward that end, the following elementary realization theorem is indispensable.

Theorem 5 (realization). *Suppose G is an upper semicontinuous decomposition of a space S and f is a closed map of S onto a space Y such that $G = \{f^{-1}(y) \mid y \in Y\}$. Then S/G is homeomorphic to Y.*

Proof. Consider the diagram

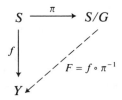

with the implicit relation $F: S/G \to Y$ defined as $F = f \circ \pi^{-1}$. The hypothesis implies f is a one-to-one and onto function. The continuity of f and the closedness of π imply that, for each closed subset C of Y, $F^{-1}(C) = \pi f^{-1}(C)$ is closed in S/G, so F is continuous. Similarly, the closedness of f and the continuity of π imply that F^{-1} is continuous. Therefore, F is a homeomorphism. ■

It is worth noting that in case f is just a map (meaning continuous function) of S onto Y for which $G = \{f^{-1}(y) \mid y \in Y\}$, then $F = f \circ \pi^{-1}$ is a one-to-one map of S/G onto Y.

Obviously the quickest applications of the realization theorem occur in the realm of compact Hausdorff spaces.

Corollary 5A. *If G is an upper semicontinuous decomposition of a compact Hausdorff space and f is a map of S onto a Hausdorff space Y such that $G = \{f^{-1}(y) \mid y \in Y\}$, then S/G is homeomorphic to Y.*

By way of application one can discern the extensive variety of spaces that arise from upper semicontinuous decompositions of some familiar spaces.

Corollary 5B. *Each Peano continuum is homeomorphic to the decomposition space associated with some upper semicontinuous decomposition of $[0, 1]$.*

Corollary 5C. *Suppose the compact Hausdorff space S contains an arc. Then each Peano continuum is homeomorphic to the decomposition space associated with some upper semicontinuous decomposition of S.*

It follows from the Tietze extension theorem that there is a map of S onto $[0, 1]$.

Corollary 5D. *Each compact metric space is homeomorphic to the decomposition space associated with some upper semicontinuous decomposition of the Cantor set.*

EXERCISES

1. Prove Proposition 1.
2. Show that if f is a closed map of the space S to the T_1-space Y, then $G = \{f^{-1}(y) \mid y \in Y\}$ is an upper semicontinuous decomposition of S.
3. A T_1-space S is regular iff for each closed subset C of S the space S/G_C is Hausdorff, where G_C is the decomposition for which $H_{G_C} = \{C\}$.
4. A T_1-space S is normal iff for each upper semicontinuous decomposition G of S, S/G is Hausdorff.
5. The decomposition space of example (8) is not Hausdorff.
6. The decomposition space of example (2) is not metrizable.
7. What are the decomposition spaces of examples (1), (3), and (5)?
8. Show that if G is an upper semicontinuous decomposition of a locally connected space S, then S/G is locally connected.
9. If G is an upper semicontinuous decomposition of a Hausdorff space S into compact subsets, then S/G is Hausdorff.

10. Suppose S is a compact connected Hausdorff space having at least two points. Then each Peano continuum is topologically the decomposition space associated with some upper semicontinuous decomposition of S.

11. Suppose G is a decomposition of a compact metric space into closed subsets such that whenever $\{g_n\}$ is a sequence of elements of G and $g \cap \liminf g_n \neq \varnothing$ $(g \in G)$, then $\limsup g_n \subset g$. Show that G is upper semicontinuous.

12. Prove the remark made following Proposition 2 about the relationship between upper semicontinuous functions and upper semicontinuous decompositions.

2. UPPER SEMICONTINUOUS DECOMPOSITIONS

From now on we shall concern ourselves only with a more restrictive notion of upper semicontinuity.

Definition (permanent). *A decomposition G of a space S is upper semicontinuous (usc) if it is upper semicontinuous in the sense of Section 1 and if, in addition, each $g \in G$ is compact.*

With this definition the unpleasant phenomena illustrated in some of the exercises of Section 1 all disappear. For instance, the Hausdorff property is preserved. More important, and ultimately justifying this definition, metrizability is preserved.

Proposition 1. *If G is an upper semicontinuous decomposition of a Hausdorff space S, then S/G is Hausdorff.*

The proof is an earlier exercise (Exercise 1.9).

Proposition 2. *If G is a usc decomposition of a metric space S, then S/G is metrizable. In particular, if G is separable metric, then S/G is separable metric.*

Proof. The case when S is separable metric is more familiar than the general case, for then S is second countable, and a general topology result implies that S/G is second countable. A basis for its topology is

$$\{\pi(U^*) \mid U \text{ any finite union of basis elements for } S\}.$$

The proof of the general case is usually attributed to A. H. Stone [1]. However, Stone himself credits the result in the situation at hand to S. Hanai [1]. At the same time, L. F. McAuley developed his own proof [1]. Proposition 2 also can be construed as a corollary to the Bing–Nagata–Smirnov metrization theorem. ∎

Another basic fact identifies a class of decompositions in metric spaces that are upper semicontinuous. A countable collection $\{A_i\}$ of subsets from a metric space is said to form a *null sequence* if, for each $\varepsilon > 0$, only finitely many of the sets A_i have diameter greater than ε.

Proposition 3. *Every decomposition G of a metric space for which H_G forms a null sequence of compact sets is usc.*

The proof is left as an exercise.

The next result applies to considerations involving the product of decomposition spaces.

Proposition 4. *If G_i is a usc decomposition of S_i ($i = 1, 2$), then $G_1 \times G_2$ is a usc decomposition of $S_1 \times S_2$ and $(S_1 \times S_2)/(G_1 \times G_2)$ is homeomorphic to $(S_1/G_1) \times (S_1/G_2)$.*

Proof. The argument showing $G_1 \times G_2$ to be usc is routine [but involves the compactness of decomposition elements! Recall example (6) of Section 1]. Let π denote the decomposition map $S_1 \times S_2 \to (S_1 \times S_2)/(G_1 \times G_2)$ and $\pi_1 \times \pi_2$ the product map $S_1 \times S_2 \to (S_1/G_1) \times (S_2/G_2)$. From the realization theorem 1.5 one knows that

$$F = (\pi_1 \times \pi_2) \circ \pi^{-1} \colon (S_1 \times S_2)/(G_1 \times G_2) \to (S_1/G_1) \times (S_2/G_2)$$

is a continuous bijection. To show that F is a homeomorphism, the easiest method is to show that it is an open map.   ∎

Given a space X, we will abuse notation a bit and also speak of X as a decomposition of itself, by which we will mean the partition into the singletons $\{x\}$, $x \in X$.

Corollary 4A. *If G is a usc decomposition of a space S and if X is a T_1-space, then $G \times X$ is usc and $(S \times X)/(G \times X)$ is homeomorphic to $(S/G) \times X$.*

Manipulations with this symbolism reveal some flaws, for we have the strange identity $X/X \approx X$. Cancellation requires brackets: $S/\{S\}$ is a point.

Here are three more notational items. Given a closed and compact (nondegenerate) subset A of a space S, we shall denote by G_A the usc decomposition of S having A as its only nondegenerate element. Given a map $f \colon S \to Y$, we shall denote by G_f the *decomposition* (of S) *induced by* f, namely, the decomposition into the point inverses $f^{-1}(y)$, $y \in Y$. Also, given a usc decomposition G of a subspace C of S, we shall denote by G^T the *trivial extension of G over S*; explicitly,

$$G^T = G \cup \{\{s\} \mid s \in S - C\}.$$

Proposition 5. *If G is a usc decomposition of a closed subspace C of a T_1-space S, then its trivial extension G^T over S is usc.*

The derivation of upper semicontinuity of G^T at $g \in G^T \cap G$ is completely straightforward and works whether or not C is closed in S. The closedness hypothesis is significant only for those $g \in G^T - G$.

EXCERCISES

1. If G is a usc decomposition of a regular space S, then S/G is regular.
2. Prove Proposition 3.
3. Prove Proposition 5.
4. Suppose G is a decomposition of a metric space (S, ρ) into compact sets and suppose there exist closed sets $C_1, C_2, \ldots$ in S and positive numbers $\varepsilon_1, \varepsilon_2, \ldots$ such that (1) each C_i is G-saturated, (2) the restriction of G to each C_i is usc, (3) $N_G \subset \bigcup C_i$, (4) $g \in G$ and $g \subset C_i$ implies diam $g < \varepsilon_i$, and (5) $\varepsilon_i \to 0$ as $i \to \infty$. Then G is usc.
5. If G_α is a usc decomposition of a space S_α for each $\alpha \in \Lambda$, then $\prod_\alpha G_\alpha$ is a usc decomposition of $\prod_\alpha S_\alpha$ and $(\prod_\alpha S_\alpha)/(\prod_\alpha G_\alpha)$ is homeomorphic to $\prod_\alpha(S_\alpha/G_\alpha)$.

3. PROPER MAPS

It follows from realization theorem 1.5 that the study of usc decompositions in compact Hausdorff spaces coincides with the study of maps between such spaces, because those maps are necessarily closed. Given a map f defined on a noncompact space S that induces a usc decomposition, we want to recognize when f is closed in order to more quickly understand the homeomorphism type of the decomposition space. To aid that, we shall say that a map $f: X \to Y$ is *proper* if for each compact subset C of Y, $f^{-1}(C)$ is compact.

Proposition 1. *If G is a usc decomposition of a space S, then $\pi: S \to S/G$ is proper.*

Proof. Consider any compact subset C of S/G and any cover $\mathcal{U} = \{U_\alpha\}$ of $\pi^{-1}(C)$ by open subsets of S. For each $g \in G$ contained in $\pi^{-1}(C)$, some finite subcollection $\{U_i^g \mid i = 1, \ldots, n(g)\}$ of $\mathcal{U}$ covers g, and the union of these sets contains a saturated open set $V_g \supset g$. Corresponding to the resulting open cover $\{\pi(V_g)\}$ of C is a finite subset Λ of G such that $\{\pi(V_g) \mid g \in \Lambda\}$ also covers C. Then the finite subcollection

$$\{U_i^g \mid g \in \Lambda; i = 1, \ldots, n(g)\}$$

of $\mathcal{U}$ must cover $\pi^{-1}(C)$. ∎

Corollary 1A.   *Every closed map with compact point inverses from a space onto a T_1-space is proper.*

This raises questions of some technical importance. Which proper maps are closed? Which proper maps induce usc decompositions?

Proposition 2.   *A map $f: X \to Y$ between locally compact Hausdorff spaces is proper if and only if its extension $f^*: X^* \to Y^*$ to the one-point compactifications that sends the point at infinity in X^* to the point at infinity in Y^* is continuous.*

The proof is straightforward.

Proposition 3.   *Each proper map $f: X \to Y$ between locally compact Hausdorff spaces is closed.*

Proof.   Apply Proposition 2 to extend f to a continuous function $f^*: X^* \to Y^*$ of one-point compactifications, preserving infinity. For any closed set B in X, $f^*(B^* = B \cup \{\infty\})$ is a closed subset of Y^*. As a result, $f(B) = f^*(B^*) \cap Y$ is a closed subset of Y.   ∎

Corollary 3A.   *Suppose G is a usc decomposition of a locally compact Hausdorff space S and f is a proper map of S onto a locally compact Hausdorff space Y such that $G = \{f^{-1}(y) \mid y \in Y\}$. Then S/G is homeomorphic to Y.*

Proposition 4.   *Each proper map $f: X \to Y$ to a first countable Hausdorff space Y is closed.*

Proof.   Suppose to the contrary that for some closed subset B of X there exists $y_0 \in \mathrm{Cl}\, f(B)$ but $y_0 \notin f(B)$. Then there exists a sequence $\{y_i \mid i = 1, 2, \dots\}$ of points in $f(B)$ converging to y_0. Since the set $Z = \{y_i \mid i = 0, 1, 2, \dots\}$ is compact, the properness of f implies $f^{-1}(Z)$ is compact as well. By choosing a point $x_i \in f^{-1}(y_i)$ for $i = 1, 2, \dots$, we produce a sequence in the compact set $B \cap f^{-1}(Z)$, which must have a limit point x_0 in B. But then $y_0 = \lim_{i \to \infty} f(x_i) = f(x_0) \in f(B)$, which is preposterous.   ∎

Corollary 4A.   *Suppose G is a usc decomposition of a space S and f is a proper map of S onto a first countable Hausdorff space Y such that $G = \{f^{-1}(y) \mid y \in Y\}$. Then S/G is homeomorphic to Y.*

The statement below summarizes and improves on the preceding analysis of proper maps.

Theorem 5.   *Suppose $f: S \to Y$ is a proper surjective map between Hausdorff spaces S and Y such that Y is either locally compact or first countable. Then the decomposition $G_f = \{f^{-1}(y) \mid y \in Y\}$ induced by f is usc and S/G_f is homeomorphic to Y.*

See Propositions 3 and 4 and their corollaries, and also Exercises 1.2 and 3.2.

As a result, we can assert that in metric spaces the study of usc decompositions coincides with the study of proper maps. The big distinction is the level of adventure undertaken: with an ordinary proper map $f: X \to Y$ one presumably has good information about the target Y, but with a proper map determined by a usc decomposition G one must venture off toward uncharted territory. Decomposition maps $\pi: X \to X/G$ are like proper maps $f: X \to ?$ with unknown range. Successful charting of that range best advances through careful survey, back in the source X, of the point preimages under f.

Warning. Conditions besides Hausdorff on the space Y in Propositions 3 and 4 are not extraneous. Let X denote an uncountable set with distinguished element Ω, endowed with the discrete topology. Let Y denote the same point set with the topology generated by the singletons $\{x\}$ where $x \neq \Omega$, together with all complements of countable sets. Then the identity function Id: $X \to Y$ is a proper continuous bijection having a locally compact Hausdorff (metric) space as its domain, but Id fails to be closed.

EXERCISES

1. If f is a map of E^1 onto E^1 such that $f^{-1}(t)$ is finite for each $t \in E^1$, then f is proper.
2. Let G denote a usc decomposition of a Hausdorff space S. Then S is locally compact iff S/G is.
3. A map f of a space S onto a locally compact Hausdorff space Y is proper iff each $y \in Y$ has a neighborhood W_y such that $f^{-1}(\text{Cl } W_y)$ is compact.
4. The identity function mentioned in the warning is proper.

4. MONOTONE DECOMPOSITIONS

Early in the twentieth century, when questions concerning the structure of continua were prevalent, it was natural for those topologists looking at decompositions to emphasize a connectedness property called monotonicity. Explicitly, a decomposition G is *monotone* if each $g \in G$ is compact and connected; similarly, a map $f: X \to Y$ is *monotone* if each inverse set $f^{-1}(y)$ is compact and connected. Currently this concept provides a valuable perspective, to be reinforced repeatedly in what lies ahead: with specific restrictions on the structure of the decomposition elements one can often establish strong conclusions about the topological type of resulting decomposition space. For example, if G is a monotone usc decomposition of E^1, then E^1/G is topologically E^1, and if G is a monotone usc decomposition of $I = [0, 1]$ ($G \neq \{I\}$), then I/G is homeomorphic to I. (Contrast these statements with Corollary 1.5B.) Another example is the theorem of R. L. Moore mentioned in the Introduction asserting that if G is a monotone usc decomposition of E^2 such that no $g \in G$ separates E^2, then E^2/G is topologically E^2.

Here is an elementary characterization of monotonicity.

Proposition 1. *Let G denote a usc decomposition of a space S. Then G is monotone if and only if $\pi^{-1}(C)$ is connected whenever C is a connected subset of S/G.*

Monotone decompositions are far from scarce. With any usc decomposition we can naturally associate a monotone one.

Proposition 2. *If G is a usc decomposition of a Hausdorff space S, then the collection M consisting of all components of elements of G is a monotone usc decomposition.*

Proof. Let U be an open subset of S containing $m \in M$, where m represents a component of $g \in G$. The compactness of g (and the fact that S is Hausdorff) enables one to find an open subset W of S containing g, with W expressed as the disjoint union of open sets W_U and W_S such that $m \subset W_U \subset U$. Since G is usc, g is contained in the G-saturated open set $W^* \subset W$. To see that M is usc, define V as $W_U \cap W^*$. Then for any $m' \in M$ that intersects V, $m' \subset W^* \subset W_S \cup W_U$, and the connectedness of m' guarantees that $m' \subset W_U \subset U$.   ∎

Corollary 2A. *In any compact Hausdorff space S the decomposition consisting of all components of S is usc.*

Proof. The degenerate decomposition $G = \{S\}$ is usc.   ∎

A primary application of Proposition 2 to abstract topology shows that certain closed maps are expressible as the composition of two maps, a monotone one and, at the other extreme, a light one. A map $f: X \to Y$ is said to be *light* if each inverse set $f^{-1}(y)$ is totally disconnected.

Theorem 3 (monotone–light factorization). *Let $f: S \to T$ be a closed map between Hausdorff spaces such that $f^{-1}(t)$ is compact for each $t \in T$. Then there exists a unique monotone usc decomposition M of S for which there is a light map $\lambda: S/M \to T$ satisfying $\lambda\pi = f$ ($\pi: S \to S/M$ being the decomposition map).*

Proof. Since f is closed and T is Hausdorff, the decomposition G_f induced by f is usc (Exercise 1.2). Let M denote the decomposition of S consisting of the components of elements from G_f. According to Proposition 2, M is monotone and usc. The map λ defined as $\lambda = f\pi^{-1}$ is light, for if C is a component of $\lambda^{-1}(t) = \pi f^{-1}(t)$, $t \in T$, then $\pi^{-1}(C) \subset f^{-1}(t)$ must be connected by Proposition 1, and $\pi\pi^{-1}(C) = C$ is necessarily a point.

The decomposition M possesses the largest possible connected elements for which $f\pi^{-1}$ is a function. The uniqueness of M follows from the observation that any decomposition having a smaller element would fail to produce a light map $\lambda = f\pi^{-1}$. ∎

A philosophical consequence of the monotone–light factorization theorem is that an understanding of all (appropriate) maps defined on a domain S can be achieved by understanding all light maps defined on monotone images of S. In case $S = I$ or $S = S^1$, where the monotone images of S coincide with S, the philosophy is particularly significant—every map from S onto a non-degenerate Hausdorff space is essentially light. In case $S = S^2$ the philosophy maintains some value, because the monotone images of S^2, 2-dimensional spaces called cactoids, were carefully analyzed in the 1920s; as a result, the passage from S^2 to its monotone images, though less routine than in the above 1-dimensional cases, does proceed along well-charted routes.

Having noted that these examples generate monotone images with dimension no larger than those of the source, the suspicious reader may wonder if monotonicity has equally nice consequences in higher-dimensional situations. It tends not to. Special features of the 1-sphere and the 2-sphere limit the possible monotone images. Monotone images of the 3-sphere, in comparison, can be highly arbitrary, as was first observed by W. Hurewicz [1].

Proposition 4. *Each compact metric space X can be embedded in the space associated with some monotone usc decomposition of S^3.*

Proof. Let τ denote a tetrahedron (3-simplex) in S^3, e_1 and e_2 a pair of nonintersecting edges of τ, and C_1 and C_2 Cantor sets in e_1 and e_2, respectively. For $i = 1, 2$ there exists a map ψ_i of C_i onto X. For each $x \in X$ let g_x denote the subset of τ consisting of all line segments jointing a point of $\psi_1^{-1}(x)$ to a point of $\psi_2^{-1}(x)$. See Fig. 4-1. The desired decomposition G is defined by $H_G = \{g_x \mid x \in X\}$.

The crucial geometric fact here is that each point of $\tau - (e_1 \cup e_2)$ lies on a unique line segment determined by a point of e_1 and a point of e_2. This implies G is a decomposition. One can readily show then that G is monotone and usc.

The desired embedding θ of X in S^3/G is given symbolically by $\theta = \pi(\psi_1)^{-1}$. Exactly as in the realization theorem, θ is an embedding. ∎

Proposition 5. *If M is a compact connected n-manifold, there is a mono-tone usc decomposition G of M having one nondegenerate element such that M/G is homeomorphic to S^n.*

Proof. Given an embedding f of E^n in M, let G denote the decomposition whose only nondegenerate element is $M - f(\text{Int } B^n)$. There is a map

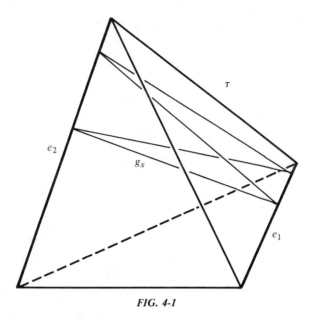

FIG. 4-1

$F: M \to S^n$ that sends $f(\text{Int } B^n)$ homeomorphically onto the complement of
some point and realizes the (monotone) decomposition G. ∎

Proposition 6. *Let N be a compact connected n-manifold and
$\psi: M \times (-1, 1) \to N$ an embedding, where M is a compact connected
$(n - 1)$-manifold, such that $N - \psi(M \times (-1, 1))$ has exactly two components. Then there exists a monotone usc decomposition G of N such that
N/G is the n-sphere and each $g \in G$ either lies in $\psi(M \times \{t\})$, for some
$t \in (-1, 1)$, or is a component of $N - \psi(M \times (-1, 1))$.*

Proof. By Proposition 5 M contains a connected set X such that M/G_X
is S^{n-1}. The decomposition G here has for its (possible) nondegenerate
elements the sets $\psi(X \times \{t\})$, $t \in (-1, 1)$, plus the two components of
$N - \psi(M \times (-1, 1))$. One readily obtains a map $F: N \to S^n$ realizing G,
which is defined in the obvious manner after Proposition 5 so as to send
$\psi(M \times (-1, 1))$ onto the complement of two points in S^n.

Example. The decomposition G of $S^3 = E^3 \cup \{\infty\}$ having nondegenerate
elements the circles C_0 and C_1 shown in Fig. 4-2, plus all figure eights as
shown whose wedge points lie on the line segment joining the two centers
of these circles, yields S^3, because its construction is patterned after the one
given for Proposition 6. To see how this works, express $S^3 - (C_0 \cup C_1)$ as
$S^1 \times S^1 \times (0, 1)$. Details are given in Bing [9].

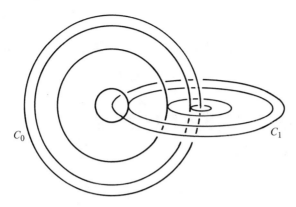

FIG. 4-2

EXERCISES

1. Every monotone map of E^1 onto itself is proper.
2. If $f: S^m \to S^n$ is a monotone surjective map, then no $f^{-1}(s)$, $s \in S^n$, separates S^m.
3. Use a topological characterization of I to show that for each monotone usc decomposition G of I ($G \neq \{I\}$), I/G is homeomorphic to I.
4. Establish the analogous result for S^1.
5. Let G be a monotone usc decomposition of a locally pathwise-connected space S and $\mathcal{V}$ a cover of S/G by connected open sets. For each map $f: I \to S/G$ there exists a partition $0 = t_0 < t_1 < t_2 < \cdots < t_k = 1$ of I such that corresponding to any choice $s_i \in \pi^{-1}(f(t_i))$ (for $i \in \{0, 1, \ldots, k\}$) is a map $F: I \to S$ satisfying (1) $F(t_i) = s_i$ and (2) both $f([t_{i-1}, t_i])$ and $\pi F([t_{i-1}, t_i])$ are contained in some $V_i \in \mathcal{V}$ for $i = 1, \ldots, k$.
6. Suppose G is a monotone usc decomposition of a locally pathwise-connected Hausdorff space S and S/G has a basis of contractible (or even simply connected) open subsets. Then $\pi: S \to S/G$ induces an epimorphism of fundamental groups.
7. There is no monotone map of S^n onto $S^1 \times S^k$ ($n \geq 2$, $k \geq 0$).
8. If G is a usc decomposition of S^n such that H_G is countable and S^n/G is the n-sphere, then G is monotone.
9. If G is a monotone usc decomposition of a *unicoherent* space S, then S/G is unicoherent. (A space X is unicoherent if, whenever X is expressed as the union of closed and connected subsets A and B, then $A \cap B$ is connected.)

II ————————————————————

THE SHRINKABILITY CRITERION

Bing's shrinkability criterion marked a redirection of decomposition theory from emphasis on properties of the decomposition space to inspection of properties held in the source space. The first task involves a study of various notions of shrinkability and their uses. With such tools Chapter II sets forth some, but far from all, of the cardinal developments from the classical period, including the fundamental notion of cellularity and its role in establishing the generalized Schönflies theorem, plus Bing's work on geometric properties implying that countable cellular decompositions of Euclidean spaces are shrinkable. On the opposite side of the coin, Chapter II displays, in Section 9, a wealth of examples of nonshrinkable cellular decompositions of E^3.

A remarkable phenomenon arose with one of these examples, Bing's dogbone space, a nonmanifold stemming from a usc decomposition G of E^3 into points and flat arcs. Although nonshrinkable, stably G is well behaved: $G \times E^1$ is a shrinkable decomposition of $E^3 \times E^1 = E^4$. Section 10 looks at related phenomena, including the result that every decomposition of E^n into points and an arc is stably shrinkable. This is used, in conjunction with a clever observation of Giffen, to reproduce a more recent example of Edwards providing noncombinatorial simplicial triangulations of S^n, $n > 4$.

Because its proof relies so heavily on tools peculiar to 3-dimensional topology, Armentrout's theorem that cellular usc decompositions of 3-manifolds are shrinkable if and only if they yield 3-manifolds, one of the pivotal facts from the classical period, is not touched on here (although its analogue for $n > 4$ eventually is). All the other loose ends concerning various notions of shrinkability are tied up in Section 13, where their equivalence is demonstrated.

5. SHRINKABLE DECOMPOSITIONS

During the 1950s R. H. Bing introduced and exploited several forms of a remarkable condition now called his *shrinkability criterion*. It prompted a major change in decomposition theory, shifting the focus from the decomposition space back to the source. The need for a fresh point of view arose in studying decompositions G of S^3 because, even when it appeared certain that S^3/G was S^3, one then had and still has no reasonable characterization of S^3 for establishing the topological equivalence. The shrinkability criterion was aimed at describing the decomposition space by means of a realization as the known source space, a realization achieved as the end result of manipulations there on the decomposition elements.

In its most general form, the criterion is expressed as follows: a usc decomposition G of a space S is *shrinkable* if and only if (shrinkability criterion) for each G-saturated open cover $\mathcal{U}$ of S and each arbitrary open cover $\mathcal{V}$ of S there is a homeomorphism h of S onto itself satisfying

(a) for each $s \in S$ there exists $U \in \mathcal{U}$ such that $s, h(s) \in U$, and
(b) for each $g \in G$ there exists $V \in \mathcal{V}$ such that $h(g) \subset V$.

In other words, the homeomorphism h called for must shrink elements of G to small size, determined by $\mathcal{V}$, under an action limited by $\mathcal{U}$ (see Fig. 5-1).

Experience suggests that the decomposition space associated with a shrinkable decomposition is likely to be homeomorphic with the source space S. To guarantee that this be true, additional restrictions, like metrizability, must be imposed on S. This section explores some of the variations on those restrictions.

A good starting point is the compact metric case. The elegance of the argument for this case, which is due to R. D. Edwards [5], compensates for the eventual duplication.

Lemma 1. *Let S be a compact metric space, G a usc decomposition of S, and ρ_G a metric on S/G. Then G is shrinkable if and only if (shrinkability criterion in the compact metric case) for each $\varepsilon > 0$ there exists a homeomorphism h of S onto itself satisfying:*

(a) $\rho_G(\pi h(s), \pi(s)) < \varepsilon$ *for each* $s \in S$, *and*
(b) $\operatorname{diam} h(g) < \varepsilon$ *for each* $g \in G$.

This lemma follows routinely from the Lebesgue covering theorem, applied in both S and S/G.

Theorem 2. *Suppose G is a usc decomposition of a compact metric space S. Then the decomposition map $\pi: S \to S/G$ can be approximated by homeomorphisms if and only if G is shrinkable.*

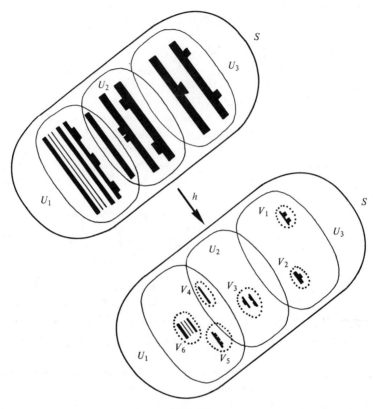

FIG. 5-1

Proof. The forward implication is the easier. Fix $\varepsilon > 0$. By hypothesis there exists a homeomorphism F of S onto S/G such that $\rho_G(F, \pi) < \varepsilon/2$. The uniform continuity of F^{-1} provides $\delta > 0$ such that the image under F^{-1} of each δ-subset of S/G has diameter less than ε. Again, there exists a homeomorphism F^* of S onto S/G for which $\rho_G(F^*, \pi) < \min\{\varepsilon/2, \delta/2\}$. For each $g \in G$, $F^*(g)$ lies in the $(\delta/2)$-neighborhood of $\pi(g)$, implying that diam $F^*(g) < \delta$. Now define h as $F^{-1}F^*$. The choice of δ guarantees that h satisfies condition (b) of the shrinkability criterion. To see that it satisfies condition (a) as well, note that for each $s \in S$

$$\rho_G(\pi(s), \pi h(s)) \leq \rho_G(\pi(s), F^*(s)) + \rho_G(F^*(s), \pi F^{-1}F^*(s))$$

$$< \varepsilon/2 + \rho_G(F(F^{-1}F^*(s)), \pi(F^{-1}F^*(s)))$$

$$< \varepsilon/2 + \varepsilon/2 = \varepsilon.$$

For the reverse implication, in the space $\mathfrak{M}$ of all maps of S to S/G, endowed with the sup-norm metric, let Z denote the closure of that subset consisting of all maps πh^{-1}, where h represents a homeomorphism of S onto itself. Also, for $n = 1, 2, \ldots$ define

$$Z_n = \{F \in Z \mid \text{diam } F^{-1}(x) < 1/n \text{ for each } x \in S/G\}.$$

The claim is that each Z_n is open and dense in Z. Openness is derived via a standard argument given by Hurewicz and Wallman [1]: for each $F \in Z_n$ the compactness of S provides $\alpha > 0$ such that $A \subset S$ and diam $A \geq 1/n$ imply diam $F(A) > \alpha$; then any $F' \in Z$ satisfying $\rho_G(F, F') < \alpha/2$ necessarily belongs to Z_n. For denseness, given $F \in Z$ and $\eta > 0$ one first of all obtains πh^{-1} (h a homeomorphism) such that $\rho_G(F, \pi h^{-1}) < \eta/2$ and then applies the shrinkability criterion to obtain another homeomorphism $H: S \to S$ for which $\rho_G(\pi, \pi H) < \eta/2$ and diam $H\pi^{-1}(x)$ is so small ($x \in S/G$) that diam $hH\pi^{-1}(x) < 1/n$. Clearly the map $\tilde{F} = \pi H^{-1}h^{-1}$ satisfies diam $\tilde{F}^{-1}(x) < 1/n$ for each $x \in S/G$. Moreover, $\rho_G(\pi, \pi H^{-1}) = \rho_G(\pi, \pi H)$ because to each $s \in S$ there corresponds $s^* \in S$ such that $H(s^*) = s$ and then

$$\rho_G(\pi(s), \pi H^{-1}(s)) = \rho_G(\pi H(s^*), \pi(s^*)).$$

As a result,

$$\rho_G(F, \tilde{F}) = \rho_G(F, \pi H^{-1}h^{-1}) \leq \rho_G(F, \pi h^{-1}) + \rho_G(\pi h^{-1}, \pi H^{-1}h^{-1})$$
$$< \eta/2 + \rho_G(\pi, \pi H^{-1}) < \eta/2 + \eta/2.$$

To conclude the argument, one observes that Z, being a closed subset of the complete metric space $\mathfrak{M}$, is itself complete. By the Baire category theorem $\bigcap_n Z_n$ is dense in Z. Any $F \in \bigcap_n Z_n$ is necessarily a homeomorphism because it is necessarily one-to-one and onto. Thus, $\pi \in Z$ can be approximated by homeomorphisms $F \in \bigcap_n Z_n$. ∎

Corollary 2A. *Let $f: S \to X$ be a surjective map between compact metric spaces. Then f can be approximated by homeomorphisms if and only if the decomposition G_f induced by f is shrinkable.*

Corollary 2B. *Let G be a usc decomposition of a compact metric space S and ρ_G a metric on S/G. Then G is shrinkable if and only if, for each $\varepsilon > 0$, there exists a map μ of S onto itself such that $G = \{\mu^{-1}(s) \mid s \in S\}$ and $\rho_G(\pi, \pi\mu) < \varepsilon$.*

Proof. First assume that μ is a map of S onto itself that realizes G and satisfies $\rho_G(\pi, \pi\mu) < \varepsilon$. Then $F = \pi\mu^{-1}$ is a homeomorphism of S onto S/G. Moreover, for each $s \in S$ there exists $s^* \in \mu^{-1}(s)$ and

$$\rho_G(\pi(s), F(s)) = \rho_G(\pi(s), \pi\mu^{-1}(s)) = \rho_G(\pi(s), \pi(s^*))$$
$$= \rho_G(\pi\mu(s^*), \pi(s^*)) \leq \rho_G(\pi\mu, \pi) < \varepsilon.$$

Thus, $\rho_G(\pi, F) < \varepsilon$ and G is shrinkable by Theorem 2.

Conversely, assume G is shrinkable. Given $\varepsilon > 0$, one can invoke Theorem 2 to produce a homeomorphism F of S onto S/G satisfying $\rho_G(\pi, F) < \varepsilon$. Define $\mu: S \to S$ as $\mu = F^{-1}\pi$. For $s \in S$,

$$\rho_G(\pi(s), \pi\mu(s)) = \rho_G(\pi(s), \pi F^{-1}\pi(s)) = \rho_G(FF^{-1}\pi(s), \pi F^{-1}\pi(s))$$

$$\leq \rho_G(F, \pi) < \varepsilon,$$

as required. ∎

For several technical reasons it is advantageous to consider further controls on the shrinking process. To that end, let G denote a usc decomposition of S and W an open subset of S containing N_G. Then G is *shrinkable fixing* $S - W$ if shrinking homeomorphisms h fulfilling the shrinkability criterion can always be obtained that keep each point of $S - W$ fixed. Furthermore, G is *strongly shrinkable* if, for every open set W containing N_G, G is shrinkable fixing $S - W$.

By restricting the action on $S - W$, one can readily adapt the proof given for Theorem 2 to establish the following, which lends itself to quick study of shrinkability in the locally compact metric case.

Theorem 3. *Suppose G is a usc decomposition of a compact metric space S and W is an open subset of S containing N_G. Then $\pi: S \to S/G$ can be approximated by homeomorphisms agreeing with π on $S - W$ if and only if G is shrinkable fixing $S - W$.*

Theorem 4. *Suppose G is a usc decomposition of a locally compact, separable metric space S and ρ_G is a metric on S/G. Then $\pi: S \to S/G$ can be approximated (in the space of maps from S to S/G, endowed with the compact-open topology) by homeomorphisms if and only if for each compact subset C of S/G and each $\varepsilon > 0$ there exists a homeomorphism h of S onto itself satisfying*

(a) $\rho_G(\pi(s), \pi h(s)) < \varepsilon$ *for each* $s \in \pi^{-1}(C) \cup h^{-1}\pi^{-1}(C)$, *and*
(b) $\operatorname{diam} h\pi^{-1}(c) < \varepsilon$ *for each* $c \in C$.

Proof. Let $S^* = S \cup \{\infty\}$ denote the one-point compactification of S, G^* the extension of G to S^* with $\{\infty\} \in G^*$, and $\pi^*: S^* \to S^*/G^*$ the obvious map, which "extends" π. Since S is locally compact and second countable, S^* is a compact metric space; the decomposition G^* is usc at $\{\infty\}$ because $\pi: S \to S/G$ is proper. The point is that π can be approximated (in the compact-open topology) by homeomorphisms if and only if $\pi^*: S^* \to S^*/G^*$ can be approximated (in the sup-norm) by homeomorphisms respecting the action on $\{\infty\}$, and that the shrinkability criterion implicit in Theorem 4 is satisfied if and only if G^* is shrinkable fixing $\{\infty\}$. This translation reduces Theorem 4 to Theorem 3. ∎

Corollary 4A. *Theorem 4 holds in case S is a locally compact and locally connected metric space.*

Proof. One can verify directly that the components of S/G are σ-compact and therefore separable (see Dugundji [1, p. 241]). The corollary follows from component-by-component application of Theorem 4, with local connectedness ensuring that if C is a component of S/G and if $h: S \to S/G$ is a homeomorphism such that $h(s_0) \in C$ for some $s_0 \in \pi^{-1}(C)$, then $h(\pi^{-1}(C)) = C$. ∎

Let $f: S \to X$ be a surjective map and $\mathcal{W}$ an open cover of X. Then a map $F: S \to X$ is said to be $\mathcal{W}$-*close* to f if for each $s \in S$ there exists $W_s \in \mathcal{W}$ such that $f(s), F(s) \in W_s$. In addition, f is said to be a *near-homeomorphism* if for each open cover $\mathcal{W}$ of X there exists a homeomorphism F of S onto X that is $\mathcal{W}$-close to f.

Suppose $\mathcal{U}$ is a cover of a space X and $A \subset X$. By the *star of A in* $\mathcal{U}$ we mean the set $\mathrm{St}(A, \mathcal{U}) = \bigcup\{U \in \mathcal{U} \mid U \cap A \neq \varnothing\}$. A cover $\mathcal{U}_1$ *star-refines* a cover $\mathcal{U}_0$ if for each $U_1 \in \mathcal{U}_1$ there exists $U_0 \in \mathcal{U}_0$ such that $\mathrm{St}(U_1, \mathcal{U}_1) \subset U_0$. The key fact to keep in mind is that for each open cover $\mathcal{U}_0$ of a paracompact Hausdorff space, there is another open cover $\mathcal{U}_1$ that star-refines $\mathcal{U}_0$ (Dugundji [1, p. 167]).

Proposition 5. *Suppose G is a usc decomposition of a paracompact Hausdorff space S such that $\pi: S \to S/G$ is a near-homeomorphism. Then G is shrinkable.*

Proof. Fix a G-saturated open cover $\mathcal{U}$ of S and an arbitrary open cover $\mathcal{V}$ of S. Since closed continuous surjections preserve paracompactness (Dugundji [1, p. 165]), S/G is paracompact.

The set $\pi(\mathcal{U}) = \{\pi(U) \mid U \in \mathcal{U}\}$ is an open cover of S/G. Hence, there exist open covers $\mathcal{W}_1$ and $\mathcal{W}_0$ of S/G, with $\mathcal{W}_1$ star-refining $\mathcal{W}_0$ and $\mathcal{W}_0$ star-refining $\pi(\mathcal{U})$. By hypothesis there exists a homeomorphism $F: S \to S/G$ that is $\mathcal{W}_1$-close to π. Let $\mathcal{W}_2$ be still another open cover of S/G, star-refining

$$\mathcal{W}_1 \cap F(\mathcal{V}) = \{W_1 \cap F(V) \mid W_1 \in \mathcal{W}_1 \text{ and } V \in \mathcal{V}\},$$

thereby simultaneously star-refining $\mathcal{W}_1$ and $F(\mathcal{V})$. Again by hypothesis there exists a homeomorphism $\tilde{F}: S \to S/G$ that is $\mathcal{W}_2$-close to π. Define $h: S \to S$ as $h = F^{-1}\tilde{F}$.

For $s \in g \in G$ both $\tilde{F}(s)$ and $\pi(s)$ lie in some $W_2 \in \mathcal{W}_2$. Hence, $\tilde{F}(g)$ is contained in $\mathrm{St}(\pi(g), \mathcal{W}_2) \subset F(V)$ for some $V \in \mathcal{V}$, implying

$$h(g) = F^{-1}\tilde{F}(g) \subset F^{-1}F(V) = V.$$

Therefore, h shrinks G $\mathcal{V}$-small.

It remains to be shown that h is $\mathcal{U}$-close to the identity. For $s \in g \in G$, $\bar{F}(s)$ is contained in $\mathrm{St}(\pi(g), \mathcal{W}_2) \subset W_1 \in \mathcal{W}_1$ and, similarly,

$$F(g) \cup \bar{F}(g) \subset \mathrm{St}(\pi(g), \mathcal{W}_1) \subset W_0 \in \mathcal{W}_0.$$

Consequently, $g \cup F^{-1}\bar{F}(g) \subset F^{-1}(W_0)$. To complete the proof it suffices to show that $F^{-1}(W_0) \subset U$ for some $U \in \mathcal{U}$. For $x \in F^{-1}(W_0)$ both $\pi(x)$ and $F(x)$ belong to some $W_x \in \mathcal{W}_0$ and $W_x \subset \mathrm{St}(W_0, \mathcal{W}_0)$; thus

$$\pi F^{-1}(W_0) \subset \mathrm{St}(W_0, \mathcal{W}_0) \subset \pi(U) \qquad \text{for some} \quad U \in \mathcal{U}. \qquad \blacksquare$$

Without some kind of completeness hypothesis on S, the converse to Proposition 5 is false.

Example.   Let C denote the Cantor set and $Q \subset E^1$ the rationals. The decomposition $G = \{C \times \{q\} \mid q \in Q\}$ is a shrinkable decomposition of $C \times Q$ but, obviously, $(C \times Q)/G$ is not homeomorphic to $C \times Q$.

Here is the central result of this chapter.

Theorem 6.   *Let G be a usc decomposition of a complete metric space S. Then $\pi: S \to S/G$ is a near-homeomorphism if and only if G is shrinkable.*

Proof.   The forward implication is Proposition 5. The reverse implication first was proved by R. D. Edwards and L. C. Glaser [1] and later was polished by A. Marin and Y. M. Visetti [1]. It is essentially Marin and Visetti's argument reproduced below.

Let ρ denote a complete metric for S. Set $\varepsilon_0 = \infty$, and let $\{\varepsilon_i\}_{i=1}^{\infty}$ denote a sequence of positive numbers approaching 0. Let $\mathcal{W}$ be an open cover of S/G. We will produce a homeomorphism $F: S \to S/G$ $\mathcal{W}$-close to π in two steps.

Part I. Analysis.   In Part II we will construct (1) a sequence $\mathcal{U}_0, \mathcal{U}_1, \mathcal{U}_2, \ldots$ of G-saturated open covers S such that

$\bar{\mathcal{U}}_0 = \{\bar{U}_0 \mid U_0 \in \mathcal{U}_0\}$ refines $\pi^{-1}(\mathcal{W})$,

$\mathcal{U}_{n+1}$ refines $\mathcal{U}_n$, and

$\mathcal{U}_n$ refines both $\{N(g; \varepsilon_n) \mid g \in G\}$ and $\{\pi^{-1}(N(\pi(g); \varepsilon_n)) \mid g \in G\}$;

as well as (2) a sequence $H_0 = \mathrm{Id}, H_1, H_2, \ldots$ of homeomorphisms of S to itself satisfying

(α_n) for each $g \in G$ there exists $U_g \in \mathcal{U}_n$ such that for every $V \in \mathcal{U}_{n+1}$ containing g,

$$H_n(V) \cup H_{n+1}(V) \subset H_n(U_g)$$

(β_n) for each $U \in \mathcal{U}_n$, diam $H_n(U) < \varepsilon_n$.

We note that (β_n) and (α_{n+i}) for $i \in \{0, 1, ..., k - 1\}$ imply

(γ_n)   for each $g \in G$ there exists $U_g^n \in \mathcal{U}_n$ such that

$$\text{diam } H_n(U_g^n) < \varepsilon_n \quad \text{and} \quad H_{n+k}(g) \subset H_n(U_g^n).$$

As a result, $\{H_n\}$ is a Cauchy sequence of homeomorphisms (measured in the sup-norm metric determined by ρ), and this sequence converges uniformly to a continuous function $\mu: S \to S$.

To show that μ is onto, we will need the following lemma.

Lemma 7. *Let $\{x_n\}$ be a sequence of points in a metric space (M, d) satisfying the hypothesis* (*):

(*)   *For each $\varepsilon > 0$ there exists a compact subset C_ε of M such that $N(C_\varepsilon; \varepsilon)$ contains all but a finite number of the points x_n.*

Then $\{x_n\}$ has a subsequence that is Cauchy with respect to d.

Proof. Noting that each subsequence of a sequence satisfying (*) also satisfies (*), we can suppose that $\text{diam}\{x_n \,|\, n = 1, 2, ...\}$ is less than some bound b. Then (*) provides a compact set C such that all but finitely many x_n belong to $N(C; b/6)$. For most n, there exists $z_n \in C$ with $\rho(x_n, z_n) < b/6$. Thus, $\{z_n\}$ has a convergent subsequence $\{z_{n(i)}\}$ for which $\text{diam}\{z_{n(i)} \,|\, i = 1, 2, ...\} < b/6$. As a result, $\text{diam}\{x_{n(i)} \,|\, i = 1, 2, ...\} < b/2$.

This procedure allows us to extract successive subsequences of $\{x_n\}$, which we denote as $\{x_n^1\}, \{x_n^2\}, ..., \{x_n^p\}, ...$, such that $\text{diam}\{x_n^p \,|\, n = 1, 2, ...\} < b/2^p$. The diagonal sequence $\{x_p^p\}$ is Cauchy.   ∎

Returning to the proof of Theorem 6, we fix $s \in S$ and consider $\{x_n\}$ where $H_{n+1}(x_n) = s$. Applying (α_n) to the element $g_n \in G$ for which $x_n \in g_n$, we obtain $U_n \in \mathcal{U}_n$ such that $g_n \subset U_n$ and $H_{n+1}(V) \cup H_n(V) \subset H_n(U_n)$ for all $V \in \mathcal{U}_{n+1}$ containing g_n. We claim that $\{U_n\}$ is a decreasing nest. To see this, if $g_{n+1} \subset W \in \mathcal{U}_{u+2}$, we have

$$s = H_{n+2}(x_{n+1}) \in H_{n+2}(g_{n+1}) \subset H_{n+2}(W) \subset H_{n+1}(U_{n+1}),$$

which implies that $x_n = H_{n+1}^{-1}(s) \in U_{n+1}$, and $g_n \subset U_{n+1}$ as well. Then by (α_n), $H_n(U_n) \supset H_n(U_{n+1})$, so $U_n \supset U_{n+1}$ for all n.

Consequently, (*) of Lemma 7 holds for the sequence $\{x_n\}$ because, given $\varepsilon > 0$ we can choose $\varepsilon_n < \varepsilon$ and can see for $k \in \{0, 1, 2, ...\}$ that

$$x_{n+k} \in U_n \subset N(g; \varepsilon_n) \subset N(g; \varepsilon),$$

for some $g \in G$. If $\{x_{n(i)}\}$ is the Cauchy subsequence promised by Lemma 7, $\{x_{n(i)}\} \to z \in S$ and

$$\mu(z) = \lim_{i \to \infty} \mu(x_{n(i)}) = \lim_{i \to \infty} H_{n(i)+1}(x_{n(i)}) = s.$$

Next, we prove that μ realizes G. Fix $g \in G$ and let $U_n \in \mathcal{U}_n$ be as in (γ_n). Then $H_{n+k}(g) \subset H(U_n)$ for $k \in \{0, 1, 2, \ldots\}$, which yields

$$\mu(g) \subset \text{Cl}\, H_n(U_n) = H_n(\bar{U}_n).$$

Since diam $H_n(\bar{U}_n) = $ diam $H_n(U_n)$, we have that diam $\mu(g) < \varepsilon_n$ for each ε_n. Thus, $\mu(g)$ is a point. Given another $g' \in G$, consider ε_n smaller than one-fourth the distance between $\pi(g)$ and $\pi(g')$, and suppose $g' \subset U_n' \in \mathcal{U}_n$. With this choice of n and ε_n it follows that $\bar{U}_n$ and $\bar{U}_n'$ are disjoint and, in turn, $\mu(g) \subset H_n(\bar{U}_n)$ and $\mu(g') \subset H_n(\bar{U}_n')$, which shows that $\mu(g) \neq \mu(g')$.

Now we show that μ is closed. Let $F \subset S$ be closed and $g \in G$ such that $g \cap \pi^{-1}\pi(F) = \varnothing$. Since π is closed, we can choose ε_n smaller than one-fourth the distance between $\pi(F)$ and $\pi(g)$. If $g \subset U_n \in \mathcal{U}_n$ as in (γ_n), then $\bar{U}_n \cap \text{Cl}(\text{St}(F, \mathcal{U}_n)) = \varnothing$. The observations

$$\mu(g) \subset H_n(\bar{U}_n) \qquad \text{and} \qquad \mu(F) \subset \text{Cl}\, H_n(\text{St}(F, \mathcal{U}_n)) = H_n(\text{Cl}(\text{St}(F, \mathcal{U}_n)))$$

reveal that $\mu(g) \in [S - \text{Cl}\, H_n(\text{St}(F, \mathcal{U}_n))] \subset S - \mu(F)$.

Finally, observe that $\pi\mu^{-1}$ is a homeomorphism of S to S/G. In particular, for $s \in S$ we have $s \in U_0 = H_0(U_0) \supset H_1(U_1) \supset \cdots$ [where $s \in g_s$ and $U_n \supset g_s$ as in (γ_n)]. Hence, $H_n(s) \in U_0$ for each n. Therefore, $\mu(s) \in \bar{U}_0$. It follows that μ is $\pi^{-1}(\mathcal{W})$-close to the identity and that $\pi\mu^{-1}$ is $\mathcal{W}$-close to π.

Part II. Construction. To get started one can easily find (using nothing deeper than the regularity of S/G) a G-saturated open cover $\mathcal{U}_0$ of S such that $\{\bar{U}_0 \mid U_0 \in \mathcal{U}_0\}$ refines $\pi^{-1}(\mathcal{W})$. Inductively, suppose we have found open covers $\mathcal{U}_0, \mathcal{U}_1, \ldots, \mathcal{U}_n$ fulfilling the appropriate refinement conditions and homeomorphisms $H_0 = \text{Id}, H_1, \ldots, H_n$ satisfying (α_i) for $i \in \{0, 1, \ldots, n-1\}$ and (β_i) for $i \in \{0, 1, \ldots, n\}$. Use the paracompactness of S/G to produce a G-saturated open cover $\mathcal{U}_n'$ of S that star-refines $\mathcal{U}_n$, and let $\mathcal{V}$ be an open cover of S by ε_{n+1}-sets. Since $H_n(G)$ is shrinkable, there exists a homeomorphism $h: S \to S$ that is $H_n(\mathcal{U}_n')$ close to the identity and shrinks elements of $H_n(G)$ to $\mathcal{V}$-size. Define H_{n+1} as hH_n.

Fix $g \in G$ and $U_n' \in \mathcal{U}_n'$ with $g \subset U_n'$. Since h is $H_n(\mathcal{U}_n')$-close to Id, both z and $h(z)$ lie in some $H_n(U_n^*) \in h_n(\mathcal{U}_n')$ implying

$$H_{n+1}(U_n') = hH_n(U_n') \subset \text{St}(H_n(U_n'), H_n(\mathcal{U}_n')) = H_n(\text{St}(U_n', \mathcal{U}_n')).$$

Since $\mathcal{U}_n'$ star-refines $\mathcal{U}_n$, there exists $U_g \in \mathcal{U}_n$ such that

(†) $$H_{n+1}(U_n') \cup H_n(U_n') \subset H_n(\text{St}(U_n', \mathcal{U}_n')) \subset H_n(U_g).$$

Clearly, for each $g \in G$ diam $H_{n+1}(g) < \varepsilon_{n+1}$. Find a G-saturated open cover $\mathfrak{M} = \{M_g \mid g \in G\}$ of S refining each of $\mathcal{U}_n'$, $\{N(g; \varepsilon_{n+1}) \mid g \in G\}$, and $\{\pi^{-1}(N(\pi(g); \varepsilon_{n+1})) \mid g \in G\}$ such that, in addition, $M_g \supset g$, diam $H_{n+1}(M_g) < \varepsilon_{n+1}$, and there exists $U_g \in \mathcal{U}_n$ such that $H_{n+1}(M_g) \cup H_n(M_g) \subset H_n(U_g)$. Then $\pi(\mathfrak{M})$ is an open cover of S/G, and we take $\mathcal{U}_{n+1}$ to be the inverse under π of an open cover star-refining $\pi(\mathfrak{M})$. Obviously H_{n+1} and $\mathfrak{U}_{n+1}$ then fulfill (β_{n+1}). To see that they fulfill (α_n) as well, observe that $\mathrm{St}(g, \mathfrak{U}_{n+1})$ is contained in some $U'_n \in \mathfrak{U}'_n$; then (α_n) follows directly from (†) above, where $U_g \in \mathfrak{U}_n$ is the open set associated with U'_n there. ∎

Corollary 6A. *Let G be a usc decomposition of a complete metric space S. Then G is shrinkable if and only if, for each G-saturated open cover $\mathfrak{U}$ of S, there exists a proper map μ of S onto itself such that μ realizes G and μ is $\mathfrak{U}$-close to the identity.*

Theorem 8. *Let G be a usc decomposition of a locally compact, separable metric space S. The following statements are equivalent:*

(a) *$\pi\colon S \to S/G$ is a near-homeomorphism;*
(b) *G is shrinkable;*
(c) *For each compact subset C of S, each $\varepsilon > 0$, and each G-saturated open cover $\mathfrak{U}$ of S, there exists a homeomorphism h of S onto itself such that h is $\mathfrak{U}$-close to the identity and $\mathrm{diam}\, h(g) < \varepsilon$ for every $g \in G$ where $g \subset C$.*

Remarks about the proof. Since S is an open subset of its one-point compactification, a compact metric space, S has a complete metric ρ. Therefore, the equivalence of (a) and (b) is Theorem 6. That (b) implies (c) is obvious. For the reverse implication, find (0) a sequence of compact G-saturated sets $C_1, C_2, \ldots$ whose union is S, (1) a sequence $\mathfrak{U}_0, \mathfrak{U}_1, \ldots,$ $\mathfrak{U}_n, \ldots$ of G-saturated open covers of S, and (2) a sequence of homeomorphisms $H_0 = \mathrm{Id}, H_1, \ldots, H_n, \ldots$ of S to itself fulfilling (α_n) of Theorem 6, as well as

(β'_n) for each $U \in \mathfrak{U}_n$ such that $U \cap C_n \neq \varnothing$, $\mathrm{diam}\, H_n(U) < \varepsilon_n$.

Then allow the analysis there to operate on this modified construction. ∎

Still stronger regulations can be imposed on the shrinking process. We say that a usc decomposition G of a space S is *ideally shrinkable* if for every open subset W of S containing N_G, every G-saturated open cover $\mathfrak{U}$ of W, and every arbitrary open cover $\mathcal{V}$ of W, there exists a homeomorphism h of S to itself such that $h \mid S - W = \mathrm{Id}$, $h \mid W$ is $\mathfrak{U}$-close to the identity, and h sends each element g of H_G into some member of $\mathcal{V}$. Obviously all ideally shrinkable decompositions are strongly shrinkable.

The limitations called for in the definition of "ideally shrinkable" can be measured in metric terms, set forth in the following result. Such measures are particularly suitable for treating the majorant topology on function spaces.

Theorem 9. *Let G be a usc decomposition of a metric space S. Then G is ideally shrinkable if and only if for every continuous function $\varepsilon\colon S \to [0, 1)$ with $\varepsilon^{-1}((0, 1)) \supset N_G$ and every continuous function $\delta\colon S/G \to [0, 1)$ with*

$\delta^{-1}((0, 1)) \supset \pi(N_G)$ *there exists a homeomorphism h of S onto itself satisfying:*

(a)   *for each $s \in S$ $\rho(\pi(s), \pi h(s)) \leq \delta(\pi(s))$ and*
(b)   *for each $g \in H_G$ there exists $s \in S$ such that $h(g) \subset N(s; \varepsilon(s))$.*

First we need a technical lemma.

Lemma 10.   *If $\mathcal{W}$ is an open cover of a metric space W, then there exists a continuous function $\varepsilon: W \to (0, 1)$ such that the cover $\{N(w; \varepsilon(w)) \mid w \in W\}$ refines $\mathcal{W}$. If, in addition, W is an open subset of a metric space S, then the map ε can be attained as the restriction of a map $\varepsilon: S \to [0, 1)$ for which $\varepsilon^{-1}(0) = S - W$.*

Proof.   For the first part, see Dugundji [1, p. 171]. Once that has been done, define a new map on all of S to be 0 on $S - W$ and $\min\{\varepsilon(w), \rho(w, S - W)\}$ on W.   ∎

Proof of Theorem 9.   Only the reverse implication will be established here. Suppose W, $\mathcal{U}$, and $\mathcal{V}$ are given. Lemma 10 provides a continuous function $\varepsilon: S \to [0, 1)$, with $\varepsilon^{-1}(0) = S - W$, such that $\{N(w; \varepsilon(w)) \mid w \in W\}$ refines $\mathcal{V}$. Lemma 10 also provides a continuous function $\delta: S/G \to [0, 1)$, with $\delta^{-1}(0) = \pi(S - W)$, such that the cover $\{N(x; \delta(x)) \mid x \in \pi(W)\}$ refines $\pi(\mathcal{U})$. With these two functions, the homeomorphism $h: S \to S$ satisfying conclusions (a) and (b) above shows G to be ideally shrinkable.   ∎

Having disposed of the prominent results, we shall now mention some properties necessarily possessed by the elements of shrinkable decompositions.

Proposition 11.   *If G is a shrinkable decomposition of a locally connected regular space S, then G is monotone.*

Proof.   Suppose to the contrary that some $g_0 \in G$ failed to be connected. Some open subset W of S containing g_0 would be expressible as the disjoint union of open sets W_1 and W_2, each of which meets g_0.

Prior to applying the shrinkability hypothesis, we construct the appropriate limiting open covers. The regularity of S/G (see Exercise 2.1) provides open sets U_1, U_2, and U_3 in S/G such that

$$\pi(g_0) \in U_{i+1} \subset \bar{U}_{i+1} \subset U_i \qquad (i = 1, 2) \qquad \text{and} \qquad \pi^{-1}(U_1) \subset W.$$

Let $\mathcal{U}$ denote the G-saturated open cover

$$\{S - \pi^{-1}(\bar{U}_2), \pi^{-1}(U_1 - \bar{U}_3), \pi^{-1}(U_2) - g_0, \pi^{-1}(U_3)\}$$

and let $\mathcal{V}$ be an open cover of S refining $\mathcal{U}$ and consisting of connected open sets.

The shrinkability of G gives rise to a homeomorphism $h\colon S \to S$ shrinking g_0 into some $V_0 \in \mathcal{V}$ while staying $\mathcal{U}$-close to Id. Then $g_0 \subset h^{-1}(V_0)$. We shall reach a contradiction by showing that $h^{-1}(V_0) \subset W$.

Suppose there exists $w \in h^{-1}(V_0) - W \subset h^{-1}(V_0) - \pi^{-1}(U_1)$. Then $S - \pi^{-1}(\bar{U}_2)$ is the only element of $\mathcal{U}$ containing w; as a result, both w and $h(w)$ lie in $S - \pi^{-1(}\bar{U}_2)$. Moreover, $S - \pi^{-1}(\bar{U}_2)$ and $\pi^{-1}(U_1 - \bar{U}_3)$ are the only elements of $\mathcal{U}$ containing $h(w) \in V_0$; thus V_0, which necessarily lies in one of these sets, is a subset of $S - \pi^{-1}(\bar{U}_3)$. On the other hand, for $s_0 \in g_0$, both s_0 and $h(s_0)$ must belong to $\pi^{-1}(U_3)$. The recollection that $h(s_0) \in h(g_0) \subset V_0$ leads to the impossibility

$$h(s_0) \in \pi^{-1}(U_3) \cap V_0 \subset \pi^{-1}(U_3) \cap [S - \pi^{-1}(\bar{U}_3)] = \varnothing. \qquad \blacksquare$$

This argument is probably more important than the result just established. Given an open cover $\mathcal{V}$ by sets with the favorite property of the moment and given any neighborhood W of $g \in G$, we produced a homeomorphism h showing $g \subset h^{-1}(V) \subset W$, for some $V \in \mathcal{V}$. The consequence merits explicit statement.

Proposition 12. *Let $\mathcal{P}$ represent a topological property applicable to subsets of a given space. Suppose G is a shrinkable decomposition of a regular space S in which each point $s \in S$ has arbitrarily small neighborhoods satisfying $\mathcal{P}$. Then each $g \in G$ has arbitrarily small neighborhoods satisfying $\mathcal{P}$.*

Examples of useful properties $\mathcal{P}$ to apply are connectedness, simple connectedness, contractibility, compactness, and being homeomorphic to E^n.

Very often in the theory of manifold decompositions a stronger notion of shrinkability arises. That notion is worth considering, for immediate contrast but eventually for the lack of contrast.

A usc decomposition G of a space S is *realized by a pseudo-isotopy* if there exists a pseudo-isotopy ψ_t of S to S such that $\psi_0 = $ identity and $G = \{\psi_1^{-1}(s) \mid s \in S\}$. By a *pseudo-isotopy ψ_t* of S to S we mean a homotopy $\psi_t\colon S \to S$ such that ψ_t is a homeomorphism for each $t \in [0, 1)$ and ψ_1 is a closed surjection. Similarly, by an *isotopy ψ_t* of S to S we mean a homotopy $\psi_t\colon S \to S$ such that ψ_t is a homeomorphism for each $t \in [0, 1]$. Throughout past history, a great deal of attention has been given to the decompositions that can be realized by pseudo-isotopies.

Proposition 13. *If G is a usc decomposition of a compact Hausdorff space S that is realized by a pseudo-isotopy, then G is monotone.*

Proof. Suppose that $g_0 \in G$ can be expressed as the disjoint union of closed sets A_0 and B_0. Then some closed set C separates A_0 and B_0, in the sense that $S - C$ can be expressed as the disjoint union of two open sets, one containing A_0 and the other B_0.

Consider $a_0 \in A_0$ and $b_0 \in B_0$. The pseudo-isotopy ψ_t implicitly defines paths $\alpha: I \to S$ as $\alpha(t) = \psi_t(a_0)$ and $\beta: I \to S$ as $\beta(t) = \psi_t(b_0)$ joining a_0 and b_0, respectively, to a common endpoint $\alpha(1) = \beta(1) = \psi_1(g_0)$. The key is that ψ_t is a homeomorphism for $t < 1$; as a result,

$$\psi_t(C) \cap [\alpha([t, 1]) \cup \beta([t, 1])] \neq \varnothing \qquad (t < 1),$$

because $\psi_t(C)$ necessarily meets the connected set $[\alpha([t, 1]) \cup \beta([t, 1])]$ joining $\psi_t(a_0)$ to $\psi_t(b_0)$. Then a typical compactness argument yields

$$\psi_1(C) \cap \alpha(1) = \psi_1(C) \cap \psi_1(g_0) \neq \varnothing.$$

which indicates $g_0 \cap \psi_1^{-1}\psi_1(C) = \psi_1^{-1}\psi_1(g_0) \cap \psi_1^{-1}\psi_1(C) \neq \varnothing$, a contradiction. ∎

The ensuing exercises make it clear that shrinkable decompositions of arbitrary metric spaces are not necessarily monotone, nor are they necessarily realized by pseudo-isotopies; however, shrinkable decompositions of manifolds possess both properties. For the first of these statements, see Exercise 2; for the second, one must look forward to Theorem 13.4.

EXERCISES

1. Suppose that G is a shrinkable decomposition of S and h is a homeomorphism of S onto itself. Show that $h(G) = \{h(g) \mid g \in G\}$ is a shrinkable decomposition of S.

2. On the Cantor set C a usc decomposition G is shrinkable iff C/G is homeomorphic to C.

3. The cone over the Cantor set C [that is, the space $(C \times [0, 1])/G_{C \times \{1\}}$] is a pathwise-connected metric space admitting a shrinkable but nonmonotone usc decomposition.

4. The decomposition $G = \{C \times \{q\} \mid q \in Q\} = \{C\} \times Q$ of $C \times Q$ is shrinkable.

5. Prove Theorem 8.

6. Complete the proof of Theorem 9.

7. Suppose G is a usc decomposition of a compact Hausdorff space X that can be realized by a pseudo-isotopy and $\mathcal{U}$ is a G-saturated open cover of X. Show that there exists a pseudo-isotopy θ_t realizing G such that to each $x \in X$ there corresponds $U_x \in \mathcal{U}$ for which $\theta_t(x) \in U_x$ for all $t \in [0, 1]$.

8. Suppose G is a usc decomposition of a compact Hausdorff space X that can be realized by a pseudo-isotopy, and suppose G_0 is the decomposition of $X \times E^1$ consisting of $\{g \times \{0\} \subset X \times E^1 \mid g \in G\}$ together with singletons from $X \times (E^1 - \{0\})$. Show that $(X \times E^1)/G_0$ is homeomorphic to $X \times E^1$. Is G_0 shrinkable?

9. Let G denote a usc decomposition of a space S. Then $\pi: S \to S/G$ is a near-homeomorphism if and only if for each G-saturated open cover $\mathcal{U}$ of S there is a closed map μ of S onto itself that realizes G and is $\mathcal{U}$-close to the identity.

6. CELLULAR SETS

Given a decomposition G of S, as a primary problem one is frequently confronted with the need for detecting the topological type of S/G. When G is shrinkable (and S is complete metric), we now know that this primary problem has a strong resolution—not only is S/G homeomorphic to S, the decomposition map itself is a near-homeomorphism. The question to face next is: under what conditions is a decomposition shrinkable?

The subsets of Euclidean spaces consistently present in shrinkable decompositions are the cellular ones. A subset X of E^n (or, more generally, of an n-manifold) is said to be *cellular* if there exists a sequence $\{B_i\}$ of n-cells in E^n such that $B_{i+1} \subset \text{Int } B_i$ $(i = 1, 2, \ldots)$ and $X = \bigcap_{i=1}^{\infty} B_i$. Alternatively, $X \subset E^n$ is cellular if and only if for each open set $U \supset X$ there exists an n-cell B such that $X \subset \text{Int } B \subset B \subset U$. As another possibility, $X \subset E^n$ is cellular if and only if X is compact and has arbitrarily small neighborhoods homeomorphic to E^n.

Clearly, cellular sets are compact and connected; however, they need not be locally connected. For example, consider the sine $(1/x)$-continuum in E^2 (see Fig. 6-1).

In manifolds cellularity of the decomposition elements is a necessary condition for shrinkability. It is not a sufficient one, as we shall see later.

Proposition 1. *If G is a shrinkable usc decomposition of an n-manifold, then each $g \in G$ is cellular.*

This follows directly from Proposition 5.12.   ∎

As a consequence of Proposition 1, we have a monotone decomposition G of an n-manifold M for which M/G is equivalent to M even though we

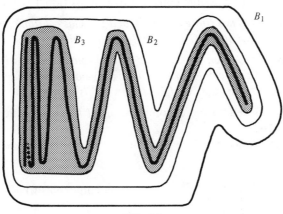

FIG. 6-1

know G fails to be shrinkable. Recall the decomposition G of S^3, whose nondegenerate elements are circles and figure eights, described at the end of Section 4. No circle C in S^3 can be cellular. Some neighborhood U of C retracts to C under a map r. If C were cellular, C would lie in a 3-cell B in U, and since there is a contraction ϕ_t of B in itself, one would obtain a contraction $r\phi_t \mid C$ of C in itself, which is impossible. Thus G cannot be shrinkable. Nevertheless, G was constructed so that S^3/G is S^3.

Proposition 2. *If X is a cellular subset of an n-manifold M, then the decomposition G_X whose only nondegenerate element is X is strongly shrinkable.*

To establish Proposition 2 we first set forth an elementary geometric fact.

Lemma. *Let X be a compact subset of the interior of an n-cell B and $\varepsilon > 0$. There is a homeomorphism h of B onto itself fixed on ∂B for which diam $h(X) < \varepsilon$.*

Briefly put, one equates B with B^n, interior to it constructs another round n-cell $B' \supset X$ centered at the origin of $B = B^n$, and radially compresses B' very near the origin while keeping ∂B pointwise fixed.

Proof of Proposition 2. Fix a G_X-saturated open cover $\mathcal{U}$ of M, another open cover $\mathcal{V}$ of M, and an open set $W \supset X$. Restricting W, if necessary, we assume Cl W is compact and lies in some $U \in \mathcal{U}$. There exists $\varepsilon > 0$ such that any subset of Cl W having diameter $< \varepsilon$ is contained in some $V \in \mathcal{V}$. The cellularity of X implies the existence of an n-cell B such that

$$X \subset \operatorname{Int} B \subset B \subset W.$$

The homeomorphism h promised by the lemma, extended over $M - B$ via the identity, shows that G satisfies the shrinkability criterion. ∎

In a sense the best possible converse to Proposition 1 is the next result.

Corollary 2A. *If G is a finite cellular (each $g \in G$ is cellular) decomposition of an n-manifold M, then G is strongly shrinkable.*

Corollary 2B. *If U is an open subset of an n-manifold and if f is a closed map of Cl U onto an n-cell R for which the only nondegenerate inverse image under f is a cellular subset X of U, then Cl U is an n-cell.*

Proof. Let G denote the decomposition of Cl U for which $H_G = \{X\}$.

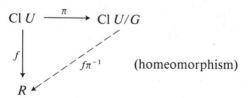

Since G is shrinkable, π is approximable by homeomorphisms. But Cl U/G is homeomorphic to R. ∎

The topic of cellularity leads to one of the major themes of this book: the intimate connection between decomposition theory and taming theory. J. W. Cannon probably first stressed this theme, but the connections themselves have been, or should have been, visible from the outset, in the work dating back to the 1950s of R. H. Bing, E. E. Moise, and Morton Brown. Brown's important generalized Schönflies theorem [1], one of the first and perhaps the most elegant taming theorem, displays an aspect of that connection through its dependence on decomposition methods.

An $(n - 1)$-sphere Σ in S^n is said to be *flat*, or *flatly embedded*, if there exists a homeomorphism of S^n to itself sending Σ to the standardly embedded sphere $S^{n-1} = \{\langle x_1, \ldots, x_{n+1} \rangle \in S^n \mid x_{n+1} = 0\}$.

A method for detecting flatness in this situation stems from the observation that S^{n-1} bounds two n-cells in S^n; accordingly, an $(n - 1)$-sphere Σ in S^n is flat if and only if it bounds two n-cells. In case Σ is known to bound, a fixed homeomorphism of Σ onto S^{n-1} can be extended over the two bounded cells, one at a time, onto the cells representing the upper and lower hemispheres of S^n.

By an *inverse set* of a map f, we mean a nondegenerate inverse image under f.

Proposition 3.   *Suppose Q is an n-cell and f is a map of Q into S^n such that $X \subset$ Int Q is the only inverse set of f and f(Int Q) is open in S^n (which must hold, by invariance of domain). Then X is cellular in Q.*

Proof.   Let U be an open subset of Int Q containing X. Then $f(U) = f(\text{Int } Q) - f(Q - U)$ is an open subset of S^n. There is a homeomorphism $\theta: S^n \to S^n$ fixed on some neighborhood V of $f(X)$ such that $\theta f(Q) \subset f(U)$. Define a function F of Q into U as the identity on $f^{-1}(V)$ and as $f^{-1}\theta f$ on $Q - X$. Since F is well defined, it is an embedding, and $F(Q)$ is an n-cell in U containing X in its interior.   ∎

Proposition 4.   *If ψ is a map of S^n onto itself with exactly two inverse sets A and B, then each of A and B is cellular.*

Proof.   We will show that B is cellular. Let Q be an n-cell in S^n containing $A \cup B$ in its interior. Then $\psi(\text{Int } Q)$ is open and contains an open set U for which $\psi(A) \in U$ but $\psi(B) \notin U$. There is a homeomorphism θ of S^n to itself carrying $\psi(Q)$ into U and fixing some neighborhood V of $\psi(A)$. Define a map f of Q into S^n as the identity on $\psi^{-1}(V)$ and as $\psi^{-1}\theta\psi$ on $Q - A$. Then $f(\text{Int } Q)$ is open and B is the only inverse set of f. By Proposition 3, B is cellular.   ∎

Proposition 5. *If f is a map of S^n onto itself with only a finite number of inverse sets, then each is cellular.*

This follows from Proposition 4 more quickly than Proposition 4 followed from Proposition 3. Later we shall be concerned with conditions under which the finiteness restriction can be removed.

Theorem 6 (generalized Schönflies theorem). *If h is an embedding of $S^{n-1} \times [-1, 1]$ in S^n, then $h(S^{n-1} \times \{0\})$ is flat. In particular, the closure of each component of $S^n - h(S^{n-1} \times \{0\})$ is an n-cell.*

Proof. Let A denote the closure of that component of $S^n - h(S^{n-1} \times \{1\})$ not containing $\Sigma = h(S^{n-1} \times \{0\})$ and B the closure of that component of $S^n - h(S^{n-1} \times \{-1\})$ not containing Σ (see Fig. 6.2). Furthermore, let D_A (D_B) denote the closure of that component of $S^n - \Sigma$ containing A (B). Let G denote the usc decomposition of $S^{n-1} \times [-1, 1]$ having as its two non-degenerate elements the sets $S^{n-1} \times \{\pm 1\}$. Then $(S^{n-1} \times [-1, 1])/G$, the suspension of S^{n-1}, is topologically S^n and there is a homeomorphism λ of $(S^{n-1} \times [-1, 1])/G$ to S^n sending the image of $S^{n-1} \times \{0\}$ to the standard sphere $S^{n-1} \subset S^n$. Extend the map $\lambda \pi h^{-1}$ from $h(S^{n-1} \times [-1, 1])$ onto S^n to a map f of S^n onto itself by defining $f(A) = \lambda \pi h^{-1}(h(S^{n-1} \times \{1\}))$ and $f(B) = \lambda \pi h^{-1}(h(S^{n-1} \times \{-1\}))$. Each of A and B is cellular (Proposition 4) and, therefore, D_A and D_B are n-cells by Corollary 2B. As an alternative way to clinch the proof, f really acts as the decomposition map for G_f; by Corollary 2A, G_f is strongly shrinkable, so there exists a homeomorphism F of S^n to itself such that F and f agree on $h(S^{n-1} \times \{0\})$. ∎

In common parlance an $(n - 1)$-manifold Σ contained in an n-manifold M is *bicollared* if there exists an embedding h of $\Sigma \times [-1, 1]$ in M such that $h(\Sigma \times \{0\}) = \Sigma$. The shorthand version of the generalized Schönflies theorem asserts that each bicollared $(n - 1)$-sphere in S^n is flat.

Applications of these techniques include a simple manifold structure theorem.

Proposition 7. *If the compact n-manifold M can be expressed as the union of two open n-cells, then M is homeomorphic to S^n.*

Proof. Let U and V denote the two open n-cells. Let f be a homeomorphism of V onto E^n, considered as S^n-point. Then f extends to a map $F: M \to S^n$ having $X = M - V$ as its only inverse set. Since $X \subset U$, X is contained in the interior of some n-cell $Q \subset U$, and Proposition 3 implies that X is cellular. By Proposition 2, the decomposition $G_F = \{F^{-1}(s) \mid s \in S^n\}$ induced by F is shrinkable. Thus, $\pi: M \to M/G_F$ is a near-homeomorphism and $F\pi^{-1}: M/G \to S^n$ is a homeomorphism, so M and S^n are topologically equivalent. ∎

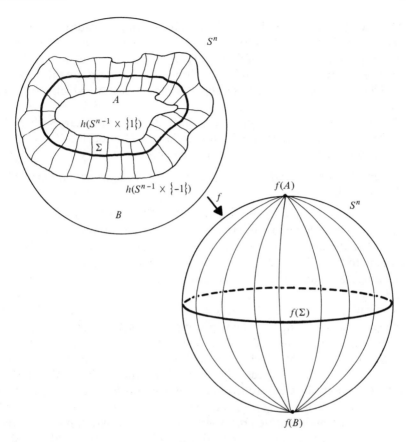

FIG. 6-2

Now consider the slightly more complicated situation where Σ_1 and Σ_2 denote disjoint bicollared $(n - 1)$-spheres in S^n. Let (Σ_1, Σ_2) denote that component of $S^n - (\Sigma_1 \cup \Sigma_2)$ having both these spheres in its frontier, $[\Sigma_1, \Sigma_2) = \Sigma_1 \cup (\Sigma_1, \Sigma_2)$ and $[\Sigma_1, \Sigma_2] = \Sigma_1 \cup \Sigma_2 \cup (\Sigma_1, \Sigma_2)$. The conjecture that $[\Sigma_1, \Sigma_2]$ is topologically $S^{n-1} \times [-1, 1]$ is a famous statement known as the annulus conjecture. It is now known in all dimensions, having been proved primarily by Kirby [1] for $n \geq 5$ and more recently by Quinn [2] for $n = 4$. Their arguments are considerably more complex than the proof of the Schönflies theorem itself. However, the comparable conjectures about the topological types of $[\Sigma_1, \Sigma_2)$ and (Σ_1, Σ_2) do follow easily from the Schönflies theorem.

One-sided variations to the concept of a bicollaring pervade the subject of embedding theory. A subset C of a space X is said to be *collared in X* provided there exists an embedding λ of $C \times [0, 1)$ onto an open subset of X such that $\lambda(\langle c, 0 \rangle) = c$ for all $c \in C$, and it is said to be *locally collared* if it can be covered by a collection of open sets (relative to C), each of which is collared in X.

A prototypical setting occurs in manifolds with boundary, where by definition the boundaries are locally collared. The local collars can be uniformly reorganized into a collar on the entire boundary.

Theorem 8. *The boundary ∂M of any n-manifold with boundary M is collared in M.*

Proof. Form a new space $\tilde{M}$ from $M \cup (\partial M \times I)$ by identifying each point p in ∂M with $\langle p, 0 \rangle$ in $\partial M \times I$. Clearly $\partial \tilde{M}$ is collared in $\tilde{M}$.

Let G be the usc decomposition of $\tilde{M}$ into singletons and the arcs corresponding to $\{p\} \times I$. It should be obvious why $\tilde{M}/G$ is equivalent to M. The remaining step is to prove G is shrinkable, indicating $\tilde{M}/G$ is also equivalent to $\tilde{M}$.

According to Theorem 5.8, it suffices to consider the case where the nondegenerate elements of G lie in a compact subset Z of $\tilde{M}$. The problem then reduces to one where $Y = Z \cap \partial \tilde{M}$ is covered by two open subsets U_1 and U_2 of $\partial \tilde{M}$, each of which is collared in $\tilde{M}$. Find closed subsets C_1 and C_2 of Y in U_1 and U_2, respectively, covering Y, and use the local collarings on the U_i's to produce controlled homeomorphisms h_1 and then h_2 of $\tilde{M}$ to itself sending each arc $\{p\} \times I$ of the appended collar into itself and, specifically, shrinking those arcs where $p \in C_1$ or, secondly, $p \in C_2$ to very small size. The composition $h_1 \circ h_2$ shrinks all of G, as required. ∎

Theorem 8 is due to M. Brown [2]. The proof was suggested by an argument of R. Connelly [1], and it actually establishes a more general result, also due to Brown.

Theorem 9. *Let C be a closed subset of a locally compact separable metric space X. Then C is locally collared in X if and only if C is collared in X.*

Returning briefly to the Schönflies theorem, one might wish to note that therefore an $(n - 1)$-sphere Σ in S^n is bicollared if it is locally collared in the closure of each component of $S^n - \Sigma$.

Prior to Brown's work, decomposition theorists tended to employ a different concept closely related to cellularity. A subset C of a space X is *pointlike* (for emphasis, in X) if $X - C$ is homeomorphic to the complement of some point of X. When the ambient space is S^n, the pointlike subsets coincide with the cellular subsets (an exercise). In an arbitrary n-manifold, each cellular subset is pointlike (Proposition 2), but the converse fails.

For example, C. O. Christenson and R. P. Osborne [1] construct a 3-manifold M by deleting from E^3 the integer points along the positive x-axis together with the circles of radius $\frac{1}{4}$ centered at integer points along the negative x-axis; the circle in M of radius $\frac{1}{4}$ centered at the origin is pointlike but noncellular. The noncompactness of M is fundamental to this and to all other examples distinguishing the twin concepts of "cellular" and "pointlike," which coincide in compact n-manifolds.

EXERCISES

1. A closed subset C of S^n is cellular iff it is pointlike.
2. A closed and connected subset C of E^n is cellular iff it is pointlike.
3. Suppose X is a cellular subset of an n-manifold M and some neighborhood U of X in M retracts to X. Show that X is contractible.
4. Prove Proposition 5.
5. Suppose Σ is an $(n - 1)$-sphere in E^n and U is the bounded component of $E^n - \Sigma$. Suppose $\psi: S^{n-1} \times [0, 1] \to E^n$ is an embedding such that $\psi(S^{n-1} \times \{0\}) = \Sigma$ and $\psi(S^{n-1} \times (0, 1]) \subset U$. Show that Cl U is an n-cell.
6. For each $n \geq 2$ there is a proper map f of a connected n-manifold M onto itself having exactly one nondegenerate inverse image X and X is noncellular.
7. Let X be a compact subset of E^n. Then E^n/G_X is a manifold iff X is cellular.
8. Suppose M is an n-manifold whose universal cover is E^n or S^n and X is a compact contractible subset of M for which M/G_X is an n-manifold. Then X is cellular.
9. If the suspension ΣX of a compact metric space X is an n-manifold, then ΣX is homeomorphic to S^n.
10. If Σ_1 and Σ_2 are disjoint bicollared $(n - 1)$-spheres in S^n, then $[\Sigma_1, \Sigma_2)$ is homeomorphic to $S^{n-1} \times [0, 1)$ and (Σ_1, Σ_2) is homeomorphic to $S^{n-1} \times (0, 1)$.

   A decomposition G (necessarily monotone) of an n-manifold M is said to be *defined by n-cells* provided there exists a sequence $\{A_j \mid j = 1, 2, ...\}$ such that

   (i)   each A_j is a finite union of pairwise disjoint n-cells;
   (ii) Int $A_j \supset A_{j+1}$;
   (iii) Cl $N_G = \bigcap_j A_j$;
   (iv) each component of $\bigcap_j A_j$ is an element of G.

11. If the decomposition G of the n-manifold M is defined by n-cells, then G is shrinkable.
12. Let A be any arc in E^3 with a designated point a_0 such that $A - \{a_0\}$ is the countable union of straight line segments (any two of which meet only in a common endpoint). Then A is cellular.

7. COUNTABLE DECOMPOSITIONS AND SHRINKABILITY

In one of his early papers about usc decompositions, R. H. Bing [2] set forth several conditions about countable cellular decompositions of E^3 implying shrinkability. His techniques depended on nothing intrinsically

3-dimensional; the arguments functioned equally well in any Euclidean space. Among the first to recognize the potential generality of Bing's methods was L. F. McAuley [2], who adapted them to nonmanifold settings by isolating a useful shrinkability property inherent in the notion of cellularity. With it we shall investigate conditions comparable to Bing's implying shrinkability for decompositions of complete metric spaces.

A compact subset C of the space S is *locally shrinkable in S* (or, simply, *locally shrinkable*) if for each open set U in S containing C and for each open cover $\mathcal{V}$ of S there exists a homeomorphism h of S onto itself such that $h(C)$ is contained in some $V \in \mathcal{V}$ and $h \mid S - U$ = identity.

Listed below are two obvious connections to previous sections. Details for their verification involve only definitions and earlier results.

Proposition 1. *Let C denote a closed and compact subset of a T_1-space S. Then the decomposition G_C of S is strongly shrinkable if and only if C is locally shrinkable in S.*

Proposition 2. *A compact subset C of an n-manifold M is cellular in M if and only if it is locally shrinkable in M.*

Not only does the next result serve as a lemma for the theorem to follow, it also has a consequence of interest in itself.

Proposition 3. *Let G be a usc decomposition of S and $\mathcal{V}$ an open cover of S. Then the union C of all $g \in G$ contained in no member of $\mathcal{V}$ is a closed subset of S.*

Proof. Each $x \in S - C$ lies in some $g \in G$ found in $V_g \in \mathcal{V}$. Upper semicontinuity gives an open neighborhood V_g^* of x, which misses C, by definition, and shows that $S - C$ is open. ∎

Corollary 3A. *Let G be a usc decomposition of a metric space (S, ρ) and $\varepsilon > 0$. Then the union of all elements of G having ρ-diameter at least ε is a closed subset of S. Consequently, N_G is an F_σ-subset of S.*

Given a decomposition G for which H_G is a countable collection of locally shrinkable sets, one might expect to be able to shrink them one at a time, in the limit effecting a total shrinking. Bing showed how to make this work in case N_G forms a G_δ-subset. One can eliminate some of the chaff from the forthcoming proof by considering the special case where N_G is closed (and G_δ). Justification of the need for some hypotheses on H_G will be given after the proof has been completed.

Theorem 4. *Suppose G is a usc decomposition of a complete metric space* *(S, ρ) such that H_G is a countable collection of locally shrinkable sets and N_G* *is a G_δ-subset of S. Then G is strongly shrinkable.*

    Proof. Consider an open set W containing N_G, a G-saturated open cover $\mathfrak{U}$ of S, and another open cover $\mathcal{V}$ of S. By hypothesis, H_G can be enumerated as $g_1, g_2, \ldots$ and N_G can be expressed as the intersection of open sets $W_1 = W, W_2, \ldots$, where $W_{i+1} \subset W_i$.

    The first step is a special replacement of $\mathfrak{U}$. Since $\pi(N_G)$ is a countable set, its cover $\pi(\mathfrak{U})$ has a refinement consisting of pairwise disjoint open sets. Pulling these back to S, for $k \in \{1, 2, \ldots\}$ we obtain a G-saturated open set U_k containing g_k and contained in some member of $\mathfrak{U}$ such that for distinct indices i and j either $U_i = U_j$ or $U_i \cap U_j = \varnothing$. Ultimately these will ensure $\mathfrak{U}$-closeness, without any further refinements. We shall produce homeomorphisms $h_0 = \mathrm{Id}, h_1, h_2, \ldots$ satisfying $h_i \,|\, S - U_i = h_{i-1} \,|\, S - U_i \, (i > 0)$, and it will follow from the construction of these open sets that each h_k is $\mathfrak{U}$-close to the identity.

    Let C_1 denote the union of all those $g \in G$ contained in no member of $\mathcal{V}$. According to Proposition 3, C_1 is closed. By normality, there exists a G-saturated open set Z_1 such that $C_1 \subset Z_1 \subset \bar{Z}_1 \subset W_1$.

    Since g_1 is locally shrinkable, there exists a homeomorphism h_1 of S to itself such that $h_1 = \mathrm{Id}$ outside $Z_1 \cap U_1$ and that $h_1(g_1)$ is $\mathcal{V}$-small. (If $g_1 \cap Z_1 = \varnothing$, g_1 is already $\mathcal{V}$-small, and $h_1 = \mathrm{Id}$ works.) Then g_1 has a G-saturated closed neighborhood Q_1 such that $h_1(Q_1)$ is contained in some $V \in \mathcal{V}$. This neighborhood Q_1 will serve as a protective buffer about g_1, in which no further motion will occur.

    Now all the components for the construction scheme have been laid out. Generally, we shall produce a sequence $h_0 = \mathrm{Id}, h_1, h_2, \ldots$ of homeomorphisms of S, another sequence $Z_1, Z_2, \ldots$ of G-saturated open sets, and a third sequence $Q_1, Q_2, \ldots$ of G-saturated closed subsets of S such that, for $i \in \{1, 2, \ldots\}$,

    (1) $Z_{i+1} \subset \bar{Z}_{i+1} \subset W_{i+1} \cap Z_i$,
    (2) $g \in G$ and $g \cap Z_i = \varnothing$ implies $h_i(g)$ is $\mathcal{V}$-small,
    (3) $h_i \,|\, S - Z_i = h_{i-1} \,|\, S - Z_i$,
    (4) $h_i \,|\, S - U_i = h_{i-1} \,|\, S - U_i$,
    (5) Q_i is a closed neighborhood of g_i such that $h_i(Q_i) \subset V_i \in \mathcal{V}$,
    (6) $h_{i+k} \,|\, Q_i = h_i \,|\, Q_i$ for every integer $k \geq 1$, and
    (7)   either $h_i(x) \neq h_{i-1}(x)$ for some $x \in g_i$ or $h_i = h_{i-1}$.

    Having constructed h_1, Z_1, and Q_1, we assume h_j, Z_j, and Q_j have been constructed. Then $h_j(G)$ is usc and, by condition (2), $h_j(g)$ is $\mathcal{V}$-small for every $g \in G$ not contained in Z_j. Hence, there exists a closed set C_{j+1}

consisting of those $g \in G$ for which $h_j(g)$ fails to be $\mathcal{V}$-small and $C_{j+1} \subset Z_j$. Furthermore, $C_{j+1} \subset W_{j+1} - \bigcup_{i=1}^{j} Q_i$ by conditions (5) and (6). Again, normality provides a G-saturated open set Z_{j+1} such that

$$C_{j+1} \subset Z_{j+1} \subset \bar{Z}_{j+1} \subset (W_{j+1} \cap Z_j) - \bigcup_{i=1}^{j} Q_i.$$

Now consider $h_j(g_{j+1})$: if $\mathcal{V}$-small, set $h_{j+1} = h_j$; if not, by repeated applications of condition (4) together with the construction of the U_i's, we see that $h_j(g_{j+1}) \subset U_{j+1}$. The local shrinkability of $h_j(g_{j+1})$ in S leads to a homeomorphism θ of S to itself fixed outside $h_j(Z_{n+1} \cap U_{j+1})$ and making $\theta h_j(g_{j+1})$ be $\mathcal{V}$-small. Define h_{j+1} as θh_j. Finally, complete the cycle by naming a G-saturated closed neighborhood Q_{j+1} of g_{j+1} such that $h_{j+1}(Q_{j+1})$ is a subset of some $V \in \mathcal{V}$. It should be noted that the definitions of Z_{j+1} and C_{j+1} plus the requirement that θ act as the identity outside $h_j(Z_{j+1})$ yield condition (2) in case $i = j + 1$. The other six conditions governing our iterative construction are readily verified.

At this point we specialize to the case that S is (locally) compact. The secret then is that the sequence of homeomorphisms $\{h_i\}$ eventually becomes (locally) stationary. For suppose to the contrary that $i(1), i(2), \ldots, i(n), \ldots$ is an infinite increasing sequence of indices such that each $h_{i(n)}$ differs from its predecessor $h_{i(n)-1}$ (on a compact neighborhood of a given point). Taking a sequence $x_n \in g_{i(n)}$, we find a limit point $y \in S$. If $y \notin N_G$, there exists an index j such that $y \notin W_j$, which implies that infinitely many of the sets $g_{i(n)}$ fall outside Z_j, forcing the homeomorphisms $h_{i(k)}$ of corresponding indices to agree with their predecessors [see condition (3)]. On the other hand, if $y \in g_j \subset N_G$, then infinitely many of the sets $g_{i(n)}$ lie inside Q_j, forcing $h_{i(k)}$ to agree with h_j on Q_j for those corresponding indices $i(k) > j$, thereby implying that $h_{i(k)} = h_{i(k)-1}$ [see condition (7)]. In either situation, we have a contradiction. Thus, there exists an index k such that $h_k = h_{k+1} = h_{k+2} = \cdots$. This means that some h_k achieves the desired shrinking. ∎

Corollary 4A. *If G is a cellular usc decomposition of an n-manifold M such that H_G is countable and N_G is a G_δ-subset of M, then G is strongly shrinkable.*

Corollary 4B. *If G is a usc decomposition of a complete metric space S such that H_G forms a countable collection of locally shrinkable sets and N_G is a G_δ set, then $\pi: S \to S/G$ is a near-homeomorphism. Moreover, for each open set W containing N_G and each open cover $\mathcal{W}$ of S/G, there exists a homeomorphism $F: S \to S/G$ with $F \mid S - W = \pi \mid S - W$ and F $\mathcal{W}$-close to π.*

If G is a usc decomposition of a metric space (S, ρ) and $\varepsilon > 0$, by $G(\geq \varepsilon)$ we mean the decomposition of S whose nondegenerate elements are those $g \in G$ having ρ-diameter at least ε. It follows from Proposition 3 and Proposition 2.5 that $G(\geq \varepsilon)$ is usc.

Corollary 4C. *If* G *is a countable, cellular usc decomposition of an* n*-manifold* M *and* $\varepsilon > 0$, *then* $G(\geq \varepsilon)$ *is strongly shrinkable.*

Proof. This is immediate because $N_{G(\geq \varepsilon)}$ is closed, hence G_δ, in M.

The corollary above may suggest too much. If M were compact and one wanted to shrink all the elements of G to small size, one could apply Theorem 4 or Corollary 4C to shrink the big elements of G, that is, those $g \in H_G$ that also lie in $G(\geq \varepsilon)$. *Caution!* Although the original big elements of G can be shrunk in this manner, there is nothing to prevent elements which are small at the outset [thus, being the union of singletons from $G(\geq \varepsilon)$] from getting stretched to very large size during the so-called shrinking process. This represents a significant problem, precluding real progress toward a total shrinking of G.

In manifolds, examples revealing the phenomenon are rather delicate. In more general metric spaces, examples are easier to dissect, and it is instructive to examine one.

Example 1. A usc decomposition G of a compact metric space S such that H_G forms a null sequence of locally shrinkable sets but G itself is not shrinkable.

The space S is a subset of E^3. The major part of it is a 2-cell B in the xy-plane consisting of all points at distance one or less from the point $\langle \frac{1}{2}, 0 \rangle$. On the interval $[0, 1] \times \{0\}$ in B one names the standard middle-thirds Cantor set $C \times \{0\}$. With each accessible point $c \in C$ (that is, a point living in the closure of one of the components of $E^1 - C$), we associate a number $r(c)$ as the length of the interval component having c in its closure, subject to the convention that $r(0) = 1 = r(1)$. For each such c, we attach to B the circle J_c in the plane perpendicular to the x-axis at c of radius $r(c)$ and centered at $\langle c, 0, r(c) \rangle$. The result is the necessarily compact space S illustrated in Fig. 7-1.

The decomposition G is the one having as its nondegenerate elements the null sequence of circles $\{J_c \mid c \text{ an accessible point of } C\}$.

The nonshrinkability of G is no surprise. The function $f: S \to B$ fixing B and sending J_c to $\langle c, 0 \rangle$ realizes G and implies that S/G is homeomorphic to B. However, S certainly cannot be homeomorphic to B. Consequently, $\pi: S \to S/G$ cannot be a near-homeomorphism or, equivalently (Theorem 5.1), G cannot be shrinkable.

Much more surprising is the claim that each $J_c \in H_G$ is locally shrinkable. To be explicit, we consider J_0 as being typical. Given $\varepsilon > 0$ we choose $c = (\frac{1}{3})^n \in C$ smaller than $\varepsilon/2$, and we name two circular disks in B, an outer one D of radius $(\frac{1}{3})^n$ centered at $\langle \frac{1}{2} \cdot (\frac{1}{3})^n, 0 \rangle$ and an inner one D' of radius $\frac{1}{2} \cdot (\frac{1}{3})^n$ centered at the same point. We rigidly rotate D' through $180°$, keeping the center point fixed and tapering the rotation off to the identity

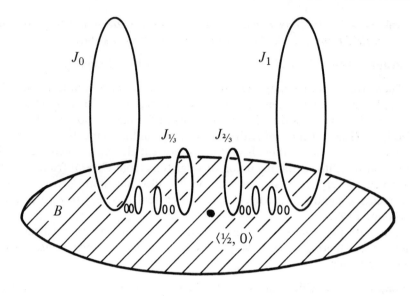

FIG. 7-1

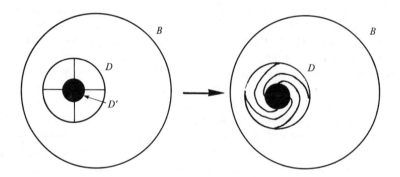

FIG. 7-2

outside D, as shown in Fig. 7-2. All that remains is to extend the homeomorphism τ of B to itself to a homeomorphism of S onto itself. Since τ
permutes the accessible points of $C \times 0$, it extends to a function τ that
permutes the curves $\{J_c\}$, whose restriction to any one of them is a homeomorphism onto another, and that reduces to the identity on those meeting
B outside D. Because these curves form a null sequence, the extended
function τ is a homeomorphism, and it shrinks J_0 to small size while keeping
points fixed outside the ε-neighborhood of J_0.   ∎

It is instructive to note that for any homeomorphism h of S to itself, $h(B) = B$. If $h(J_0)$ is small, $J_0 = h(J_c)$ for some $c \in C$, probably $c \neq 1$. In other words, no matter how one attempts to shrink G, some $J_c \in H_G$ will be sent onto J_0. This points to a rigidity property of the decomposition G: every homeomorphism of S to itself permutes the decomposition elements, precluding any possibility of shrinking G.

With a minor variation to G we can discern further limitations to the usefulness of shrinkability for establishing whether a decomposition of an arbitrary space is homeomorphic to its source. In spaces other than complete metric ones, the example of Section 5 reveals that shrinkability is not sufficient for doing this. Even in compact metric spaces, however, shrinkability is not a necessary condition; the example of Section 4 provides a monotone usc decomposition G of S^3 such that S^3/G is topologically S^3 but G is nonshrinkable, because its elements are not cellular (locally shrinkable). The local shrinkability of decomposition elements is not enough to rectify the matter.

Example 2. A usc decomposition G' of a compact metric space S such that $H_{G'}$ forms a null sequence of locally shrinkable sets and $S/G' \approx S$, but G' fails to be shrinkable.

The space S coincides with the one in Example 1. The only difference between G there and G' is that G' involves just "half" of the other's nondegenerate elements. Explicitly, the nondegenerate elements of G' consist of those $J_c \in H_G$ where $c \geq \frac{1}{2}$. ■

Given a countable usc decomposition into locally shrinkable sets, how then should one hope to shrink the individual nondegenerate elements successively so as to produce a shrinking of the entire collection? While shrinking any particular element, one must be able to monitor the elements nearby so that their sizes do not expand excessively. The next result (also from Bing [2]) furnishes a suitable monitoring condition.

Theorem 5. *Suppose S is a locally compact metric space and G is a countable usc decomposition of S such that, for each $g_0 \in H_G$ and each $\varepsilon > 0$, there exists a homeomorphism h of S onto itself satisfying*

(a) $h \mid S - N(g_0; \varepsilon) = \text{Id}$,
(b) $\text{diam}\, h(g_0) < \varepsilon$, *and*
(c) *for $g \in G$, either* $\text{diam}\, h(g) < \varepsilon$ *or* $h(g) \subset N(g; \varepsilon)$.

Then G is strongly shrinkable.

Proof. Note first that given a homeomorphism f of S to itself, $g_0 \in G$, and $\varepsilon > 0$ such that $\text{Cl}\, N(g_0; \varepsilon)$ is compact, one can use the uniform continuity of $f \mid \operatorname{Cl} N(g_0 ; \varepsilon)$ to obtain another homeomorphism $f' = fh$ of S satisfying

(a') $\quad f' \mid S - N(g_0 ; \varepsilon) = f \mid S - N(g_0 ; \varepsilon)$,
(b') $\quad \operatorname{diam} f'(g_0) < \varepsilon$,
(c') $\quad$ for each $g \in G$, $\operatorname{diam} f'(g) < \varepsilon + \operatorname{diam} f(g)$.

We use this modified version of the hypothesis to prove Theorem 5. We shall consider only the case in which S is compact; the more general result follows exactly as in the hints for the locally compact case dropped during the proof of Theorem 4.

Let W be an open set containing N_G and $\varepsilon > 0$. By Proposition 3, the set $N_{G(\geq \varepsilon/2)}$ is compact. Enumerate the elements of $H_{G(\geq \varepsilon/2)}$ as $g_1, g_2, \ldots$.

As in the proof of Theorem 4, we can determine G-saturated open sets $U_1, U_2, \ldots$ in W such that U_i contains g_i, U_i and U_j either coincide or do not intersect, and $\pi(U_i)$ has diameter less than ε. Then, just as before, we shall encounter the desired shrinking homeomorphism h_k at the end of a finite sequence $h_0 = \operatorname{Id}$, h_1, h_2, ..., h_k of homeomorphisms satisfying $h_i \mid S - U_i = h_{i-1} \mid S - U_i$ ($i \in \{1, 2, \ldots, k\}$), which will imply that h_k agrees with the identity outside W and that h_k is ε-close to π.

By hypothesis, there exists a homeomorphism h_1 of S onto itself such that

(a$_1$) $\quad h_1 \mid S - U_1 = \operatorname{Id}$,
(b$_1$) $\quad \operatorname{diam} h_1(g_1) < \varepsilon/4$,
(c$_1$) $\quad \operatorname{diam} h_1(g) < (\varepsilon/4) + \operatorname{diam} g$, for each $g \in G$.

As a result, g_1 has a G-saturated closed neighborhood Q_1 such that, for each $g \in G$ with $g \subset Q_1$, $\operatorname{diam} h_1(g) < \varepsilon$.

Recursively, we aim to produce a sequence of homeomorphisms $h_0 = \operatorname{Id}$, h_1, ..., h_j, ... and a sequence Q_1, ..., Q_j, ..., where Q_j denotes a G-saturated closed neighborhood of g_j, satisfying (for $j \in \{1, 2, \ldots\}$),

(a$_j$) $\quad h_j \mid S - U_j = h_{j-1} \mid S - U_j$,
(b$_j$) $\quad \operatorname{diam} h_j(g_j) < \varepsilon$,
(c$_j$) $\quad \operatorname{diam} h_j(g) < (1 - 1/2^j)(\varepsilon/2) + \operatorname{diam} g$, for each $g \in G$,
(d$_j$) $\quad h_{j+1} \mid Q_1 \cup \cdots \cup Q_j = h_j \mid Q_1 \cup \cdots \cup Q_j$,
(e$_j$) $\quad g \in G$ and $g \subset Q_j$ implies $\operatorname{diam} h_j(g) < \varepsilon$, and
(f$_j$) $\quad h_j = h_{j-1}$ if $\operatorname{diam} h_{j-1}(g_j) < \varepsilon$.

Assuming that $h_0, h_1, \ldots, h_j$ and $Q_1, \ldots, Q_j$ have been constructed subject to the conditions above, we consider the element g_{j+1}. If $\operatorname{diam} h_j(g_{j+1}) < \varepsilon$, we define h_{j+1} as h_j. If not, it follows from the various possible conditions (d$_i$) and (e$_i$) that $g_{j+1} \notin Q_1 \cup \cdots \cup Q_j$. Application of the modified version of the hypothesis yields a homeomorphism h_{j+1} such that

$$h_{j+1} \mid S - (U_{j+1} \cup [Q_1 \cup \cdots \cup Q_j]) = h_j \mid S - (U_{j+1} \cup [Q_1 \cup \cdots \cup Q_j]),$$

$$\operatorname{diam} h_{j+1}(g_{j+1}) < \varepsilon/2^{j+2},$$

and for each $g \in G$,

$$\operatorname{diam} h_{j+1}(g) < (\varepsilon/2^{j+2}) + \operatorname{diam} h_j(g).$$

The third of these combines with condition (c_j) to give

$$\operatorname{diam} h_{j+1}(g) < (1/2^{j+1})(\varepsilon/2) + (1 - 1/2^j)(\varepsilon/2) + \operatorname{diam} g$$

$$= (1 - 1/2^{j+1})(\varepsilon/2) + \operatorname{diam} g.$$

In either case, it is possible to find a G-saturated closed neighborhood Q_{j+1} of g_{j+1} for which every $g \in G$ in Q_{j+1} satisfies $\operatorname{diam} h_{j+1}(g) < \varepsilon$.

The compact set $N_{G(\geq \varepsilon/2)}$ is covered by the interiors of the sets Q_j. Hence, for some k that set is contained in $Q_1 \cup \cdots \cup Q_k$. Repeated applications of conditions (d_j) and (e_j) establish that $\operatorname{diam} h_k(g) < \varepsilon$ for all $g \in G$ found in $Q_1 \cup \cdots \cup Q_k$. Conditions (f_{k+i}) force all the later homeomorphisms h_{k+i} to agree with h_k. Finally, for those $g \in G$ that were not large originally, that is, for those of diameter less than $\varepsilon/2$, condition (c_k) reveals that $\operatorname{diam} h_k(g) < \varepsilon$. This means that h_k has shrunk everything ε-small, and it fulfills the requirements of the shrinkability criterion in the compact metric case. ■

The argument just given actually yields the following variation to Theorem 5.

Theorem 6. *Suppose S is a locally compact metric space and G is a countable usc decomposition of S such that for each homeomorphism f of S onto itself, each $g_0 \in G$, and each $\varepsilon > 0$, there exists a homeomorphism f' on S satisfying*

$$f' \mid S - N(g_0; \varepsilon) = f \mid S - N(g_0; \varepsilon),$$

$$\operatorname{diam} f'(g_0) < \varepsilon,$$

and for each $g \in G$,

$$\operatorname{diam} f'(g) < \varepsilon + \operatorname{diam} f(g).$$

Then G is strongly shrinkable.

EXERCISES

1. Suppose G is a strongly shrinkable, countable usc decomposition of a locally connected metric space S. Then each $g_0 \in H_G$ is locally shrinkable in S. In particular, show that for each homeomorphism f of S onto itself, each $g_0 \in H_G$, and each $\varepsilon > 0$, there exists a homeomorphism f' on S such that

$$f' \mid S - N(g_0; \varepsilon) = f \mid S - N(g_0; \varepsilon),$$

$$\operatorname{diam} f'(g_0) < \varepsilon,$$

and for each $g \in G$,

$$\operatorname{diam} f'(g) < \varepsilon + \operatorname{diam} f(g).$$

2. Prove Theorem 4 in case S is a complete metric space.

3. Prove Theorems 5 and 6 in case S is a locally compact metric space.

4. Prove Theorem 4 in case S is a (locally) compact normal space.

5. Is Theorem 5 valid in case S is a complete metric space? (Apparently this question is unsolved.)

6. Show that there exists a nonshrinkable decomposition G of a compact metric space C such that H_G is a null sequence of 2-cells, each locally shrinkable in C, and C/G is homeomorphic to C.

8. COUNTABLE DECOMPOSITIONS OF E^n

This section applies the work from Section 7, mainly Theorem 7.5, to countable decompositions of Euclidean space. Based on results developed first, in spirit if not in deed, by R. H. Bing [2], the hypotheses here invariably include special geometric properties possessed by the decomposition elements. Such hypotheses are necessary, since according to examples to be presented in the next section not all countable, cellular decompositions of E^n are shrinkable.

Given a space X with some preferred embedding e in E^n, one says that a subspace X' of E^n homeomorphic to X is *flat* (or is *flatly* embedded in E^n) provided there exists a homeomorphism θ of E^n onto itself such that $\theta(X') = e(X)$. In particular, one says that an arc A in E^n is *flat* provided there is a self-homeomorphism θ of E^n carrying A to a straight line segment.

Theorem 1. *If G is a countable usc decomposition of E^n into points and flat arcs, then G is strongly shrinkable.*

Any flat arc can be considered as a line segment. Invoking uniform continuity of the inverse of the flattening homeomorphism, restricted to a compact neighborhood of the segment, one quickly reduces the verification that the hypotheses of Theorem 7.5 are satisfied to the lemma below.

Lemma 2.   *Let G be a monotone usc decomposition of E^n, $g_0 \in G$ a straight line segment, and $\delta > 0$. Then there exists a homeomorphism H of E^n onto itself satisfying*

(a) $H \,|\, E^n - N(g_0; \delta) = \mathrm{Id}$,
(b) $\mathrm{diam}\, H(g_0) < \delta$, *and*
(c)   *for $g \in G$ either $H(g) \subset N(g; \delta)$ or $\mathrm{diam}\, H(g) < \delta$.*

Proof.   Without loss of generality, g_0 is the line segment $\{\langle 0, \ldots, 0, x_n \rangle \in E^n \,|\, 0 \leq x_n \leq b\}$. Choose an integer $k > 0$ large enough that $b/k < \delta/8$, and let P_i denote the $(n - 1)$-dimensional hyperplane of E^n perpendicular to the x_n-axis at the point $(ib)/k$ $(i \in \{0, 1, \ldots, k\})$.

For each $\varepsilon > 0$ the set of points at distance exactly ε from g_0 can be seen to be a bicollared $(n - 1)$-sphere; details, for a much more general setting, are provided in Lemma 3. Hence, for $i \in \{0, 1, \ldots, k\}$ there exists $\varepsilon(i) > 0$, with

$$\delta/8 > \varepsilon(0) > \varepsilon(1) > \cdots > \varepsilon(i) > \varepsilon(i + 1) > \cdots > \varepsilon(k),$$

such that the sets S_i of points at distance $\varepsilon(i)$ from g_0 are bicollared $(n - 1)$-spheres and

(∗)   whenever $g \in G$ and $g \cap S_i \neq \varnothing$,   then $g \cap S_{i+1} = \varnothing$.

The $\varepsilon(i)$'s (and S_i's) are obtained in successive order; upper semicontinuity of G makes condition (∗) possible.

Three additional notational matters: for $i \in \{1, 2, \ldots, k\}$ T_i will denote the points of S_i below the plane P_{i-1}, B_{i-1} will denote the open $(n - 1)$-cell $P_{i-1} \cap N(g_0; \varepsilon(i))$, and C_i (a chamber) will denote the closure of that part of $N(g_0; \varepsilon(0))$ between the planes P_{i-1} and P_i while C_0 will denote the closure of that part of $N(g_0; \varepsilon(0))$ below P_0 (see Fig. 8-1). An important aspect of these chambers is their size, which, by construction, is less than $\delta/2$.

The homeomorphism H will be defined so that $H(g_0) \subset C_k$, forcing $H(g_0)$ to have small size. Control on the other elements will be derived from the following: if $x \in N(g_0; \varepsilon(i - 1)) - N(g_0; \varepsilon(i + 1))$ for $i \in \{1, \ldots, k - 1\}$ and if $H(x) \neq x$, then

$$H(x) \in C_{i-1} \cup C_i.$$

For those $g \in G$ meeting $N(g_0; \varepsilon(0))$, condition (∗) implies the existence of an integer i such that

$$g \in N(g_0; \varepsilon(i - 1)) - N(g_0; \varepsilon(i + 1))$$

[or $g \in E^n - N(g_0; \varepsilon(1))$; or $g \in N(g_0; \varepsilon(k))$]. It will follow then that

$$H(g) \subset g \cup C_{i-1} \cup C_i$$

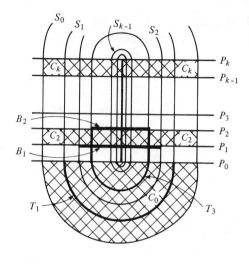

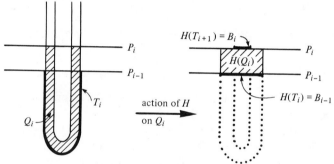

FIG. 8-1

[or $H(g) \subset g \cup C_0$; or $H(g) \subset g \cup C_k$]. The monotonicity hypothesis coupled with the fact $\mathrm{diam}(C_{i-1} \cup C_i) < \delta$ yields that either $H(g) \subset N(g; \delta)$ [which occurs in case $H(g) \cap g \neq \varnothing$] or $\mathrm{diam}\, H(g) < \delta$.

Here is one way to define H. Require that it be the identity everywhere outside S_0 [that is, on $E^n - N(g_0; \varepsilon(0))$], also on those points of E^n above P_i and outside S_i ($i \in \{1, ..., k - 1\}$), and finally on all points of E^n above P_k. Define $H \mid T_i$ so that $H(T_i) = B_{i-1}$ for $i \in \{1, ..., k - 1\}$. Extend over the remaining parts of $N(g_0; \varepsilon(0))$ in a cell-by-cell fashion. This is straightforward because if Q_i denotes the closure of the region below P_i and between S_i and S_{i+1}, then Q_i is an n-cell and H is defined so as to send ∂Q_i onto the boundary of another cell in E^n. ∎

The next topic is a generalized convexity property. A compact set X in E^n is *starlike with respect to a point* $x_0 \in X$ if for any other point $x \in X$ the line segment determined by x and x_0 is a subset of X; equivalently, each geometric ray in E^n emanating from x_0 intersects X is a connected set. Generally, a compact subset X of E^n is *starlike* if it is starlike with respect to some point $x_0 \in X$. It should be obvious that every compact, convex subset of E^n is starlike with respect to each of its points.

Lemma 3. *Let X be a compact subset of E^n that is starlike with respect to $x_0 \in X$ and Z the set of all points at distance exactly some fixed $\delta > 0$ from X. Then each ray emanating from x_0 intersects Z in a single point, and Z is a bicollared $(n - 1)$-sphere.*

Corollary 3A. *Each starlike subset of E^n is cellular.*

Proof. The complement of Z can be expressed as the union of $N(X; \delta)$ and $E^n - \mathrm{Cl}(N(X; \delta))$. Since each ray emanating from x_0 meets both of these, it must also meet Z at least once. We argue that it meets Z exactly once.

Suppose to the contrary that on some ray R there are two points z_1 and z_2 of $R \cap Z$. Without loss of generality we can assume z_1 lies between x_0 and z_2 on R. By definition of Z and by compactness of X, there exists a point x_2 of X whose distance from z_2 equals δ. Let $\alpha = \rho(x_0, z_1)/\rho(x_0, z_2)$ and note that $\alpha < 1$.

The connectedness of $X \cap R$ ensures that the ray R' from x_0 through x_2 is not collinear with R. On R' let $\tilde{x}_1$ denote the point such that $\rho(\tilde{x}_1, x_0) = \alpha\rho(x_2, x_0)$. (See the accompanying diagram.) Then the triangles determined by x_0, x_2, and z_2 and by x_0, $\tilde{x}_1$, and z_1 are similar, implying that $\rho(z_1, \tilde{x}_1) = \alpha\rho(z_2, x_2)$. As a result,

$$\rho(z_1, X) \leq \rho(z_1, \tilde{x}_1) = \alpha \cdot \delta < \delta,$$

which is impossible.

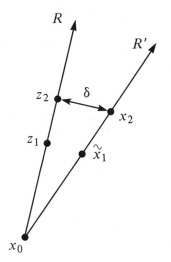

The fact that rays from x_0 meet Z in unique points implies that radial projection of $E^n - \{x_0\}$ to the unit $(n-1)$-sphere centered at x_0, when restricted to Z, gives a one-to-one and onto (continuous) function. To the reader we leave the verification that this homeomorphism can be extended to a ray-preserving homeomorphism of E^n, thereby establishing that Z is a flat $(n-1)$-sphere. ∎

Like flat arcs, starlike sets admit carefully controlled shrinkings; a particularly nice shrinking is described in what follows.

Lemma 4. *Suppose X is a compact subset of E^n that is starlike with respect to $x_0 \in X$, U is an open subset of E^n containing X, $\delta > 0$, and B is a round n-cell centered at x_0 such that $X \subset N(B; \delta)$. Then there exists a homeomorphism f of E^n onto itself satisfying*

(a) $f \mid E^n - (U \cap N(B; \delta)) = \mathrm{Id}$,
(b) $f(X) \subset \mathrm{Int}\, B$,
(c) $f(X)$ *is starlike with respect to* x_0, *and*
(d) $\rho(f, \mathrm{Id}) < \delta$.

Proof. First choose $\alpha_1 > 0$ such that $N(X; \alpha_1) \subset U \cap N(B; \delta)$, set $\alpha_2 = \alpha_1/2$, and define $(n-1)$-spheres Z_j as

$$Z_j = \{z \in E^n \mid \rho(z, X) = \alpha_j\} \qquad (j = 1, 2).$$

Let r represent the radius of the n-cell B.

The homeomorphism f will send each ray emanating from x_0 onto itself, implying conclusion (c). On each such ray R three points hold special significance: the point $q = R \cap Z_1$, the point $p = R \cap Z_2$, and the point p' at distance r from x_0 ($p' = R \cap \partial B$). Certainly p lies between x_0 and q on R; however, where p' is to be found depends on circumstances like the length of $X \cap R$ and the sizes of α_1 and α_2.

Of course, f is required to send x_0 to itself. In addition, it is to act as the identity on q and on all other points of $R \cap (E^n - N(X; \alpha_1))$. What f does to p breaks down into two cases: in case p' lies between x_0 and p, $f(p) = p'$; in case $p = p'$ or p lies between x_0 and p', $f(p) = p$. Finally, f is defined to take the interval $[x_0, p]$ in R linearly onto $[x_0, f(p)]$ and the interval $[p, q]$ linearly onto $[f(p), q]$. (See the accompanying diagram.)

Several features of f are worth attention. In case $f(p) \ne p$, $\rho(f(p), p) < \delta$, because p is trapped between $f(p) = p'$ and the point at distance $r + \delta$ from x_0 on R. In the other case $\rho(f(p), p) < \delta$ for trivial reasons. Thus, in either case the linearity properties of f lead to conclusion (d). Moreover, since $X \cap R \subset [x_0, p)$,

$$f(X \cap R) = f(X) \cap R \subset [x_0, p') = R \cap \mathrm{Int}\, B,$$

which implies conclusion (b). ∎

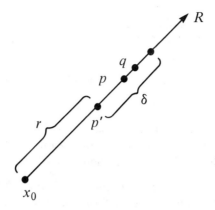

The next argument supplies a fundamental technique, a shrinking of one element by means of a composition of finitely many small moves, with an interspersed monitoring of other displaced decomposition elements to prevent undue stretching.

Lemma 5. *Suppose G is a usc decomposition of E^n, $g_0 \in G$ is a starlike set, and H_G forms a null sequence (or the elements of G in some neighborhood of g_0 form a null sequence). Then for each $\delta > 0$ there exists a homeomorphism F of E^n to itself satisfying*

(a) $F \mid E^n - N(g_0; \delta) = \mathrm{Id}$,
(b) $\mathrm{diam}\, F(g_0) < \delta$, *and*
(c)   *for $g \in G$ either* $\mathrm{diam}\, F(g) < \delta$ *or* $F(g) = g$.

Proof. Fix $x_0 \in g_0$ such that g_0 is starlike with respect to x_0, and find the least positive integer k for which $g_0 \subset N(x_0; k\delta/3)$.

The homeomorphism F will be expressed as $F = f_{k-1} \circ \cdots \circ f_1$, where the f_i's are determined from Lemma 4 and where they produce, in stages, sets $f_j \circ \cdots \circ f_1(X)$ starlike with respect to x_0 and contained in $N(x_0; (k - j) \cdot \delta/3)$. At the first stage one uses the nullity of H_G to insist that f_1 move no point outside an open set U_1 obtained so near to g_0 that any other $g \in G$ meeting U_1 has diameter less than $\delta/3$; after $f_{j-1}, \ldots, f_1$ have been defined, one insists that f_j move no point outside an open set $U_j \subset U_{j-1}$ so near $f_{j-1} \circ \cdots \circ f_1(g_0)$ that, for any other $g \in G$ with $f_{j-1} \circ \cdots \circ f_1(g) \cap U_j \neq \varnothing$, $\mathrm{diam}\, f_{j-1} \circ \cdots \circ f_1(g) < \delta/3$. In the application of Lemma 4, the f_j's will move no point more than $\delta/3$. Thus, for $g \in G$ ($g \neq g_0$) and $j \in \{2, 3, \ldots, k - 1\}$, either

$$f_j \circ \cdots \circ f_1(g) = f_{j-1} \circ \cdots \circ f_1(g)$$

or

$$\mathrm{diam}\, f_j \circ \cdots \circ f_1(g) < \delta.$$

In particular, the above holds for $F = f_{k-1} \circ \cdots \circ f_1$, and if at the jth stage, $\delta/3 \leq \operatorname{diam} f_j \circ \cdots \circ f_1(g) < \delta$, then $F(g) = f_j \circ \cdots \circ f_1(g)$, which implies conclusion (c). Finally, since

$$F(g_0) = f_{k-1} \circ \cdots \circ f_1(g_0) \subset N(x_0; \delta/3),$$

F shrinks g_0 to sufficiently small size. ∎

A subset X of E^n is said to be (ambiently) *starlike-equivalent* if there exists a homeomorphism θ of E^n onto itself sending X onto a starlike set. Once this terminology is at hand, the theorem should be transparent.

Theorem 6. *If G is a usc decomposition of E^n such that H_G forms a null sequence of starlike-equivalent sets, then G is strongly shrinkable.*

Theorem 6 officially was proved by R. J. Bean [1], who attributed credit for much of the technique to R. H. Bing. It can be derived from Lemma 5 just as Theorem 1 was derived from Lemma 2.

Corollary 6A. *Every usc decomposition of E^n into points and a null sequence of flat cells is strongly shrinkable.*

The kind of argument around which Theorem 6 revolves, given here to prove Lemma 5, plays a central role in many shrinking theorems, including R. D. Edwards's cell-like approximation theorem and M. H. Freedman's analysis of Casson 2-handles in 4-space. For that reason Theorem 6 probably represents the most significant result of this section.

Whether Theorem 6 is valid when H_G forms a countable collection (rather than a null sequence) of starlike-equivalent sets is unknown for $n \geq 4$, although it is known for $n = 3$ (Denman–Starbird [1]); it is valid when H_G is a countable collection of rectilinearly starlike sets. Demonstrating this fact is our next aim. The published proof, again due to Bing [2], is rather complicated; the one given below, which was described orally by him in response to a complaint about intricacies of his published argument, more clearly displays the rigidly geometric controls imposed on the shrinking process. Denman [1] presents a slightly different proof of this fact.

Theorem 7. *If G is a countable usc decomposition of E^n into starlike sets, then G is strongly shrinkable.*

Globally, the proof is exactly like that of Theorem 7.6. For a given $\varepsilon > 0$ we enumerate the elements of $H_{G(\geq \varepsilon/2)}$ as $g_1, g_2, \ldots$ and construct a sequence $\{h_i\}$ of homeomorphisms on E^n satisfying, first of all, conditions to ensure that each h_i is some preassigned G-saturated open cover $\mathcal{U}$-close to the identity and equals the identity off some preassigned open neighborhood W of N_G, and second,

(1)   diam $h_i(g_i) < \varepsilon$,
(2)   diam $h_i(g) < $ diam $g + \sum_{j=1}^{i} (\varepsilon/2^{j+2})$ for all $g \in G$,
(3) $h_i \,|\, Q_1 \cup \cdots \cup Q_{i-1} = h_{i-1} \,|\, Q_1 \cup \cdots \cup Q_{i-1}$, and
(4) $h_i = h_{i-1}$ if diam $h_{i-1}(g_i) < \varepsilon$,

where Q_j, as usual, denotes a G-saturated closed neighborhood of g_j ($j = 1, 2, \ldots$) such that diam $h_j(Q_j) < \varepsilon$. Ultimately, the point of the argument will be that the rule $h = \lim h_i$ is actually a homeomorphism because locally it is just h_k, for some large integer k. With such a rule, almost before one expects it, the clutch of this mechanism engages, terminating the motion. The role of starlikeness in all of this is to enable the construction of the shrinking homeomorphisms h_i satisfying conditions (1) and (2) simultaneously. Condition (2) prevents undue stretching of the surrounding elements.

A controlled shrinking of any one decomposition element, which combines techniques from Theorem 1 and Theorem 6, is isolated in the next result.

Lemma 8. *Suppose G is a usc decomposition of E^n, $g_0 \in G$ is a starlike set, and δ is a positive number. Then there exist an integer $K > 0$ and positive numbers $\delta_0, \delta_1, \ldots, \delta_K$, with*

$$\delta/4 = \delta_0 > \delta_1 > \cdots > \delta_K,$$

such that to each $g \in G$ corresponds $i \in \{0, 1, \ldots, K\}$ for which

$$g \subset N(g_0; \delta_{i-1}) - N(g_0, \delta_{i+1})$$

(here interpret δ_{-1} as ∞ and $N(g_0; \delta_{k+i})$ as $\varnothing$) and there exists a homeomorphism f of E^n onto itself satisfying

(a) $f \,|\, E^n - N(g_0; \delta_1) = $ Id,
(b)   diam $f(g_0) < \delta$, and
(c)   diam $f(X) < \delta + $ diam X for each set $X \subset N(g_0; \delta_{i-1}) - N(g_0; \delta_{i+1})$.

Proof. Name a point $x_0 \in g_0$ for which g_0 is starlike with respect to x_0. Choose K as the least integer for which $g_0 \subset N(x_0; K \cdot \delta/4)$. Find $\delta_1 \in (0, \delta/4)$ such that any $g \in G$ meeting $N(g_0; \delta_1)$ is contained in $N(g_0; \delta_0)$ and Cl $N(g_0; \delta_1) \subset N(x_0; K \cdot \delta/4)$. Then determine the numbers δ_i ($i \in \{2, \ldots, K\}$) successively so that any $g \in G$ meeting $N(g_0; \delta_i)$ is contained in $N(g_0; \delta_{i-1})$.

Let S_i denote the frontier of $N(g_0; \delta_i)$; on any ray R emanating from x_0 let $s_i = R \cap S_i$ (see Lemma 3) and let t_i be the point of R at distance $(K - i + 1) \cdot \delta/4$ from x_0 ($i \in \{0, 1, \ldots, K\}$). Define $f \,|\, R$ as the identity on x_0 as well as on points of R beyond s_0;

$$f(s_i) = \begin{cases} s_i & \text{if } s_i \text{ is between } x_0 \text{ and } t_i \\ t_i & \text{otherwise;} \end{cases}$$

and so as to be linear on the intervals between s_{i+1} and s_i and also on the one between x_0 and s_K. Because $g_0 \subset N(x_0; K \cdot \delta/4)$, $\operatorname{Cl} N(g_0; \delta_0) \subset N(x_0; (K + 1) \cdot \delta/4)$, which implies that s_0 lies between x_0 and t_0; by construction of δ_1, $\operatorname{Cl} N(g_0; \delta_1) \subset N(x_0; K \cdot \delta/4)$, which implies that s_1 lies between x_0 and t_1. As a result, $f(s_0) = s_0$ and $f(s_1) = s_1$. Furthermore,

$$f(g_0 \cap R) \subset R \cap N(x_0; \delta/2).$$

These observations indicate that f satisfies conclusions (a) and (b).

To see why it satisfies conclusion (c) as well, let ψ_i denote the radial projection of E^n onto $\operatorname{Cl} N(x_0; (K - i + 1)\delta/4)$ sending all points of R beyond t_i to t_i ($i \in \{0, 1, ..., K\}$). Geometric considerations reveal that ψ_i does not increase distances between pairs of points, so that for any set X, $\operatorname{diam} \psi_i(X) \leq \operatorname{diam} X$. In particular, if X is a compact subset of $N(g_0; \delta_{i-1}) - N(g_0; \delta_{i+1})$, then $f(X) \subset N(\psi_i(X); \delta/2)$ (here $i > 0$), which yields

$$\operatorname{diam} f(X) < \delta + \operatorname{diam} \psi_i(X) \leq \delta + \operatorname{diam} X. \quad \blacksquare$$

Proof of Theorem 7. *Step 1.* If $\operatorname{diam} g_1 < \varepsilon$, set $h_1 = \operatorname{Id}$. If not, apply Lemma 8 for the decomposition G, starlike set g_1, and positive number $\delta^1 < \varepsilon/8$, to obtain positive numbers $\delta_0^1, \delta_1^1, ..., \delta_{K(1)}^1$ and a homeomorphism $h_1 = f_1$ satisfying those conclusions. Afterwards, find a G-saturated closed neighborhood Q_1 of g_1 such that $\operatorname{diam} h_1(Q_1) < \varepsilon$.

Step 2. If $\operatorname{diam} h_1(g_2) < \varepsilon$, set $h_2 = h_1$ (and, implicitly, define $f_2 = \operatorname{Id}$). If not, $g_2 \cap Q_1 = \varnothing$, and there is some neighborhood U_2' of g_2 in $E^n - Q_1$ such that, for some index $i \in \{0, 1, ..., K(1)\}$,

$$U_2' \subset N(g_1; \delta_{i-1}^1) - N(g_1; \delta_{i+1}^1).$$

Pick $\delta^2 \in (0, \varepsilon/16)$ so small that $N(g_2; \delta^2) \subset U_2'$ and that, for any $A \subset U_2'$ having diameter less than δ^2, $\operatorname{diam} h_1(A) < \varepsilon$. Apply Lemma 8 again for the starlike set g_2 and positive number δ^2 to obtain positive numbers $\delta_0^2, \delta_1^2, ...,$ $\delta_{K(2)}^2$ and a homeomorphism f_2. Define h_2 as $h_1 f_2$. Certainly $\operatorname{diam} h_2(g_2) < \varepsilon$. For any set X in $N(g_2; \delta_{j-1}^2) - N(g_2; \delta_{j+1}^2)$ where $j > 0$,

$$\operatorname{diam} f_2(X) < \delta^2 + \operatorname{diam} X < \varepsilon/16 + \operatorname{diam} X;$$

since $f_2(X) \subset N(g_1; \delta_{i-1}^1) - N(g_1; \delta_{i+1}^1)$,

$$\operatorname{diam} h_2(X) = \operatorname{diam} h_1(f_2(X)) < \varepsilon/8 + \operatorname{diam} f_2(X) < \varepsilon/8 + \varepsilon/16 + \operatorname{diam} X.$$

Of course, for $X \subset E^n - N(g_2; \delta_1^2)$, $f_2 \mid X = \operatorname{Id}$, and thus $h_2 \mid X = h_1 \mid X$, which indicates that X is no more distorted after Step 2 than after Step 1. Name, as usual, a G-saturated closed neighborhood Q_2 of g_2 for which $\operatorname{diam} h_2(Q_2) < \varepsilon$.

Step 3. If diam $h_2(g_3) < \varepsilon$ or if $g_3 \cap U_2^i = \varnothing$, this step proceeds like Step 2. If not, there is some neighborhood U_3^i of g_3 in $E^n - (Q_1 \cup Q_2)$ such that, for some index $j \in \{0, 1, \ldots, K(2)\}$,

$$U_3^i \subset N(g_2; \delta_{j-1}^2) - N(g_2; \delta_{j+1}^2).$$

Pick $\delta^3 \in (0, \varepsilon/2^5)$ so small that $N(g_3; \delta^3) \subset U_3^i$ and that diam $h_2(A) < \varepsilon/2^5$ for any subset A of U_3^i having diameter $< \delta^3$. Apply Lemma 8 again, for the set g_3 and number δ^3, to obtain a homeomorphism f_3 and define h_3 as $h_2 f_3$. Now check that for any subset X of some $N(g_3, \delta_{k-1}^3) - N(g_3; \delta_{k+1}^3)$ where $k > 0$,

$$\text{diam } f_3(X) < \delta^3 + \text{diam } X < \varepsilon/2^5 + \text{diam } X,$$

and since $f_3(X) \subset N(g_2, \delta_{j-1}^2) - N(g_2; \delta_{j+1}^2)$,

$$\text{diam } h_3(X) = \text{diam } h_2 f_3(X) < \varepsilon/2^3 + \varepsilon/2^4 + \text{diam } f_3(X)$$

$$< \varepsilon/2^3 + \varepsilon/2^4 + \varepsilon/2^5 + \text{diam } X;$$

on the other hand, for $X \subset E^n - N(g_3; \delta_1^3)$, $f_3(X) = X$, so

$$\text{diam } h_3(X) = \text{diam } h_2 f_3(X) = \text{diam } h_2(X).$$

Name Q_3.

Later steps. The above process is continued, as with Steps 2 and 3, but with more cases to consider. If diam $h_m(g_{m+1}) < \varepsilon$, set $h_{m+1} = h_m$. If $g_{m+1} \cap U_m' = \varnothing$, Step $m + 1$ proceeds exactly like Step m. If $g_{m+1} \subset U_m'$, one can repeat the constructions and analyses as above to obtain f_{m+1} and $h_{m+1} = h_m f_{m+1}$, with size controls applying not simply to $g \in G$ but more broadly to subsets of the various sets $N(g_i; \delta_{j(i)-1}^i) - N(g_i; \delta_{j(i)+1}^i)$, which makes possible analogous size controls under later homeomorphisms h_{m+k}. ∎

In most of the earlier work, in order to construct a sequence $\{h_i\}$ of homeomorphisms successively shrinking more decomposition elements $g_i \in G$ to small size, once we had obtained h_i we turned to the usc decomposition $h_i(G)$ and, by whatever means available, we compressed $h_i(g_{i+1})$ to produce h_{i+1}. In Theorem 8 that approach would not be so fruitful, because an arbitrary homeomorphism on E^n is likely to destroy the starlikeness of decomposition elements. To circumvent the potential difficulty, we obtain h_{i+1} from h_i instead by constructing a homeomorphism f_{i+1} affecting g_{i+1} rather than $h_i(g_{i+1})$ and setting $h_{i+1} = h_i f_{i+1}$. Each approach has its merits, and occasionally one functions smoothly when the other resists.

We exploit radial shrinking techniques still another time in the final result of this section, originally established by T. M. Price [1].

Theorem 9.   *Suppose G is a countable usc decomposition of an n-manifold M such that, for each $g_0 \in H_G$ and each open set $U \supset g_0$, there exists an n-cell B such that $g_0 \subset B \subset U$ and $\partial B \cap N_G = \emptyset$. Then G is strongly shrinkable.*

Proof.   All that is required here is a verification that G satisfies the conditions of Theorem 7.5. With that in mind, fix $g_0 \in H_G$ and $\varepsilon > 0$.

The hypothesis provides an n-cell B in M with $g_0 \subset B \subset N(g_0; \varepsilon)$ and $\partial B \cap N_G = \emptyset$. Let θ be a homeomorphism of B onto the standard n-cell B^n. Choose $\delta > 0$ so that diam $\theta^{-1}(A) < \varepsilon$ whenever $A \subset B^n$ has diameter less than 4δ, and let $\theta(G)$ represent the decomposition of B^n given by $\{\theta(g) \mid g \in G$ and $g \subset B\}$.

Determine an integer $k > 0$ such that $1/k < \delta$. Since $N_{\theta(G)} \cap \partial B^n = \emptyset$, there exists $\delta_0 \in (0, 1)$ so small that, for any $g \in \theta(G)$ intersecting $N(\partial B^n; \delta_0)$, its radial projection to ∂B^n has diameter less than 2δ. Find a sequence of numbers $\delta_0 > \delta_1 > \delta_2 > \cdots > \delta_k > 0$ such that every $g \in \theta(G)$ intersecting $N(\partial B^n; \delta_{i+1})$ is contained in $N(\partial B^n; \delta_i)$.

For any radius R of B^n, let s_i denote the point of R at distance δ_i from $R \cap \partial B^n$ and let t_i denote the point of R at distance i/k from the origin. Define a homeomorphism f of B^n onto itself as the identity on the endpoints of R, so as to send s_i onto t_i, and so as to be linear on the intervals in between. Now consider $g' \in \theta(G)$: if g' meets $B^n - N(\partial B^n; \delta_0)$, then $g' \subset B^n - N(\partial B^n; \delta_1)$ and $f(g') \subset N(\text{origin}; \delta)$, which has diameter less than 2δ; otherwise, as long as $g' \cap \partial B^n = \emptyset$, g' lies in some $N(\partial B^n; \delta_{i-1}) - N(\partial B^n; \delta_{i+1})$, and then $f(g')$ is trapped in that sector of the cone over its own projection onto ∂B^n between the spheres of radius $(i - 1)/k$ and of radius $(i + 1)/k$ centered at the origin, which implies that diam $f(g') < 4\delta$, because the sector itself is that small.

Finally, define $h: M \to M$ as $\theta^{-1} f \theta$ on B and as the identity elsewhere.   ∎

The following is a corollary to each of the theorems given in this section.

Corollary 10.   *Every monotone usc decomposition G of E^1 or S^1, except $G = \{S^1\}$, is strongly shrinkable.*

SOME UNSOLVED QUESTIONS

1. If G is a usc decomposition of E^n with points and countably many flat cells, is G shrinkable? [Yes, if $n = 3$ (Starbird–Woodruff [1], Everett [2]).]
2. If G is a usc decomposition of E^n into points and countably many starlike-equivalent sets, is G shrinkable? [Yes, if $n = 3$ (Denman–Starbird [1]). T. L. Lay [1] has obtained a partial result, which appears as Exercise 4.]

3. Suppose G is a usc decomposition of E^n such that for each $g \in H_G$ and each open neighborhood W_g of g there exists a neighborhood U_g with $g \subset U_g \subset W_g$, where the frontier of U_g is an $(n - 1)$-sphere missing N_G. Is G shrinkable? [Yes, if $n = 3$ (Woodruff [1]).]

4. Is there a usc decomposition of E^n into points and straight line segments that is not shrinkable? [Yes, if $n = 3$ (Armentrout [4]). See also Bing [9] and Eaton [3].]

EXERCISES

1. If G is a usc decomposition of an n-manifold M such that H_G forms a countable collection of essentially flat arcs, then G is strongly shrinkable. [An arc A in an n-manifold is said to be *essentially flat* if there exist a neighborhood U of A and a homeomorphism ψ of U onto E^n such that $\psi(A)$ is flat.]

2. Suppose G is a usc decomposition of E^n such that H_G forms a null sequence and, for each $g_0 \in H_G$ and each open set U containing g_0, there exists a starlike-equivalent set X_0 with $g_0 \subset X_0 \subset U$ and $X_0 \cap N_G = g_0$. Then G is strongly shrinkable.

3. Suppose G is a usc decomposition of E^n for which H_G has at most two elements. Then E^n/G can be embedded in E^{n+1}.

4. Suppose G is a countable monotone usc decomposition of E^n such that for each $g \in H_G$ there exists a homeomorphism θ_g of E^n onto itself sending g into the cone from the origin over a compact 0-dimensional subset of S^{n-1}. Then G is strongly shrinkable.

5. If G is a usc decomposition of E^n such that H_G forms a null sequence of collapsible (tame) polyhedral subsets, then G is strongly shrinkable.

   A monotone usc decomposition G of an n-manifold M is said to be *constrained by n-cells* if for each $\varepsilon > 0$ and each open set $W \supset N_{G(\geq \varepsilon)}$ there exist pairwise disjoint n-cells $B_1, \ldots, B_k$ in M such that $N_{G(\geq \varepsilon)} \subset \bigcup_i B_i \subset W$ and $\partial B_i \cap N_G = \varnothing$ $(i = 1, \ldots, k)$.

6. If G is a monotone usc decomposition of an n-manifold such that (1) G is constrained by n-cells and (2) $\pi(N_G)$ is 0-dimensional, then G is strongly shrinkable.

9. SOME CELLULAR DECOMPOSITIONS OF E^3

The classical concept of a defining sequence provides a prefabricated form for building decompositions of manifolds. A specific instance where this already has been put to work occurs in the exercises at the end of Section 6. Generally, a *defining sequence for a decomposition* of an n-manifold M is a sequence $\mathcal{S} = \{C_i \mid i = 1, 2, \ldots\}$, where each C_i is a compact n-manifold with boundary and Int $C_i \supset C_{i+1}$; the *decomposition G associated with $\mathcal{S}$* consists of the components of $\bigcap C_i$ and the singletons from $M - \bigcap C_i$.

As a more inclusive way of treating noncompact manifolds M it is permissible, in place of requiring that each C_i in a defining sequence be compact, to demand instead that each of its components be compact. We will not strive to be fully comprehensive, however, until Section 37, where we will lay out an all-inclusive notion of defining sequence.

Proposition 1. *The decomposition G associated with any defining sequence S of an n-manifold is usc; moreover,* Cl $\pi(N_G)$ *is compact and 0-dimensional.*

Proof. According to Corollary 4.2A and Proposition 2.5, G is usc. By Proposition 4.1, Cl $\pi(N_G)$ is totally disconnected (and compact). ∎

Often the elementary result below is perfectly tailored for identifying a cellular decomposition.

Proposition 2. *Suppose* $S = \{C_1, C_2, \ldots\}$ *is a defining sequence in an n-manifold such that for each index i and each component K of C_{i+1} there exists an n-cell B_K with*

$$K \subset \text{Int } B_K \subset B_K \subset C_i.$$

Then the decomposition G associated with S is cellular.

The aims now are to delineate some historically important cellular (in one case, merely cell-like) decompositions of E^3, all of which crop up from defining sequences, and then to examine their properties.

A general remark about the forthcoming descriptive procedure is in order at the outset. The examples will be specified largely by a single figure depicting the initial (first two) stages of a defining sequence. The way to "see" later stages is to replicate the pictured pattern. Making such replication possible is the fact that in every example the largest object, the first stage C_1 of the defining sequence, is a compact 3-manifold with boundary T and each of the smaller objects T_j, a component of the second stage C_2, is homeomorphic to T. The third stage C_3 can be produced by choosing a homeomorphism f_j of T onto T_j and setting C_3 equal to the union of all the various sets $f_j(C_2)$. Successive stages can be reproduced in exactly the same fashion. As a result, like the old Quaker Oats box with the picture of the elderly Quaker holding another box, on which is a picture of the elderly Quaker holding still another box, etc., the initial structure is reproduced repeatedly, so that the part of C_{i+1} inside a given component of C_i looks just like C_2 inside C_1. Unless additional requirements are explicitly imposed, the choice of homeomorphisms f_j arranging C_{i+1} inside C_i will not be significant.

Example 1. The decomposition G_1 has a defining sequence resulting from infinite iteration of the structures illustrated in Fig. 9-1.

It should be clear from Fig. 9-1 and Proposition 2 that G_1 is a cellular decomposition. Even though defined by solid tori rather than 3-cells, G_1 is shrinkable. As we shall learn, $\pi(\bigcap C_i)$ is a Cantor set strangely embedded in the decomposition space, since its complement fails to be simply connected. If $\{C_i\}$ is constructed with care, $\pi \,|\, P$ can be made 1-1. With such a construction, R. H. Bing [1] established the shrinkability of G_1. Trading on

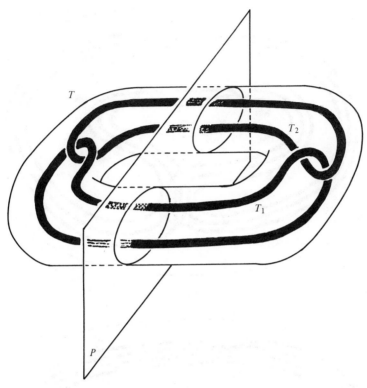

FIG. 9-1

symmetries of $\{C_i\}$ about the plane P, he devised thereby an involution of the decomposition space, which coincides with E^3, interchanging its left and right halves by reflecting through the wild plane $\pi(P)$; the involution is not equivalent to any standard involution, because the fixed point set is wild.

Example 2. The decomposition G_2 has a defining sequence resulting from infinite iteration of the structures illustrated in Fig. 9-2.

As before, G_2 is cellular. We shall show in Proposition 7 that it is not shrinkable.

Example 3. The decomposition G_3 is suggested in the usual way by the structures illustrated in Fig. 9-3.

A more complicated version of Example 2, G_3 is another cellular, nonshrinkable decomposition. The same would be true if Fig. 9-3 were modified to position an arbitrary number $k > 4$ of solid tori T_j in similar linked fashion inside T.

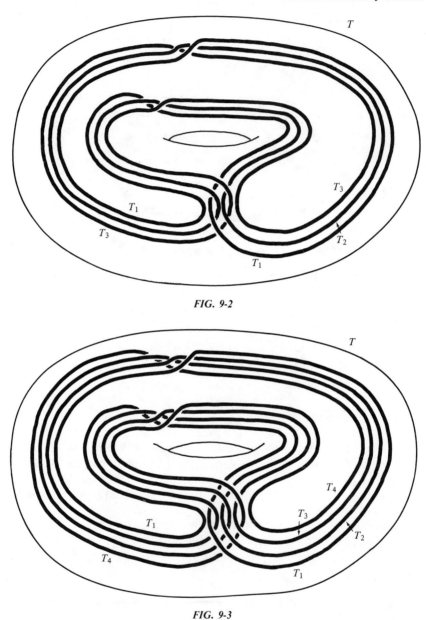

FIG. 9-2

FIG. 9-3

Example 4. The decomposition G_4 has a defining sequence suggested by Fig. 9-4 and gives rise to the famous dogbone decomposition space of Bing [3].

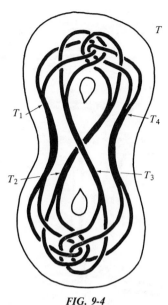

FIG. 9-4

Like the two previous examples, G_4 is nonshrinkable, but the argument for this case is formidable and we will not retrace it.

The defining sequence can be aligned so the nondegenerate elements are all flat arcs. That punctuates a fine distinction with Theorem 8.1. As an even more significant attribute, E^3/G_4 is a manifold factor, the first ever to be discovered. Bing [5] proved in 1959 that $(E^3/G_4) \times E^1$ is homeomorphic to E^4. In particular, his work on this, on Example 1, and on countable decompositions of E^3 supplied the original evidence exposing the usefulness of the shrinkability criterion.

Example 5. The decomposition G_5 has a defining sequence determined by the entangled sets pictured in Fig. 9-5.

With careful regulation G_5 can be constructed so all its nondegenerate elements are flat arcs. Consequently, it represents another cellular, nonshrinkable dogbone decomposition, comparable to Example 4. A close relative of an example discovered by W. T. Eaton [1], this particular manifestation was found by D. G. Wright [2]. It is cited here because G_5, rather than G_4, functions as the model for higher-dimensional analogues— cellular, nonshrinkable decompositions of E^n—and also, a matter directly pertinent to this section, because the forthcoming examination of linking patterns in E^3 lends itself to a more efficient analysis of G_5 than of G_4. Part of the reason is that the solid tori forming the upper and lower ends of T

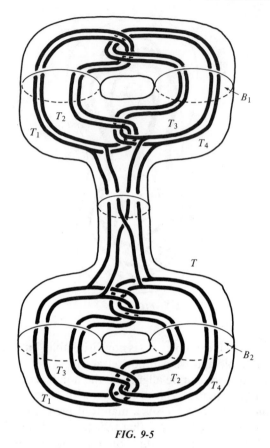

FIG. 9-5

in Fig. 9-5 contain several subsets from

$$C_2 = T_1 \cup T_2 \cup T_3 \cup T_4$$

reminiscent of structures depicted in Fig. 9-1. An important by-product of the shrinkable decomposition G_1 is the wild Cantor set it engenders. Here G_5 ramifies that wildness at both the upper and lower ends of T and, we shall see, joins the pieces together in a manner designed specifically to preclude shrinkability.

Example 6. The decomposition G_6 has a defining sequence suggested by Fig. 9-6.

To incorporate the more remarkable aspects of G_6, the most delicate among these examples, one must be exceptionally meticulous about the iterative construction of its defining sequence. Two features of the initial

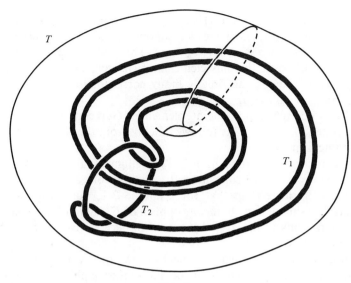

FIG. 9-6

structures deserve notice. First, each T_i has a core simple closed curve that lies in one of two 2-dimensional planes, either a horizontal plane P_H or a vertical one P_V, and T_i can be regarded as a slight thickening of this core. Second, T_1 appears nearly as large as T itself while T_2 can be made as small as desired; specify it with diameter less than $\frac{1}{4}$.

Observe that T also can be regarded as a slight thickening of a core circle in P_H. Here then is a recursive description of an optimal defining sequence $\{C_i\}$ for G_6: in forming C_i from C_{i-1}, presume each component T' of C_{i-1} has a core curve either in P_H or in P_V; determine solid tori T_1' and T_2' in T', with the pair embedded there as T_1 and T_2 are embedded in T, but subject to two additional conditions—first, that T_1' and T_2' both have a core simple closed curve in P_H or in P_V, one in each plane, with each T_k' included setwise in the 2^{-i}-neighborhood of its own core ($k = 1, 2$), and second, that T_2' has diameter less than 2^{-i}.

With such a defining sequence $\{C_i\}$, the associated decomposition G_6 has only a null sequence of nondegenerate elements, by the second of the two conditions above. Components of $\bigcap C_i$ can be uniquely specified by infinite sequences $\{e_j \mid j = 2, 3, \ldots\}$, where $e_j \in \{1, 2\}$. The correspondence comes about naturally, by associating with each initial segment $\{e_2, \ldots, e_i\}$ of the sequence a component $A(e_2, \cdots e_i)$ of C_i, defined by the rules (1) that

$$A(1) = T_1 \subset C_2 \quad \text{and} \quad A(2) = T_2 \subset C_2$$

and generally (2) that $A(e_2 \cdots e_i 1)$ is the large component of C_{i+1} in $A(e_2 \cdots e_i)$ while $A(e_2 \cdots e_i 2)$ is the small one; a component X of $\bigcap C_i$ is paired with the sequence $\{e_i\}$ for which $X = \bigcap A(e_2 \cdots e_i)$. This makes it easy to enumerate the big elements, since the only elements of G_6 having diameter as large as 2^{-n} $(n > 1)$ are among the finitely many corresponding to sequences $\{e_i\}$ where e_i is 1 for all $i \geq n$.

Furthermore, each nondegenerate element of G_6 is contained in one of two planes, either P_H or P_V. This follows quickly from the first fact governing the defining sequence.

By construction and Proposition 2, G_6 is cellular. It also is nonshrinkable, a fact to be explored in Proposition 10. The rigorous accounting promises to be more intricate than that for Example 7.1, but the philosophical explanation is similar: any homeomorphism of E^3 submitting to mild cover controls and shrinking the large elements of G_6 necessarily must expand some of the small ones.

Example 6 was developed, as one by now must expect, by Bing [8]. Interesting not merely by virtue of the null sequence phenomenon, it arose as the first nonshrinkable cellular decomposition involving just countably many nondegenerate elements. Its additional properties permit it to stand as the minimal nonshrinkable example.

Example 7. The decomposition G_7 results by iterating the structures pictured in Fig. 9-7.

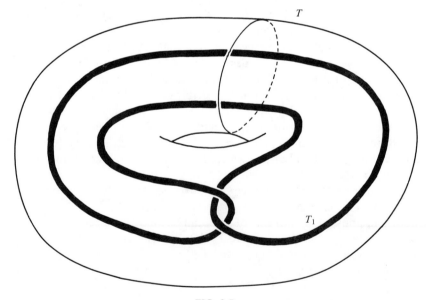

FIG. 9-7

This differs radically from the previous examples since it has only a single nondegenerate element, a noncellular set, causing G_7 to be nonshrinkable for trivial reasons. When the iterative nest is constructed efficiently to force (among other things) 1-dimensionality of the intersection Wh, then Wh is called the *Whitehead continuum*, after J. H. C. Whitehead [1].

What makes this unusual? After all, if C is a simple closed curve in E^3, G_C cannot be shrinkable, nor can it yield a 3-manifold (Exercise 6.7). Example 7 plays a foreshadowing role, hinting that ultimately the paramount concern about decompositions might not be cellularity. In addition, it is mentioned here because (1) like Example 5, but for simpler reasons, it is a nonmanifold that is a manifold factor (indeed, all of the decomposition spaces described in this section are factors of E^4), (2) the noncellularity of Wh follows from properties employed to expose the nonshrinkability of the other examples given in this section, and (3) these Whitehead continua play a major part in M. H. Freedman's [1] notable 4-manifold decomposition work, culminating in his result that Casson 2-handles are real 2-handles and, consequently, in the solution of the 4-dimensional Poincaré conjecture.

Next, some explanations of the properties possessed by these examples. Briefly, about an exception among them, the shrinkable decomposition G_1: we intend to present a pivotal clue about its shrinkability but will leave the details as an instructive exercise in manipulative techniques. Looking deep enough into the defining sequence $\{C_i\}$ for G_1, one should compress the various components T of some C_i close to a core curve and then should dice up each such T by a finite collection of planes P_j, as shown in Fig. 9-8, spaced close enough together that the part of T between any two planes P_{j-1} and P_{j+1} has small size. One achieves the desired shrinking by rearranging those elements of G_1 in T so that none meets more than one of the planes P_j. To be successful, according to Bing, one must avoid the trap of being too greedy by attempting to accomplish too much at once; instead, the tactics are to diligently reduce, a little bit at a time, the maximal number of planes P_j intersecting any one component of the successive stages C_{i+m} in T. The crux of this suggestion is the rearrangement pictured in Fig. 9-8.

The remainder of this section is devoted primarily to an investigation of nonshrinkability in the other examples, excluding Example 4. To that end we shall use (without proof) the following elementary transversality result. When both manifolds present are PL, the conclusion can be achieved most easily by adjusting the given map f to one in general position with respect to Σ.

Proposition 3. *Suppose M is an n-manifold, Σ is an $(n-1)$-manifold embedded in M as a closed and bicollared subset, $f: B^2 \to M$ is a map, U is an open subset of M containing Σ, and $\varepsilon > 0$. Then there exists a map $F: B^2 \to M$ satisfying*

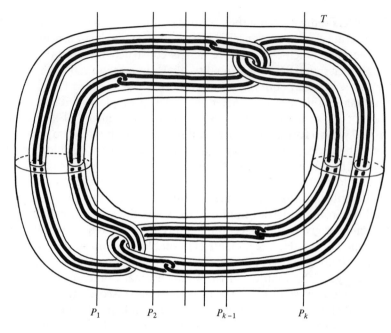

FIG. 9-8

(a) $F \mid B^2 - f^{-1}(U) = f \mid B^2 - f^{-1}(U)$,
(b) $\rho(F, f) < \varepsilon$,
(c) $F^{-1}(\Sigma)$ *is a* 1-*manifold in* B^2 *with* $\partial F^{-1}(\Sigma) = F^{-1}(\Sigma) \cap \partial B^2$, *and*
(d) *for any point* $p \in F^{-1}(\Sigma)$, $F(B^2)$ *pierces* Σ *at* $F(p)$ [*that is, for each neighborhood* W *of* p *in* B^2, $F(W)$ *touches both sides of the bicollar on* Σ *near* $F(p)$].

Given a map $F: B^2 \to M$ *satisfying* (c) *and* (d), *we say that* $F(B^2)$ *is transverse to* Σ.

Consider a defining sequence $\{C_i\}$ for a usc decomposition G of an n-manifold such that each ∂C_i is bicollared, a property holding in Examples 1–7 here. Consider also a map $f: B^2 \to M$ with $C_2 \cap f(\partial B^2) = \varnothing$. The foremost benefit of Proposition 3 is to supply a map $F: B^2 \to M$ close to f, agreeing with f on ∂B^2, with $F^{-1}(\partial C_i)$ being a collection of pairwise disjoint simple closed curves in Int B^2, for all $i > 1$. Of course, the planar Schönflies theorem ensures that every component of $F^{-1}(\partial C_i)$ bounds a 2-cell in Int B^2.

To fully comprehend the nonshrinkability of these decompositions, in effect one must come to terms with the wildness of Antoine's necklace, a Cantor set with an unusual embedding in E^3 (Antoine [1]). It is generated by an infinite iteration of the objects pictured in Fig. 9-9, the construction of which can be controlled so that $\bigcap C_i$ is totally disconnected, in which event

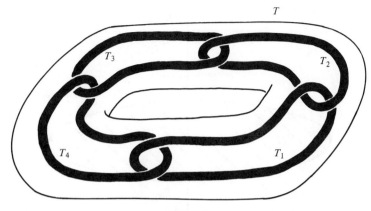

FIG. 9-9

$A = \bigcap C_i$ is Antoine's necklace. Although the construction gives a defining sequence somewhat like that of Example 1, the associated decomposition, being trivial, is not of interest.

For some geometric tricks to study A, refer first to Fig. 9-10.

1. $E^3 - L$ retracts to $D - \{x\}$.

Retract $E^3 - L$ vertically into the plane of D and then radially to $D - \{x\}$.

Compactify E^3 and then decompactify again by removing another point, not on L. This transforms the line L into the circle J of Fig. 9-11. Thus we have:

2. $E^3 - J$ retracts to $D - \{x\}$.

As a simple consequence, the inclusion-induced homomorphism $\pi_1(D - \{x\}) \to \pi_1(E^3 - J)$ is an injection. Split E^3 by means of a vertical plane separating the curves J_1 and J_2, shown in Fig. 9-12, and apply the Seifert–van Kampen theorem to derive:

3. $\pi_1(D - \{x, y\}) \to \pi_1(E^3 - (J_1 \cup J_2))$ is an injection.

Lemma 4. *For the structures pictured in Fig. 9-9, any loop in* $E^3 - T$ *that contracts in* $E^3 - \bigcup T_i$ *is null homotopic in* $E^3 - \text{Int } T$.

Proof. Let J_i denote a core of T_i; explicitly, the pair (T_i, J_i) is homeomorphic to the standard pair $(S^1 \times B^2, S^1 \times \{0\})$. Install planar disks D_2 and D_4, bounded by J_2 and J_4, respectively, so that each of these disks D_i meets the two adjacent circles as in Fig. 9-12.

The set $J_1 \cup D_2 \cup J_3 \cup D_4$ contains a simple closed curve C which is a core of T. Hence, $T - C$ retracts to ∂T, and the smaller $T - (J_1 \cup D_2 \cup J_3 \cup D_4)$ retracts to ∂T as well.

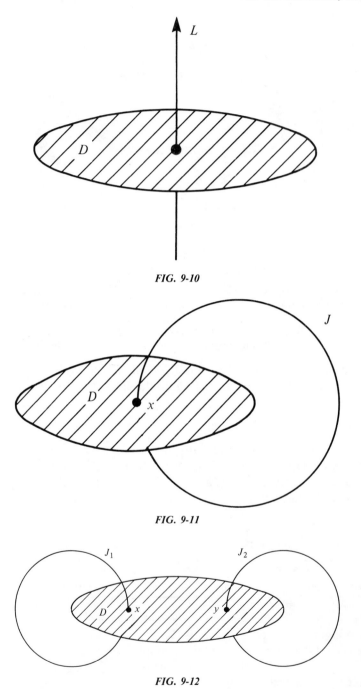

FIG. 9-10

FIG. 9-11

FIG. 9-12

Let L denote a loop in $E^3 - T$ that contracts in $E^3 - \bigcup J_i$. It suffices to prove L contracts in $E^3 - (J_1 \cup D_2 \cup J_3 \cup D_4)$. Name a map f of B^2 into $E^3 - \bigcup J_i$ with $f \mid \partial B^2$ defining L.

By Proposition 3, f can be approximated by a new map (still called f) with the properties above such that, in addition, $f^{-1}(D_2 \cup D_4)$ is a 1-manifold K. Let H denote the component of $B^2 - K$ containing ∂B^2. Each component S of ∂H, other than ∂B^2, is mapped by f to either $D_2 - \bigcup J_i$ or $D_4 - \bigcup J_i$. Trick 3 ensures that f can be redefined on the disk B_S in B^2 bounded by S to send it into the part of D_2 or D_4 missing $\bigcup J_i$. The new map f can be further adjusted, near each B_S, to send the image to the appropriate side of a disk D_i, causing $f(B^2)$ to avoid D_2 and D_4 completely, thereby demonstrating that L contracts in the complement of $(J_1 \cup D_2 \cup J_3 \cup D_4)$.   ∎

As a tool for sorting out nonshrinkability properties as well as detecting the failure of simple connectedness, we shall exploit 2-planes and 2-cells embedded or, more generally, mapped in a given n-manifold. The ensuing terminology helps identify certain decisive intersections. Let Q denote an n-manifold with boundary and H a 2-manifold with boundary. A map $f: H \to Q$ for which $f(\partial H) \subset \partial Q$ is said to be *I-inessential* (an abbreviation of *interior-inessential*) if there exists a map $f': H \to \partial Q$ satisfying $f' \mid \partial H = f \mid \partial H$; when no such map f' exists, f is said to be *I-essential*.

What are the I-essential maps of, say, B^2 to Q? Then $f(\partial B^2)$ is an essential loop in ∂Q but an inessential loop in Q. In particular, in case Q is a solid torus T, this indicates $f \mid \partial B^2$ is homotopic in ∂T to some multiple of a meridian.

A technical variation of the notion of I-essential map is even more useful in the situations to be studied ahead. Let H denote a compact 2-manifold in a 2-cell B. A map f of H to an n-manifold Q with $f(\partial H) \subset \partial Q$ is said to be *virtually I-essential* if f extends to an I-essential map $F: B_H \to Q$ satisfying $F(B_H - H) \subset \partial Q$, where B_H denotes the unique 2-cell in B containing H for which $\partial B_H \subset \partial H$. The value of this variation becomes apparent when a covering map $p: Q' \to Q$ is at hand, for, due to the simple connectivity of B_H, f then admits a lift $f': H \to Q'$ (with $pf' = f$).

Remark. A map $f: H \to Q$ such that $f(\partial H) \subset \partial Q$ is virtually I-essential if and only if f sends the outermost component of ∂H to a nontrivial loop in ∂Q while sending all other components of ∂H to trivial curves there.

Proposition 5. $E^3 - A$ *fails to be simply connected.*

Proof. Given a map f of ∂B^2 to a homotopically nontrivial loop in $E^3 - T$, we shall show that f does not extend to a map f of B^2 into $E^3 - A$.

Suppose the contrary. Use Proposition 3 to alter f so that $f^{-1}(\bigcup T_i)$ is a compact 2-manifold K. By Lemma 4 K is nonvoid. It must contain some component H on which $f \mid H$ is I-essential, for otherwise f could be redefined on K so that $f(B) \cap (\bigcup J_i) = \varnothing$, which would contradict Lemma 4. Among all such components H, select an innermost one H' (H' is contained in a 2-cell in B^2 missing the other components on which f is I-essential).

Now we claim that $f \mid H'$ is virtually I-essential. There are two types of boundary components in H': the special outermost boundary component S', which coincides with the boundary of the minimal 2-cell $B_{H'}$ containing H', and the others, individually denoted as S, which are characterized by the property of bounding 2-cells D in B^2 whose interiors meet no component of K on which f is I-essential. As argued in the preceding paragraph, $f \mid D$ can be modified to satisfy $f(\text{Int } D) \cap (\bigcup J_i) = \varnothing$. Assuming T_1 to be the second stage torus containing $f(S)$, $f(S)$ is some combination of meridian and longitude on ∂T_1. Nontrivial multiples of the meridian, plus any multiple of the longitude, link the core of T_1 in E^3, and nontrivial multiples of the longitude plus meridians link the core of T_2, but $f(S)$ links neither, since $f(D)$ provides a contraction of $f(S)$ in their complement. Accordingly, $f(S)$ must be homotopically trivial on ∂T_1. With this information we see that $f(S')$ must be nontrivial there, for otherwise $f \mid H$ would not be I-essential. Consequently, $f \mid H$ is virtually I-essential.

Since it misses A, $f(H)$ also misses some stage C_k of the defining sequence for A. (Here equate C_1 with T and C_2 with $\bigcup T_i$.) The argument given above translates, with little change, to one showing inductively for each $j > 2$ the existence of a component H_j in $f^{-1}(C_j)$ [after suitable adjustments to f, maintaining the property that $f(H)$ misses C_k] on which $f \mid H_j$ is virtually I-essential. This leads to a contradiction when $j = k$, and, therefore, no such contraction of $f(\partial B^2)$ in E^3 can exist. ∎

Before scrutinizing the examples, we pause to state more of the terminology from embedding theory. A Cantor set X embedded in E^n is said to be *tame* if there exists a homeomorphism θ of E^n to itself such that $\theta(X)$ lies in a straight line segment; if X is not tame, it is said to be *wild*.

Corollary 5A.   *Antoine's necklace is wild.*

It should be clear that every tame Cantor set in E^n, $n \geq 3$, has simply connected complement.

The proof of Proposition 5 yields the following:

Lemma 6.   *Let H be a compact 2-manifold with boundary in a 2-cell B, and f a virtually I-essential map of H into the solid torus T of Fig. 9.9. Then $f(H)$ intersects $\bigcup T_i$. Moreover, on some component of $f^{-1}(\bigcup T_i)$, f is virtually I-essential.*

Proposition 7. *The decomposition G_2 of Example 2 is nonshrinkable.*

Proof. Let B_1 and B_2 be two disjoint, meridional disks in the solid torus T of Fig. 9-2 (that is, each B_j is equivalent in $T = S^1 \times B^2$ to a 2-cell of the form $\{p\} \times B^2$). Assume to the contrary that G_2 is strongly shrinkable. Then there exists a homeomorphism of E^3 to itself, fixed outside T, shrinking the elements of G_2 to such small size that, for some stage k, no component of C_k meets both B_1 and B_2. Its inverse h satisfies:

(∗) for any component P of C_k, P does not intersect both $h(B_1)$ and $h(B_2)$.

Using Proposition 3 we find a map f on $B_1 \cup B_2$ near h such that for each component Q of C_j, $0 < j < k$, $f^{-1}(Q)$ is a 2-manifold with boundary.

Remark. The constructions required for demonstrating the failure of shrinkability are only slightly more elaborate than the ones given about strong shrinkability. In order to prove directly that G_2 is nonshrinkable, one should enlarge the 2-cells B_1, B_2 to disjoint 2-cells B_1', B_2' with

$$B_j \subset \operatorname{Int} B_j' \qquad \text{and} \qquad B_j' \cap T = B_j.$$

Exercising standard controls, one can obtain a homeomorphism h, satisfying (∗) as above, such that each $h(\partial B_j') \subset E^3 - T$, where it is not null homotopic. Applying earlier arguments, one can find a map f defined on $B_1' \cup B_2'$ such that $f^{-1}(C_i)$ is a 2-manifold with boundary $(1 \leq i \leq k)$ and $(f \mid B_j')^{-1}(T)$ contains a component P_j' on which $f \colon P_j' \to T$ is virtually I-essential. This ensures the correct sort of intersection with the first stage, which automatically occurred in the strongly shrinkable situation. Now one can engage the remainder of the forthcoming proof, with only minor changes.

Lemma 8. *Let H_i denote a compact 2-manifold in B_i such that $f \colon H_i \to T$ is a virtually I-essential map $(i = 1, 2)$. Then there exists $j \in \{1, 2, 3\}$ for which H_1 and H_2 contain (nonempty) components M_1 and M_2, respectively, of $f^{-1}(T_j)$, such that $f \colon M_i \to T_j$ is virtually I-essential.*

Proof. One key here is the observation that any two of the second stage tori T_i in Fig. 9-2 are embedded in T exactly like the two second stage tori found in Fig. 9-1. In the standard 2–1 cover $p \colon T' \to T$ of T, these two second stage tori have preimages $p^{-1}(T_1 \cup T_2)$ lying in T' just like the first stage picture given to define Antoine's necklace in Fig. 9-9. The other key is that $f \mid H_1 \cup H_2$ lifts to $F \colon H_1 \cup H_2 \to T'$ because, being virtually I-essential, $f \mid H_i$ can be extended to map f' of a 2-cell D_i to T. According to Lemma 6, $F(H_1)$ and $F(H_2)$ each hit $p^{-1}(T_1 \cup T_2)$ in virtually I-essential fashion. Thus, either T_1 or T_2 is hit the same way by $f(H_1)$, and one or the other is also hit by $f(H_2)$. The same holds true of the other two pairs of second stage tori given in Fig. 9-2. Lemma 8 follows by the pigeonhole principle. ∎

The contradiction that will establish Proposition 7 should be obvious now. One of the second stage tori, say T_1, has nontrivial intersection with both $f(B_1)$ and $f(B_2)$, where the nontriviality is measured in a form allowing repeated application of Lemma 8. As a result, some component of the third stage in T_1 has similar nontrivial intersection with both singular disks, and so on through successive stages. The impossibility of this occurring is fully transparent at the kth stage. ∎

A duplicate analysis can be employed to verify the nonshrinkability of Example 3. As an alternative, one could determine a closed subset C of the decomposition space E^3/G_3 such that the decomposition $G_3(C)$ induced over C coincides with G_2; if G_3 were shrinkable, homeomorphisms shrinking G_3 could be modified somewhat to do the same for $G_3(C)$.

Turning next to Example 7, we already have employed all the techniques needed for showing each meridional disk B in T intersects the nondegenerate element of G_7. The proof of the following is left as a review exercise.

Proposition 9. *The decomposition G_7 of Example 7 is noncellular.*

Proposition 10. *The decomposition G_6 of Example 6 is nonshrinkable.*

Proof. Let B_1 and B_2 denote disjoint, meridional disks in the solid torus T of Fig. 9-6. As in the proof of Proposition 7, suppose the contrary and obtain a homeomorphism h of E^3 to itself, fixed outside T, such that no component P of some stage C_k of the defining sequence for G_6 meets both $h(B_1)$ and $h(B_2)$. We shall consider only the technically simple case in which $h(B_1)$ and $h(B_2)$ are transverse to $T_1 \cup T_2$ (h can be adjusted to a PL homeomorphism in general position to ensure this).

The crux of the argument is given in the following lemma.

Lemma 11. *There exists $j \in \{1, 2\}$ such that B_1 and B_2 contain nonempty components H_1 and H_2, respectively, of $h^{-1}(T_j)$ for which $h: H_i \to T_j$ is virtually I-essential ($i = 1, 2$).*

Proof. Let $p: T' \to T$ denote the standard 2–1 covering, and let T_1' and T_2' denote the lifts of T_1 and T_2 to T' shown in Fig. 9-13 (the other pair of such lifts to T' has no role to play in Lemma 11). These two are embedded in T' just as the two second stage tori pictured in Fig. 9-1 sit in the torus T there.

Assume for the moment that $h(B_1)$ misses T_2. Name the two lifts F and F' of $h \mid B_1$ to T'. By the first part of the proof of Lemma 8, $F(B_1)$ has a virtually I-essential intersection with T_1' or T_2', and it must be with T_1'. On the other hand, only one of the components W of

$$T' - (F(B_1) \cup F'(B_1))$$

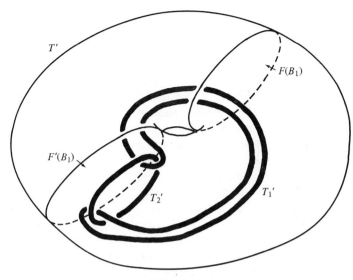

FIG. 9-13

contains T_2', and only one of the two lifts of $h \mid B_2$ to T' goes into W; the other lift must have a virtually I-essential intersection with T_1', since it has none with T_2'. Consequently, both $h(B_1)$ and $h(B_2)$ have the same type of intersection with T_1.

In any event, $h(B_1)$ has some virtually I-essential intersection with T_1 or T_2, say with T_1, and $h(B_2)$ has the same with the other. To draw closer to the situation where $h(B_1)$ misses T_2, we do disk trading between $h(B_i)$ and $\partial(T_1 \cup T_2)$. Whenever J is a component of $h^{-1}(\partial T_1 \cup \partial T_2)$ in B_i such that $h(J)$ bounds a disk in $\partial T_1 \cup \partial T_2$, we redefine h on the disk D_J in B_i bounded by J to send it (homeomorphically) onto a disk in ∂T_i and then adjust slightly, using the collar structure on the boundary to push the image off that T_i. By continually working with curves J whose images bound innermost disks in $\partial T_1 \cup \partial T_2$ [i.e., disks E_J there whose interiors intersect neither $h(B_1)$ nor $h(B_2)$], we modify h, without introducing any singularities, so that no component of $h(B_i) \cap \partial T_j$ bounds a 2-cell in ∂T_j.

Because there exists a virtually I-essential component of intersection between $h(B_2)$ and T_2, the intersection contains a meridional disk. If after disk trading $h(B_1)$ still meets ∂T_2, there exists an innermost component K in B_1 of $h^{-1}(\partial T_2)$, and $h(K)$ bounds a meridional disk in T_2, since it necessarily is parallel in ∂T to each of the components of $h(B_2) \cap \partial T_2$. If D_K denotes the disk in B_1 bounded by K, $h(D_K)$ must be contained in T_2, for otherwise $h(D_K)$ would give a null homotopy of $h(K)$ in E^3-Int T_2, which is ruled out by trick 2. This implies $h: D_K \to T_2$ is virtually I-essential and proves that T_2 satisfies the desired conclusion. ∎

Rephrasing Lemma 11 to expedite iteration, one can complete the proof of Proposition 10 by imitating the way Lemma 8 was used to prove Proposition 7.   ■

Finally, we study Example 5. At this point detecting its nonshrinkability is fairly simple—the example is far less delicate than Example 6. The structures comprising the defining sequence, cubes with two handles, differ from the less complicated solid tori employed in the other examples. Nonetheless, ignoring the stems joining the stuff in the upper part to the stuff in the lower part, we see ramified, or duplicated, versions of the more familiar, shrinkable structures pictured in Fig. 9-1. The extra ingredient in this example, and the source of its nonshrinkability, is the connection between nondegeneracy in the upper and lower parts given by these stems. We begin the analysis by exploring that connection in an abstract setting.

Prescribe a Cantor set K as the countable product of copies of the two-point space $\{0, 1\}$. Think of K as arising from a defining sequence $\{C_k\}$, so that its points are naturally and uniquely expressed as $\bigcap D(i_1, i_2, ..., i_k)$, where $i_k \in \{0, 1\}$ and, in addition, for all k

$$D(i_1, ..., i_k) \text{ is a component of } C_k$$

and

$$D(i_1, ..., i_k, i_{k+1}) \subset D(i_1, ..., i_k).$$

Say that a (nonempty) compact subset A of K is *admissible* if, whenever A intersects $D(i_1, ..., i_k)$ and k is odd, then A intersects both $D(i_1, ..., i_k, 0)$ and $D(i_1, ..., i_k, 1)$.

Name the switching homeomorphism s of the space $Z = \{0, 1\} \times \{0, 1\}$ to itself sending $\langle x, y \rangle$ to $\langle y, x \rangle$, and determine a mixing homeomorphism $S: K \to K$ using s as the action on each copy of $Z = Z_i$ in the representation $K = \prod_i Z_i$.

Lemma 12. *For every two admissible subsets A and A' of K, $S(A)$ intersects A'.*

Proof. Being nonempty, A' meets either $D(0)$ or $D(1)$, say $D(1)$. By definition of admissibility, A' must meet both $D(1, 0)$ and $D(1, 1)$. Similarly, A intersects both $D(i_1, 0)$ and $D(i_1, 1)$. The definition of S ensures that $S(A)$ and A' intersect the same component $D(1, i_1)$ of C_2.

This represents the inductive step of the argument eventually showing both $S(A)$ and A' intersect a common component of C_{2k}, from which Lemma 12 follows.   ■

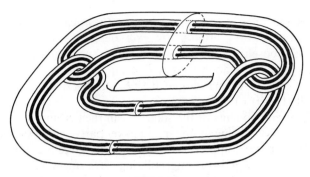

FIG. 9-14

Displayed in Fig. 9-14 are the first three stages of a defining sequence, regarded as stages 0, 1, and 2 to correspond with the abstract setup given prior to the statement of Lemma 12. With such structures a single application of the typical iteration procedure to the kth stage produce stages $k + 1$ and $k + 2$. Assuming the resulting defining sequence generates a Cantor set $\bigcap C_k$, label the components of C_k as $D(i_1, ..., i_k)$, where $i_k \in \{0, 1\}$ and

$$D(i_1, ..., i_k, i_{k+1}) \subset D(i_1, ..., i_k),$$

and say that a nonempty compact subset A of $\bigcap C_k$ is *admissible* if it has the intersection property previously specified.

Lemma 13. *Suppose H is a compact 2-manifold in a 2-cell B and $f: H \to T$ is a virtually I-essential map. Then $f(H)$ contains an admissible subset of $\bigcap C_k$.*

Proof. Invoke Proposition 3 to approximate f, fixing ∂H and $f^{-1}(\bigcap C_k)$, so that each component of $f^{-1}(C_k)$ is a 2-manifold. Let R_k denote the union of all components of $f^{-1}(C_k)$ on which f is virtually I-essential. According to the rephrased Lemma 11 (see also Exercise 3), R_1 is nonvoid. The concentricity of elements at the next stage guarantees that every component H_1 of R_1 contains components H_2 and H_2' from R_2 such that

$$f(H_2) \subset D(i_1, 0) \quad \text{and} \quad f(H_2') \subset D(i_1, 1).$$

The preceding argument serves as the inductive step establishing that each component of R_k contains a component of R_{k+1} and, whenever k is odd, that each component H_k of R_k contains at least two components of R_{k+1}, one mapped to each of the two components of C_{k+1} interior to the part of C_k touching $f(H_k)$. Consequently, $f(\bigcap R_k)$ is an admissible subset of $\bigcap C_k$. ∎

Now for Example 5.

Proposition 14. *The decomposition G_5 of Example 5 is nonshrinkable.*

Proof. Suppose the contrary and obtain a homeomorphism h of E^3 to itself, fixed outside the double torus T of Fig. 9-5, such that no element of G_5 intersects each of two meridional disks B_1 and B_2, spanning the upper and lower toroidal parts of T.

Each of the components of the stages C_k in the defining sequence for G_5 is a double torus consisting of upper and lower parts, with a stem connecting the two. The set of all upper parts (lower parts) gives a defining sequence quite like the one in Fig. 9-14; interpolating extra stages whose components in any piece of C_k consist of two solid tori, each incorporating a pair of the parallel, ramifying objects from C_{k+1}, makes the correspondence with Fig. 9-14 exact. Moreover, if the upper and lower parts are regarded as two copies of the same defining sequence, then the connection between components provided by the stems is precisely that specified by the mixing homeomorphism S of Lemma 12.

We plar. to adjust $h \mid B_1 \cup B_2$ to a map f transverse to certain parts of the defining sequence structures. The adjustment can easily be limited so no element of G_5 intersects both $f(B_1)$ and $f(B_2)$. Subject to that stipulation, modify $h \mid B_1$ to obtain $f \mid B_1$ transverse to just the upper parts. Nothing confines $f(B_1)$ to the upper part; like $h(B_1)$, it can wander anywhere through T. Temporarily inserting a 3-cell plug to fill the hole of the lower part, we enlarge T to a solid torus T' in which the upper defining sequence is topologically that of Fig. 9-14, and into which f maps B_1 in virtually I-essential fashion (because h does). By Lemma 13, $f(B_1)$ contains an admissible subset of the upper part. Similarly, $f(B_2)$ contains an admissible subset of the lower part. Combined with the observation above that the stems of the defining sequence geometrically realize the mixing homeomorphism S, Lemma 12 establishes the desired contradiction: some element of G_5 *does* intersect both $f(B_1)$ and $f(B_2)$. ∎

EXERCISES

1. Verify that G_1 is shrinkable.
2. Prove Proposition 9.
3. If $f\colon H \to T$ is a virtually I-essential map into the solid torus T of Fig. 9-1, then $f^{-1}(T_1 \cup T_2)$ contains a component H' such that $f\colon H' \to T_1 \cup T_2$ is virtually I-essential.
4. Suppose $\mathbb{S} = \{C_i\}$ is a defining sequence for a strongly shrinkable decomposition of an n-manifold M. If Q is a closed subset of M/G and $\mathcal{G}$ is the trivial extension of the decomposition $\{\pi^{-1}(q) \mid q \in Q\}$ induced over Q, then $\mathcal{G}$ is strongly shrinkable.

10. PRODUCTS OF DECOMPOSITIONS WITH A LINE

For every known cellular decomposition G of E^n, not only is the product of E^n/G with E^1 homeomorphic to E^{n+1} but, in addition, the decomposition $G \times E^1$ of $E^n \times E^1 = E^{n+1}$ is (strongly) shrinkable. What is more, the cellularity of G functions as an unnecessarily strong hypothesis. This section is devoted to an exploration of these facts in some specific cases; a more general and extensive exploration is reserved for later.

First, a result supplying a convenient method for detecting shrinkability in certain product decompositions. It is a direct consequence of Theorem 5.4.

Theorem 1. *Let G denote a usc decomposition of E^n associated with a defining sequence $\{C_1, C_2, ...\}$ and let $\pi': E^{n+1} \to (E^n \times E^1)/(G \times E^1)$ denote the related decomposition map. Then π' can be approximated by homeomorphisms if, for each $\varepsilon > 0$ and each integer k, there exists a homeomorphism h of E^{n+1} onto itself satisfying (for all $t \in E^1$ and $g \in G$)*

(a) $h \,|\, (E^n - C_k) \times E^1 = \mathrm{Id}$,
(b) $h(C_k \times \{t\}) \subset C_k \times [t - \varepsilon, t + \varepsilon]$, *and*
(c) $\mathrm{diam}\, h(g \times \{t\}) < 3\varepsilon$.

The initial goal here is to demonstrate how, even though G may be nonshrinkable, the product $G \times E^1$ might be shrinkable. Bing credits A. Shapiro (see Bing [7] and also Freedman [1, p. 422]) with discovery of this phenomenon for the decomposition of Example 9.7 involving the Whitehead continuum. The argument below, due to J. J. Andrews and L. R. Rubin [1], is as slick as any found in this area.

Lemma 2. *Suppose the interior of the solid torus $T = B^2 \times S^1$ contains a finite polyhedron Q such that the inclusion $Q \hookrightarrow T$ is null homotopic. Then there exists $\alpha > 0$ and there exists a homeomorphism f of $T \times E^1$ onto itself satisfying (for all $t \in E^1$)*

(a) $f \,|\, \partial B^2 \times S^1 \times E^1 = \mathrm{Id}$,
(b) $f(T \times \{t\}) \subset T \times [t - \alpha, t + \alpha]$, *and*
(c) $f(Q \times \{t\}) \subset B^2 \times \{\text{point}\} \times [t - \alpha, t + \alpha]$.

Proof. Prescribe coordinates of S^1 as e^{it} and adjust the situation so that $Q \subset B^* \times S^1$, where B^* denotes the subdisk of B^2 consisting of those points having norm $\leq \frac{1}{2}$.

Let $p: B^2 \times E^1 \to B^2 \times S^1$ denote the universal covering, explicitly given as $p(\langle b, t \rangle) = \langle b, e^{it} \rangle$, and let proj: $T \times E^1 \to T$ denote the projection. Define an embedding k of the total space $B^2 \times E^1$ into $B^2 \times S^1 \times E^1 = T \times E^1$ as $k(\langle b, t \rangle) = \langle b, e^{it}, t \rangle$, and note that proj $\circ\, k = p$.

According to the homotopy lifting theorem, the inclusion $j: Q \to T$ lifts to a map (embedding) $J: Q \to B^2 \times E^1$ such that $pJ = j$. Following this map J by projection to the E^1-factor gives a map $w: Q \to E^1$ such that for points $q = \langle b, s \rangle \in Q$

$$J(\langle b, s \rangle) = \langle b, w(\langle b, s \rangle) \rangle \in B^2 \times E^1.$$

The requirement $pJ = j$ leads to the crucial identity

(†) $e^{iw(\langle b, s \rangle)} = s.$

Moreover, $w: Q \to E^1$ satisfies the additional identity: for $q \in Q$

(††) $kJ(q) = \langle q, w(q) \rangle \in T \times E^1,$

because proj $\circ\, kJ = pJ = j$.

Certainly there exists some $\alpha > 0$ such that $w(Q) \subset [-\alpha, \alpha]$. By Tietze's extension theorem, w extends to a map $\psi: T \to [-\alpha, \alpha]$ such that $\psi(u) = 0$ for every $u \in \partial T = \partial B^2 \times S^1$.

The desired homeomorphism f is expressed as the composition of a lift and a twist. The lift $\lambda: T \times E^1 \to T \times E^1$ is defined as

$$\lambda(\langle u, t \rangle) = \langle u, \psi(u) + t \rangle.$$

To describe the twist, for each $t \in E^1$ specify a homeomorphism τ_t of T onto itself as

$$\tau_t(\langle b, s \rangle) = \begin{cases} \langle b, se^{-2it(1-|b|)} \rangle & \text{if} \quad \langle b, s \rangle \notin B^* \times S^1 \\ \langle b, se^{-it} \rangle & \text{if} \quad \langle b, s \rangle \in B^* \times S^1 \end{cases}$$

What τ_t does is to twist the inner torus $B^* \times S^1$ rigidly an amount that depends purely on t, with the twisting action tapering off on the region $T - (B^* \times S^1)$, reducing to the identity on ∂T. Then the twist τ of $T \times E^1$ is defined as

$$\tau(\langle u, t \rangle) = \langle \tau_t(u), t \rangle.$$

It should be obvious that both λ and τ are homeomorphisms; continuity is the only concern, for λ and τ have two-sided inverses given by $\langle u, t \rangle \to \langle u, t - \psi(u) \rangle$ and $\langle u, t \rangle \to \langle \tau_t^{-1}(u), t \rangle$, respectively.

Finally, it must be shown that the composition $f = \tau\lambda$ satisfies the conclusions of Lemma 2. First, given $\langle b, s \rangle \in \partial B^2 \times S^1$, we have $|b| = 1$, and we see that

$$f(\langle b, s, t \rangle) = \tau\lambda(\langle b, s, t \rangle) = \tau(\langle b, s, t \rangle)$$

$$= \langle \tau_t\langle b, s \rangle, t \rangle = \langle b, s, t \rangle.$$

Second, since τ preserves E^1-coordinates,

$$f(T \times \{t\}) = \tau\lambda(T \times \{t\}) \subset \tau(T \times [t - \alpha, t + \alpha]) = T \times [t - \alpha, t + \alpha].$$

Third, for $\langle b, s \rangle \in Q$

$$\begin{aligned}
\text{proj} \circ f(\langle b, s, t \rangle) &= \text{proj} \circ \tau\lambda(\langle b, s, t \rangle) \\
&= \text{proj} \circ \tau\langle b, s, t + \psi(\langle b, s \rangle)\rangle \\
&= \langle b, se^{-i(\psi(\langle b, s \rangle)+t)}\rangle \\
&= \langle b, se^{-i\psi(\langle b, s \rangle)}e^{-it}\rangle \\
&= \langle b, se^{-iw(\langle b, s \rangle)}e^{-it}\rangle \\
&= \langle b, e^{-it}\rangle, \qquad \text{by Condition } (\dagger).
\end{aligned}$$

In other words, $\text{proj} \circ f(Q \times \{t\}) \subset B^* \times \{e^{-it}\}$ (see Fig. 10-1). This establishes Conclusion (c) and completes the proof of Lemma 2. ∎

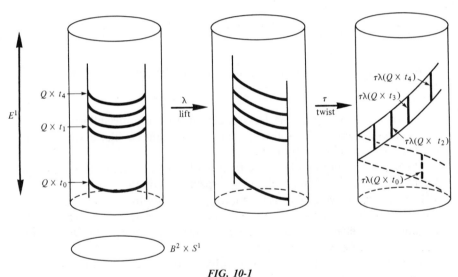

FIG. 10-1

Theorem 3. *Suppose that G is a usc decomposition of E^3 defined by solid tori (that is, G has a defining sequence $\{C_k\}$ and each component of each C_k is a solid torus) such that every inclusion of one solid torus into its predecessor in the nest is null homotopic. Then $(E^3/G) \times E^1$ is homeomorphic to E^4.*

Proof. Fix the defining sequence $\{C_k\}$ for G. We shall appeal to Theorem 1 to show that the map $\pi': E^3 \times E^1 \rightarrow (E^3 \times E^1)/(G \times E^1)$ can be uniformly approximated by homeomorphisms. The desired conclusion will then follow from Corollary 2.4A.

In order to verify that Theorem 1 applies, consider $\varepsilon > 0$ and an integer k. Focus on one component T of C_k. Let $Q = T \cap C_{k+1}$. Since, for each component T' of C_{k+1} contained in T, the inclusion $T' \hookrightarrow T$ is null homotopic, it follows that $Q \hookrightarrow T$ is null homotopic. Now inside $T = B^2 \times S^1$ we perform one modification: we find a slightly smaller copy $B^* \times S^1$ of T in Int T, with $B^* \times S^1 \supset Q$ and shrink this, via a homeomorphism θ fixed outside T, so close to a core (point $\times S^1$) that each diam $\theta(B^* \times \{s\}) < \varepsilon$. From now on we assume this situation prevailed at the onset and suppress θ.

Lemma 2 provides a positive number α and a homeomorphism f_T of $T \times E^1$ to itself. It is important to note that α could have been chosen small, had we redescribed the covering map $p: B^2 \times E^1 \to B^2 \times S^1$ at the beginning of the proof for Lemma 2. Alternatively, we can reparameterize the E^1-coordinate to accomplish this now, yielding $[t - \alpha, t + \alpha] \subset [t - \varepsilon, t + \varepsilon]$ for all $t \in E^1$. The required homeomorphism h is defined as f_T on each $T \times E^1$ and as the identity elsewhere. ∎

Corollary 3A. *For the decomposition G_6 of Section 9, $(E^3/G_6) \times E^1$ is homeomorphic to E^4.*

Corollary 3B. *For the decomposition G_7 of Section 9, $(E^3/G_7) \times E^1$ is homeomorphic to E^4.*

A striking property of the latter stems from the noncellularity of G_7: noncellular decompositions in some sense can become cellular upon forming the Cartesian product with a line.

The prototype among the exotic factors (that is, nonmanifold factors) of Euclidean space is Bing's dogbone space, Example 5 of Section 9. J. J. Andrews and M. L. Curtis [1] adapted Bing's argument establishing this fact to show that the product of E^1 with any decomposition of E^n whose only nondegenerate element is an arc yields a space homeomorphic to E^{n+1}. Our next goal is to derive their result.

An important step in the proof involves a special case of the following clever rearrangement device, originally given by V. L. Klee [1]. It reveals, among other things, why certain objects in $E^n \times \{0\}$, no matter how badly embedded there, are cellularly embedded in $E^n \times E^1$.

Proposition 4 (flattening). *Suppose $e_1: X \to E^m$ and $e_2: X \to E^n$ are both embeddings of a compact space X. Define embeddings $f_1, f_2,$ and d of X in $E^{m+n} = E^m \times E^n$ as $f_1(x) = \langle e_1(X), 0 \rangle$, $f_2(x) = \langle 0, e_2(x) \rangle$, and $d(x) = \langle e_1(x), e_2(x) \rangle$, respectively. Then $f_1, f_2,$ and d are equivalent embeddings; that is, there is a homeomorphism θ_i of E^{m+n} onto itself such that $\theta_i f_i = d$ $(i = 1, 2)$.*

Proof. The proof is an elementary consequence of Tietze's extension theorem. To see, for instance, that f_1 and d are equivalent, determine an n-cube C in E^n large enough to contain $e_2(X)$, and extend $e_2(e_1)^{-1} | e_1(X)$ to a map $\mu: E^m \to C$. Define $\theta_1: E^m \times E^n \to E^m \times E^n$ as

$$\theta_1(\langle z_1, z_2 \rangle) = \langle z_1, z_2 + \mu(z_1) \rangle,$$

where addition in E^n is defined coordinatewise. Certainly θ_1 is continuous, and it is a homeomorphism because it has for an inverse the (continuous) function $\langle z_1, z_2 \rangle \to \langle z_1, z_2 - \mu(z_1) \rangle$. For each $x \in X$,

$$\theta_1 f_1(x) = \theta_1(\langle e_1(x), 0 \rangle) = \langle e_1(x), \mu e_1(x) \rangle$$
$$= \langle e_1(x), e_2 e_1^{-1} e_1(x) \rangle = d(x).$$

Corollary 4A. *Every arc in $E^n \times \{0\} \subset E^{n+1}$ is flat in E^{n+1}.*

Corollary 4B. *Every k-cell in $E^n \times \{0\} \subset E^{n+k}$ is flat in E^{n+k}.*

Corollary 4C. *Every Cantor set in $E^n \times \{0\} \subset E^{n+1}$ is tame in E^{n+1}.*

What comes next pertains to the Andrews and Curtis result. We suppose A to be an arc in E^n and $\varepsilon > 0$. We specify a homeomorphism α of I onto A and then partition I by points $\{t_i \mid i = 0, 1, \dots, m + 1\}$, with $\alpha([t_{i-1}, t_i])$ small enough to be contained in an open subset U_i of E^n ($i = 1, \dots, m + 1$) such that

$$U_i \cap U_j \neq \varnothing \text{ iff } |i - j| \leq 1$$

and

$$\text{diam } (U_i \cup U_{i+1} \cup U_{i+2} \cup U_{i+3}) < \varepsilon \qquad (i = 1, \dots, m - 2).$$

To provide control in the E^1-direction, we identify some intervals and related sets: for $i \in \{1, \dots, m - 1\}$ let $J_i = [i, 2m - i]$ and then let $L_i = J_i - \text{Int } J_{i+1}$ when $i < m - 1$.

It is a consequence of the Klee trick that each level arc $A \times \{t\} \subset E^{n+1}$ can be shrunk to small size. To support the work of shrinking all level arcs simultaneously, we begin by showing how to combine a vertical compression with the pinching of a certain level arc to bring about a shrinking of the product of J_{i+1} and a subarc of A.

Lemma 5. *Let V_i be a neighborhood of $\alpha([0, t_i])$ in E^n ($i = 1, 2, \dots, m - 2$). Then there exists a homeomorphism h_i of $E^{n+1} = E^n \times E^1$ onto itself satisfying*

(a) $h_i | E^{n+1} - (V_i \times J_i) = \text{Id}$,
(b) $h_i | \alpha([t_i, 1]) \times E^1 = \text{Id}$, *and*
(c) $h_i(\alpha([0, t_i]) \times J_{i+1}) \subset U_{i+1} \times J_i$.

Proof. By Corollary 4A the arc $A \times \{i + 1\} = \alpha(I) \times \{i + 1\}$ is flat in E^{n+1}. This makes it possible to squeeze the flat subarc $\alpha([0, t_i]) \times \{i + 1\}$ close to the point $\alpha(t_i) \times \{i + 1\}$, by means of a homeomorphism μ of E^{n+1} such that

$$\mu \,|\, E^{n+1} - (V_i \times J_i) = \mathrm{Id},$$

$$\mu \,|\, \alpha([t_i, 1]) \times E^1 = \mathrm{Id},$$

and

$$\mu(\alpha([0, t_i]) \times \{i + 1\}) \subset U_{i+1} \times J_i.$$

It follows that $\mu^{-1}(U_{i+1} \times \mathrm{Int}\, J_i) \supset (\alpha([0, t_i]) \times \{i + 1\}) \cup (\alpha(t_i) \times J_{i+1})$. As a result, there exists $\delta > 0$ such that $\mu^{-1}(U_{i+1} \times \mathrm{Int}\, J_i)$ also contains $\alpha([t_i - \delta, t_i]) \times J_{i+1}$. Now one can produce a homeomorphism v of E^{n+1} itself, which acts there as a vertical compression, affecting only the E^1-coordinate and satisfying

$$v \,|\, E^{n+1} - (V_i \times J_i) = \mathrm{Id},$$

$$v \,|\, \alpha([t_i, 1]) \times E^1 = \mathrm{Id},$$

and

$$v(\alpha([0, t_i]) \times J_{i+1}) \subset \mu^{-1}(U_{i+1} \times \mathrm{Int}\, J_i).$$

It can be obtained as follows: name a positive number $d < 1$ such that

$$\mu^{-1}(U_{i+1} \times \mathrm{Int}\, J_i) \supset \alpha([0, t_i]) \times [i + 1, i + 1 + d];$$

define a continuous function $s: E^n \to [i + 1 + d, 2m - i - 1]$ sending points of $E^n - V_i$ and points of $\alpha([t_i, 1])$ to $2m - i - 1$ while sending points of $\alpha([0, t_i - \delta])$ to $i + 1 + d$; finally, define $v: E^{n+1} \to E^{n+1}$ as the identity above $E^n \times \{2m - i\}$ and below $E^n \times \{i + 1\}$, so as to send each point $\langle p, 2m - i - 1 \rangle$ to $\langle p, s(p) \rangle$, and so as to be the obvious linear map on the remaining vertical intervals $\{p\} \times J_{i+1}$ and $\{p\} \times [2m - i - 1, 2m - i]$. The effect of v is illustrated in Fig. 10-2.

In order to complete this proof, simply define h_i as μv. Then

$$h_i(\alpha([0, t_i]) \times J_{i+1}) = \mu v(\alpha([0, t_i]) \times J_{i+1}) \subset \mu \mu^{-1}(U_{i+1} \times J_i),$$

as desired. The other properties required of h_i should be transparent. ∎

Lemma 6. *There exists a homeomorphism λ of E^{n+1} onto itself satisfying*

(a) $\lambda \,|\, E^{n+1} - \bigcup_{i=1}^{m-2}(U_i \times J_i) = \mathrm{Id}$,

(b) $\lambda(\alpha([0, t_{i+1}]) \times L_i) \subset (U_i \cup U_{i+1}) \times J_{i-1}$ *for $i \in \{1, ..., m - 2\}$*,

(c) $\lambda(\alpha([0, t_m]) \times J_{m-1}) \subset (U_{m-1} \cup U_m) \times J_{m-2} \subset (U_{m-1} \cup U_m) \times J_1$.

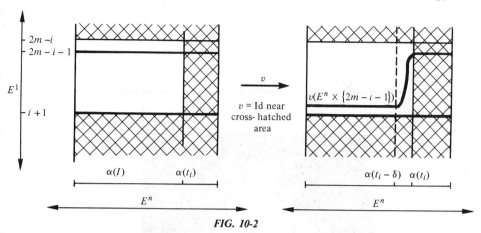

FIG. 10-2

Proof. Here λ will arise as a composition $h_1 h_2 \cdots h_{m-2}$ of homeo-morphisms resulting from Lemma 5. To begin, one obtains h_1 that way, using the neighborhood $V_1 = U_1$ of $\alpha([0, t_1])$.

Since h_1 acts as the identity on $\alpha([t_1, t_2]) \times E^1$ and sends $\alpha([0, t_1]) \times J_2$ into $U_2 \times J_1$, there exists a neighborhood V_2 of $\alpha([0, t_2])$ in $U_1 \cup U_2$ such that $h_1(V_2 \times J_2) \subset U_2 \times J_1$. One applies Lemma 5 again, with this neighborhood V_2, to obtain h_2.

The iterative step retraces the pattern just laid down for obtaining h_2. After h_{i-1} has been procured subject to the conditions

$$h_{i-1} \,|\, \alpha([t_{i-1}, 1]) \times J_{i-1} = \mathrm{Id}$$

and

$$h_{i-1}(\alpha([0, t_{i-1}]) \times J_i) \subset U_i \times J_{i-1},$$

[conclusions (b) and (c) of Lemma 5], one determines a neighborhood V_i of $\alpha([0, t_i])$ in $U_1 \cup \cdots \cup U_i$ such that $h_{i-1}(V_i \times J_i) \subset U_i \times J_{i-1}$ and then applies Lemma 5 with this neighborhood V_i to obtain h_i.

It should be obvious from the choices of V_i in $U_1 \cup \cdots \cup U_i$ and conclusion (a) of Lemma 5 that $\lambda = h_1 h_2 \cdots h_{m-2}$ satisfies conclusion (a) above (see Fig. 10-3). In analyzing conclusions (b) and (c) it is useful to recall that $(U_1 \cup \cdots \cup U_i)$ and $(U_{i+2} \cup \cdots \cup U_{m+1})$ are disjoint. Based on the choices of V_j, (a) of Lemma 5 then yields

(†) $h_j \,|\, (U_{i+2} \cup \cdots \cup U_{m+1}) \times E^1 = \mathrm{Id}$ whenever $j \le i$, as well as

$$h_j \,|\, E^n \times (E^1 - J_i) = \mathrm{Id} \text{ whenever } j \ge i.$$

Since $L_i \subset \mathrm{Cl}(E^1 - J_{i+1})$, the latter implies

(††) $$h_j \,|\, E^n \times L_i = \mathrm{Id} \text{ whenever } j > i.$$

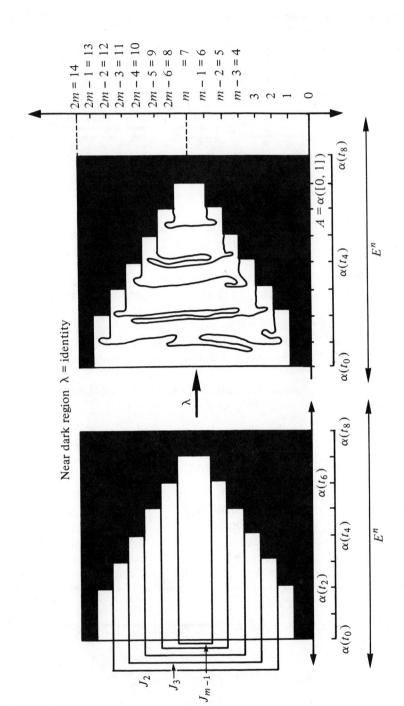

FIG. 10-3

To see why Conclusion (c) holds, note that

$$\lambda(\alpha([0, t_{m-2}]) \times J_{m-1}) = h_1 h_2 \cdots h_{m-2}(\alpha([0, t_{m-2}]) \times J_{m-1})$$

$$\subset h_1 h_2 \cdots h_{m-3}(U_{m-1} \times J_{m-2}) \quad \text{by (c) of Lemma 5}$$

$$\subset U_{m-1} \times J_{m-2} \quad \text{by (†).}$$

In addition, by conclusion (b) of Lemma 5,

$$\lambda(\alpha([t_{m-2}, t_m]) \times J_{m-1}) = \alpha([t_{m-2}, t_m]) \times J_{m-1}$$

$$\subset (U_{m-1} \cup U_m) \times J_{m-1},$$

and these two inclusions readily combine to yield (c).

In order to verify conclusion (b), it is advantageous to first observe, by conclusion (a) of Lemma 5, that $h_i(\alpha([0, t_i]) \times J_i) \subset V_i \times J_i$. Then (for $i > 1$)

$$\lambda(\alpha([0, t_{i+1}]) \times L_i) = h_1 h_2 \cdots h_{m-2}(\alpha([0, t_{i+1}]) \times L_i),$$

$$= h_1 h_2 \cdots h_i(\alpha([0, t_{i+1}]) \times L_i) \quad \text{by (††),}$$

$$= h_1 h_2 \circ \cdots \circ h_i(\alpha([0, t_i]) \times L_i)$$

$$\cup h_1 h_2 \circ \cdots \circ h_i(\alpha([t_i, t_{i+1}]) \times L_i)$$

$$\subset h_1 h_2 \circ \cdots \circ h_i(\alpha([0, t_i]) \times J_i) \cup (U_{i+1} \times J_i)$$

$$\text{by (b) of Lemma 5,}$$

$$\subset h_1 h_2 \circ \cdots \circ h_{i-1}(V_i \times J_i) \cup (U_{i+1} \times J_i) \quad \text{as above,}$$

$$\subset h_1 h_2 \circ \cdots \circ h_{i-2}(U_i \times J_{i-1}) \cup (U_{i+1} \times J_i)$$

$$\text{by choice of } V_i,$$

$$\subset (U_i \times J_{i-1}) \cup (U_{i+1} \times J_i) \quad \text{by (†)}$$

$$\subset (U_i \cup U_{i+1}) \times J_{i-1}.$$

Why the condition also holds when $i = 1$ should be clear to anyone who has read this far. ■

Theorem 7. *For each arc A in E^n, $(E^n/G_A) \times E^1$ is homeomorphic to E^{n+1}.*

Proof. This will follow from Corollary 2.4A and Theorem 1, once we construct a homeomorphism h of E^{n+1} to itself satisfying the three conditions:

$$h \,|\, (E^n - N(A; \varepsilon)) \times E^1 = \text{Id},$$

$$h(E^n \times \{t\}) \subset E^n \times [t - \varepsilon, t + \varepsilon],$$

and

$$\text{diam } h(A \times \{t\}) < 3\varepsilon, \quad \text{for each } t \in E^1.$$

Such a homeomorphism can be fabricated by exploiting the structures named for Lemmas 5 and 6, craftily pieced together. To explain formally, we let k range over the integers and we set

$$D_k = \bigcup_{i=1}^{m=2} (U_i \times [2mk + i, 2mk + 2m - i]) \quad \text{and}$$

$$D_k' = \bigcup_{i=1}^{m=2} (U_i \times [2mk + m + i, 2mk + 3m - i]).$$

These domains have been designed so no D_k meets any D_j'. This is best understood by studying Fig. 10-4. For details in a typical instance, consider $\langle x, t \rangle \in D_0$ where $t \le m$. Choose the least integer i such that $x \in U_i$. Then $t \ge i$. The only set D_k' that could possibly contain $\langle x, t \rangle$ is D_{-1}'. If that were the case, since $x \in U_{m+2-j}$ where $j = m + 2 - i$ and perhaps where $j = m + 1 - i$ as well, the definition of D_{-1}' would force

$$t \le m - (m + 1 - i) = i - 1 < i,$$

a contradiction.

For $k \in \mathbb{Z}$ and $i \in \{1, 2, \ldots, m - 1\}$ define

$$J_i' = \bigcup_k [2mk + i, 2mk + 2m - i]$$

$$P_i' = \bigcup_k [2mk + m + i, 2mk + 3m - i],$$

and for $i < m - 1$ let $L_i' = J_i' - \text{Int } J_{i+1}'$ and $Q_i' = P_i' - \text{Int } P_{i+1}'$. In addition, let $D = \bigcup_k D_k$ and $D' = \bigcup_k D_k'$. According to Lemma 6 there exist homeomorphisms λ_R for the J_i' and L_i' and λ_L for the P_i' and Q_i', each given as a translate of the Lemma 6 homeomorphism λ to the appropriate components, satisfying

(1) $\lambda_R \,|\, E^{n+1} - D = \text{Id}$ and
     $\lambda_L \,|\, E^{n+1} - D' = \text{Id}$;

(2) $\lambda_R(\alpha([0, t_{i+1}]) \times L_i') \subset (U_i \cup U_{i+1}) \times J_i'$ and
     $\lambda_L(\alpha([t_{m-i}, 1]) \times Q_i') \subset (U_{m-i+1} \cup U_{m-i+2}) \times P_i'$;

(3) $\lambda_R(\alpha([0, t_m]) \times J_{m-1}') \subset (U_{m-1} \cup U_m) \times J_i'$ and
     $\lambda_L(\alpha([t_1, 1]) \times P_{m-1}') \subset (U_2 \cup U_3) \times P_i'$.

Now define h as $\lambda_R \lambda_L$. The resultant shrinking depends on the explicit juxtaposition of right and left stacks. For example, if $t \in \text{Int } L_i'$, then $t \in Q_j'$ where $j = m - 1 - i$. Thus, by (2)

$$h(A \times \{t\}) = \lambda_R \lambda_L(\alpha([0, t_{i+1}]) \times L_i') \cup \lambda_R \lambda_L(\alpha([t_{m-j}, 1]) \times Q_j')$$

$$\subset [(U_i \cup U_{i+1}) \times E^1] \cup [(U_{i+2} \cup U_{i+3}) \times E^1].$$

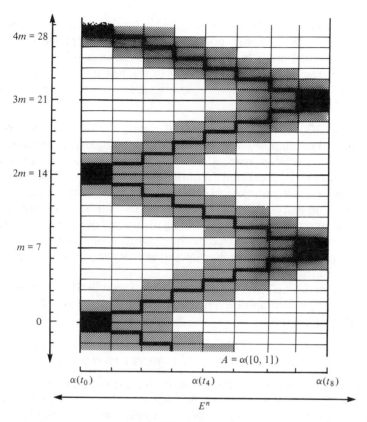

$A = \alpha([0, 1])$

$\alpha(t_0)$ $\alpha(t_4)$ $\alpha(t_8)$

E^n

FIG. 10-4

Of course, neither λ_R nor λ_L moves points vertically more than $2m$, which, because D and D' are disjoint, implies

$$h(A \times \{t\}) \subset (U_i \cup U_{i+1} \cup U_{i+2} \cup U_{i+3}) \times [t - 2m, t + 2m].$$

Initially the U_i's were chosen so that the diameter of any four successive ones was small. To complete the proof, all that remains is to rescale the E^1-coordinate with $2m < \varepsilon$. ■

Corollary 7A. *For each arc A in S^n, $(S^n/G_A) \times E^1$ is homeomorphic to $S^n \times E^1$.*

Corollary 7B. *For each arc A in S^n, the suspension of (S^n/G_A) is homeomorphic to S^{n+1}.*

Corollary 7C. *For each $n \geq 3$ S^n contains a wild arc α such that $S^n - \alpha$ fails to be simply connected.*

Proof. Such an arc in S^3 was described in Section 9. Generally, given an arc A in S^{n-1} ($n > 3$) with nonsimply connected complement, one can produce a related arc α in S^n, considered as the suspension of S^{n-1}/G_A, where α corresponds to the suspension of the bad point. Then $S^n - \alpha$ is topologically equivalent to $(S^{n-1} - A) \times (-1, 1)$, and neither space is simply connected. ∎

Corollary 7C, due to Morton Brown [3], clearly displays how pathological, nonmanifold decomposition spaces can lead to pathological embeddings of nice objects in Euclidean space. Conversely, when the topic of cellularity was introduced and the generalized Schönflies theorem was established, it should have been clear that pathological embeddings of cells do lead to pathological decomposition spaces. As a consequence, we have further data revealing the intimate connections between embedding theory and decomposition theory. Those connections are not limited to the production of pathology—after all, the concept of cellularity evolved as the empowering device for detecting flatly embedded $(n - 1)$-spheres in E^n.

Suspending a k-cell in S^n to obtain a $(k + 1)$-cell in S^{n+1} combines with the suspending a decomposition space technique of Corollary 7C to yield the improvement below.

Corollary 7D. *For $n \geq 3$ and each integer k with $1 \leq k \leq n$, S^n contains a k-cell K such that $S^n - K$ fails to be simply connected.*

The proof of Theorem 7 develops uniform controls on a shrinking process to bring about the desired conclusion but falls short of incorporating the cover controls required by the shrinkability criterion. Those regulations can be managed with just a little extra effort.

Theorem 8. *If A is any arc in E^n, then the decomposition $G_A \times E^1$ of E^{n+1} is strongly shrinkable.*

Outline of the Proof. Name a saturated open cover $\mathcal{U}$ of E^{n+1}, an arbitrary open cover $\mathcal{V}$, and an open set W containing $A \times E^1$. Find another saturated open cover $\mathcal{U}'$ star-refining $\mathcal{U}$, but exercise care so every $U' \in \mathcal{U}'$ intersecting $A \times E^1$ lies in W; moreover, construct $\mathcal{U}'$ so as to have an increasing collection $\{t_k \mid k \in \mathbb{Z}\}$ of real numbers for which each $A \times [t_{k-1}, t_k]$ is contained in some $U_k \in \mathcal{U}'$.

Since each arc $A \times \{t_k\}$ is flat, there exists a homeomorphism f_1 of E^{n+1} to itself shrinking each $A \times \{t_k\}$ into some $V_k \in \mathcal{V}$ while moving no point outside $\bigcup(U'_k \cap U'_{k+1})$ (the collection $\{U'_k \cap U'_{k+1}\}$ should be restricted so as to be pairwise disjoint). For each $k \in \mathbb{Z}$ determine $\delta_k > 0$ and a neighborhood W_k of A in E^n such that

$$W_k \times [t_{k-1}, t_k] \subset U'_k \quad \text{and} \quad f_1(W_k \times [t_k - 3\delta_k, t_k + 3\delta_k]) \subset V_k.$$

Let $\alpha_k = \min\{\delta_{k-1}, \delta_k\}$. Now use a variation to Theorem 7, obtained after identifying (t_{k-1}, t_k) with E^1 by then allowing the homeomorphisms λ_R and λ_L to operate on just a finite number of the domains D_k and D_k' named therein, to produce a homeomorphism h_k of $E^n \times (t_{k-1}, t_k)$ to itself satisfying

$$h_k \,|\, (E^n - W_k) \times (t_{k-1}, t_k) = \mathrm{Id},$$

$$h_k \,|\, E^n \times ((t_{k-1}, t_{k-1} + \alpha_k) \cup (t_k - \alpha_k, t_k)) = \mathrm{Id},$$

$$h_k(E^n \times \{t\}) \subset E^n \times [t - \alpha_k, t + \alpha_k]$$
$$\text{for all} \quad t \in (t_{k-1}, t_k),$$

$$\operatorname{diam} h_k(A \times \{t\}) < \varepsilon \qquad \text{for all}$$
$$t \in (t_{k-1} + 2\alpha_k, t_k - 2\alpha_k)$$

where ε is chosen to be very small relative to the cover of $f_1(\mathrm{Cl}(W_k)) \times [t_{k-1} + \alpha_k, t_k - \alpha_k])$ by $\mathcal{V}$. The combination of all such h_k, extended elsewhere via the identity, is a homeomorphism f_2 of E^{n+1}, and $f_1 f_2$ achieves the desired shrinking. ∎

As a corollary to the proof, one can derive an analogous result with S^1 replacing E^1.

Corollary 8A. *If A is an arc in E^n (or S^n), then the decomposition $G_A \times S^1$ of $E^n \times S^1$ (or $S^n \times S^1$) is strongly shrinkable.*

Theorem 8 enhances interest in the questions: given a compact subset X of E^n, under what conditions is $(E^n/G_X) \times E^1$ homeomorphic to E^{n+1}? When is the product decomposition $G_X \times E^1$ of E^{n+1} shrinkable? We know that the cellularity of X in E^n is a sufficient condition in the first case and that the cellularity of $X \times \{0\}$ in E^{n+1} is a necessary condition in the second. These stand as fundamentally important questions, to be readdressed in what lies ahead.

Historical Remarks. The exotic factors of E^4 resulting from Theorem 3 and 7 are unlike anything known in E^3. The absence of such phenomena in dimension 3 has a classical explanation: if $X \times E^1$ is homeomorphic to E^n, then X is what is called a generalized $(n - 1)$-manifold, and the only generalized 2-manifolds are genuine 2-manifolds—in case X is a noncompact, simply connected generalized 2-manifold, it must be E^2.

J. L. Bryant [1,3] has shown that, for each k-cell K in E^n, $(E^n/G_K) \times E^1$ is homeomorphic to E^{n+1}. In essence, his proof proceeds much like the one described here, squeezing out one coordinate of the k-cell at a time (this synopsis does depend on our current knowledge, not available at the time his proof was conceived, that each k-cell in $E^n \times \{0\}$ is flat in E^{n+1}).

There is an extensive history to the study of other exotic factors of E^{n+1} arising from decompositions of E^n. Recounted at length in (Daverman [6]), it is a subject of Section 26.

Concerning factors of cells, Bing [7] showed that a space X is a 3-cell if its product with I is a 4-cell. One can easily see that, given an arc A in the interior of I^n, $(I^n/G_A) \times I$ is I^{n+1} iff $I^n/G_A = I^n$. Kwun and Raymond [1] have proved that the product of I^n modulo an arc in its interior and I^2 is I^{n+2}. P. W. Harley [1] has shown that the product of I^n modulo an arc in its boundary and I is I^{n+1}.

In the manifold setting, V. Poenaru [1] and B. Mazur [1] separately have given examples of PL 4-manifolds that are not 4-cells whose products with I are 5-cells. Similar examples (in reality, compact, contractible n-manifolds with nonsimply connected boundary) have been given in case $n \geq 5$ by M. L. Curtis [1] and L. C. Glaser [1]; the product of any such space with I must be an $(n + 1)$-cell.

EXERCISES

1. If A represents an arc in an n-manifold M, show that $(M/G_A) \times E^1$ is homeomorphic to $M \times E^1$.
2. Prove Corollary 7D.
3. Prove Corollary 8A.
4. Let X be a compact subset of E^n such that $G_X \times S^1$ is a shrinkable decomposition of $E^n \times S^1$. Show that $(E^n/G_X) \times E^1$ is homeomorphic to E^{n+1}.
5. Let G be a usc decomposition of S^n such that H_G forms a null sequence of arcs. Show that $G \times S^1$ is a strongly shrinkable decomposition of $S^n \times S^1$. Show also that $(S^n/G) \times E^1 \approx S^n \times E^1$.

11. SQUEEZING A 2-CELL TO AN ARC

For any positive integer k let B^k denote the standard k-cell in E^k consisting of all points at distance no more than 1 from the origin, and let S^{k-1} denote its boundary. When $m < k$, clearly B^m can be regarded as a subset of B^k; in that case let $p \colon B^k \to B^m$ denote the vertical projection sending $\langle x_1, \ldots, x_k \rangle$ to $\langle x_1, \ldots, x_m \rangle$.

Suppose K is a k-cell in some n-manifold M. A map f of M to itself is said to *squeeze K to an m-cell* if there exist homeomorphisms h of B^k onto K and H of B^m onto $f(K)$ such that f carries $M - K$ homeomorphically onto $M - f(K)$ and $fh = Hp$. (See the accompanying diagram.) In particular, we say that f squeezes K to an m-cell [or to the image cell $f(K)$] *along h*. From the perspective of decomposition theory, whenever some map squeezes a k-cell K in M to an m-cell along the guiding homeomorphism h, the appropriate

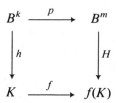

$$B^k \xrightarrow{\quad p \quad} B^m$$

(with vertical maps h and H, bottom row $K \xrightarrow{\quad f \quad} f(K)$)

decomposition of M to examine is G_f, the one induced by f; to accentuate the role of h we prefer to denote this decomposition as $h(G_p)^{\mathrm{T}}$, the trivial extension over M of the decomposition on K induced by ph^{-1}—in other words, the decomposition of M consisting of the singletons from $M - K$ and the sets $hp^{-1}(x)$, where $x \in B^m$.

The situation where the fountainhead cell K is two-dimensional will attract the bulk of our consideration. Under a mild assumption that arcs in K can be altered to arcs in K that are locally flat in M (which, in actuality, involves no restriction whatsoever), with an argument inspired by the one used in Theorem 10.7 and developed in Daverman and Eaton [1], we will show how such a 2-cell in M can be squeezed to a 1-cell (arc). As a corollary, we will obtain an alternate proof, in the case $k = 2$, of a result due to Bryant alluded to in Section 10, namely that for such 2-cells K, $(M/G_K) \times E^1$ is a manifold.

An enormous payoff accrues in the subsequent section, where work of C. H. Giffen, coupled with the aforementioned corollary, is capitalized on to demonstrate the existence of a noncombinatorial simplicial triangulation of S^n $(n > 4)$.

Now, about the flatness assumption, we say that a 2-cell K in a manifold M satisfies Axiom LF in M if, for each arc A in K with $A \cap \partial K \subset \partial A$ and each $\varepsilon > 0$, there is an ε-homeomorphism λ of K onto itself, fixed off the ε-neighborhood of A, such that $\lambda(A)$ is locally flat in M. Although we will have no cause to use the fact, every 2-cell in an n-manifold M satisfies Axiom LF; this was proved by R. H. Bing [10] for $n = 3$, by R. B. Sher [1] for $n = 4$, and by C. L. Seebeck III [1] for $n > 4$.

An explicit statement of the central result, around which this entire section revolves, is given below.

Theorem 1. *If D is a 2-cell satisfying Axiom LF in an n-manifold M, then D can be squeezed to an arc. Specifically, if h is a homeomorphism of B^2 onto D and $\varepsilon > 0$, then there exist a map $f: M \to M$ and another homeomorphism $h: B^2 \to D$ such that $p(h', h) < \varepsilon$, $h' \mid S^0 = h \mid S^0$, and f squeezes D to an arc along h'.*

Corollary 1A. *If D is a 2-cell satisfying Axiom LF in an n-manifold M, then $(M/G_D) \times E^1$ is an $(n + 1)$-manifold.*

See also Theorem 10.7 and Exercise 10.1.

A bizarre phenomenon comes up when we attempt to prove Theorem 1. Conferred upon us is a decomposition $h(G_p)^T$ of M into points and arcs. Unexpectedly, our instructions are not to show it is shrinkable (which it might not be), but instead to establish an existence statement: near $h(G_p)^T$ is another decomposition $h'(G_p)^T$ yielding M as the decomposition space [indeed, $h'(G_p)^T$ is shrinkable, as we could show later, but that is not a pressing issue here]. The ensuing lemma presents a suitable analog of the shrinkability criterion for the rather nebulous decomposition problem at hand.

Lemma 2. *Suppose D is a 2-cell satisfying Axiom LF in M, h is a homeomorphism of B^2 onto D, U is an open subset of M containing $h(B^2 - S^0)$, and $\varepsilon > 0$. Then there exist a homeomorphism $h_1: B^2 \to D$ and a homeomorphism F of M onto itself such that*:

(a) $F \,|\, M - U = \text{Id}$,

(b) $h_1 \,|\, S^0 = h \,|\, S^0$,

(c) $p(h_1, h) < \varepsilon$,

(d) $\text{diam } Fh_1(g) < \varepsilon$ *for all g in G_p*,

(e) $Fh_1(g) \subset N(h(g); \varepsilon)$, *and*

(f)   *to each $x \in M$ for which $F(x) \neq x$ there corresponds $g \in G_p$ such that* $\{x, F(x)\} \subset N(h(g); \varepsilon)$.

One-half the argument for Lemma 2 involves an arc-shrinking technique something like what was employed in Lemma 8.2. The other half entails a rearrangement of the homeomorphism from B^2 onto D, identifying the new decomposition $h'(G_p)^T$, which, when the rearrangement is properly meshed with the first technique, has been shrunk to small size. The next result focuses on the first half.

Lemma 3. *Suppose $r > 0$ and $H: [-r, r] \times I \to M$ is an embedding such that $H(\{0\} \times I)$ is locally flat, V is an open subset of M containing $H(\{0\} \times (0, 1])$,*

$$0 = t_0 < t_1 < t_2 < \cdots < t_{m-1} < t_m = 1,$$

is a partition of I, and $\delta > 0$. Then there exist a homeomorphism F of M to itself, fixed off $M - V$, and there exist numbers

$$0 = r_0 < r_1 < r_2 < \cdots < r_{m-1} < r_m < r$$

such that, for $j = 1, \ldots, m$

(a)   *F is fixed on*

$$\bigcup_j H((([r_j, r] \times [t_j, 1]) \bigcup ([-r, -r_j] \times [t_j, 1])),$$

(b) $FH([r_{j-1}, r_j] \times [t_{j-1}, 1])$ *is contained in the δ-neighborhood of* $H([r_{j-1}, r_j] \times [t_{j-1}, t_j])$, *and*

(c) $FH([-r_j, -r_{j-1}] \times [t_{j-1}, 1])$ *is contained in the δ-neighborhood of* $H([-r_j, -r_{j-1}] \times [t_{j-1}, t_j])$.

Proof. The idea is to shrink the locally flat arc $H(\{0\} \times I)$ as one would a telescope having m chambers. To that simple idea one must append regulators to curb the appropriate sets. Figure 11-1 suggests how they might appear.

Choose $r_m \in (0, r)$ so that

(i) $H([0, r_m] \times [t_{m-1}, 1])$ lies in the δ-neighborhood of $H([r_m, r] \times [t_{m-1}, 1])$, and

(ii) $H([-r_m, 0] \times [t_{m-1}, 1])$ lies in the δ-neighborhood of $H([-r, -r_m] \times [t_{m-1}, 1])$.

Let $W_m = N(H([-r_m, r_m] \times [t_{m-1}, 1]); \delta)$. Use the flatness of $H(\{0\} \times I)$ to find a homeomorphism ψ_m of M to itself, fixed outside $V \cap W_m$ and on

$$H([-r, r] \times I) - H([-r_m, r_m] \times [t_{m-1}, 1]),$$

shrinking $H(\{0\} \times [t_{m-1}, 1])$ into the δ-neighborhood of $H(\{0\} \times \{t_{m-1}\})$.

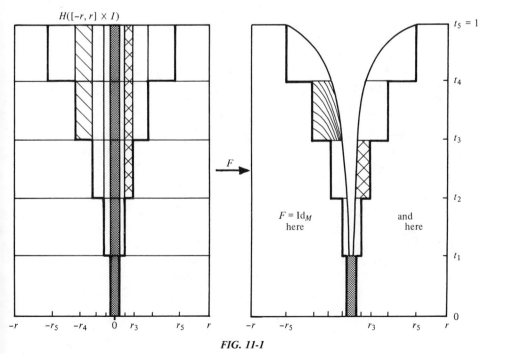

FIG. 11-1

Next, obtain $r_{m-1} \in (0, r_m)$ such that

(i) $\psi_m H([0, r_{m-1}] \times [t_{m-2}, 1])$ lies in the δ-neighborhood of $H([r_{m-1}, r] \times [t_{m-2}, t_{m-1}])$, and

(ii) $\psi_m H(\{-r_{m-1}, 0] \times [t_{m-2}, 1])$ lies in the δ-neighborhood of $H([-r, -r_{m-1}] \times [t_{m-2}, t_{m-1}])$.

Let $W_{m-1} = N(\psi_m H([-r_{m-1}, r_{m-1}] \times [t_{m-2}, 1]); \delta)$. Use flatness again to obtain a homeomorphism ψ_{m-1} shrinking the flat arc $\psi_m H(\{0\} \times [t_{m-2}, 1])$ into the δ-neighborhood of $H(\{0\} \times \{t_{m-2}\})$ while moving no point outside $V \cap W_{m-1}$ or on

$$H(([-r, r] \times I) - ([-r_{m-1}, r_{m-1}] \times [t_{m-2}, 1])).$$

Continuing in this fashion, one determines successive points

$$r_m > r_{m-1} > \cdots > r_{m-k}$$

and homeomorphisms $\psi_m, \psi_{m-1}, \ldots, \psi_{m-k}$ (for $k = 0, 1, \ldots, m-1$) of M to itself, each fixed outside V, such that

(i) $\psi_{m-k+1} \cdots \psi_m H([0, r_{m-k}] \times [t_{m-k-1}, 1])$ lies in the δ-neighborhood of $H([r_{m-k}, r] \times [t_{m-k-1}, t_{m-k}])$,

(ii) $\psi_{m-k+1} \cdots \psi_m H([-r, -r_{m-k}, 0] \times [t_{m-k-1}, 1])$ lies in the δ-neighborhood of $H([-r, -r_{m-k}] \times [t_{m-k-1}, t_{m-k}])$,

(iii) $\psi_{m-k} \cdots \psi_m H(\{0\} \times [t_{m-k-1}, 1])$ lies in the δ-neighborhood of $H(\{0\} \times \{t_{m-k-1}\})$, and

(iv) ψ_{m-k} is fixed on $\psi_{m-k+1} \cdots \psi_m H(([-r, r] \times I) - ([-r_{m-k}, r_{m-k}] \times [t_{m-k-1}, 1]))$ and outside $N(\psi_{m-k+1} \cdots \psi_m H([-r_{m-k}, r_{m-k}] \times [t_{m-k-1}, 1]); \delta)$.

Then the desired F is just the composition $\psi_1 \cdots \psi_m$.   ■

Proof of Lemma 2.   The argument is given in two steps. In Step 1 the disk D is sectioned into thin, "vertical" strips by arcs, which, after alteration to flat arcs, are squeezed to small size using Lemma 3. In Step 2 the resulting image disk is sectioned into small disks. Care must be exercised in Step 1 to ensure that the arcs of Step 2 can be realized as images of approximately vertical segments from B^2.

For what will be done a rectangular model for the 2-cell is preferable to the round model; hence, we choose $s \in (0, 1)$ and a first coordinate-preserving homeomorphism d of $[-s, s] \times I$ onto $p^{-1}([-s, s]) \subset B^2$, where s is so close to 1 that each component of $D - hd([-s, s] \times I)$ has diameter less than $\varepsilon/8$.

Step 1.   Applying Axiom LF, we find an integer $k > 0$, a partition

$$-s = s_0 < s_1 < \cdots < s_{2k} = s$$

of $[-s, s]$ and a homeomorphism λ of B^2 onto D satisfying

(1) λ agrees with h on $p^{-1}([-1, 1] - (-s, s))$,
(2) $\rho(\lambda, h) < \varepsilon/2$,
(3) $\lambda d(\{s_{2j}\} \times I)$ is flat, for $j = 1, ..., k - 1$, and
(4) diam $\lambda d([s_{i-1}, s_i] \times [(j - 1)/2k, j/2k]) < \varepsilon/8$ for $i, j = 1, 2, ..., 2k$.

For $j = 1, 2, ..., k - 1$, let $B_j = \{s_{2j}\} \times I$. Choose pairwise disjoint, open subsets $V_1, V_2, ..., V_{k-1}$ of U with

(5) $\lambda d(B_j) \subset V_j \subset N(d(B_j); \varepsilon/8)$ and
(6) $V_j \cap \lambda d(([-s, s_{2j-1}) \cup (s_{2j+1}, s]) \times I) = \varnothing$.

Reparameterize the various intervals $[s_{2j-1}, s_{2j+1}]$ as $[-r, r]$, with s_{2j} corresponding to 0, in order to put Lemma 3 into operation on the associated flat arcs $\lambda d(B_j)$, open sets V_j, partition

$$0 < 1/k < 2/k < \cdots < (k - 1)/k < 1$$

of I, and positive number $\gamma = \varepsilon/4$. The arcs are to be squeezed in alternating directions, from top toward bottom on one and from bottom toward top on those adjacent to it. Precisely, on the arc determined by B_j, when j is even use Lemma 3 with the embedding λd, and when j is odd, with λdu, where $u: [-s, s] \times I \to [-s, s] \times I$ is the inversion sending $\langle x, t \rangle$ to $\langle x, 1 - t \rangle$ (see Fig. 11-2). The resulting shrinkings in the mutually disjoint open sets V_j combine to give the desired homeomorphism F on M.

Step 2. At this stage the specific arcs $F\lambda d(B_j)$ have size less than ε, but the same is not necessarily valid for an arbitrary arc $F\lambda d(\{x\} \times I)$. In order to procure the required rearrangement, use Lemma 3 for $j = 0, 1, ..., k - 1$ to extract numbers

$$s_{2j} < r(j, 1) < r(j, 2) < \cdots < r(j, k - 1) < s_{2j+1} <$$

$$v(j, 1) < v(j, 2) < \cdots < v(j, k - 1) < s_{2j+2}$$

for which

(7) diam $F\lambda d(C) < \varepsilon$,

where C is a component of $([s_{2j}, s_{2j+2}] \times I) - \bigcup_i A_i$ and where A_i denotes the arc

$$(\{r(j, i)\} \times [i/k, 1]) \cup ([r(j, i), v(j, i)] \times \{i/k\}) \cup (\{v(j, i)\} \times [0, i/k])$$

in case j is an odd integer and the arc

$$(\{r(j, i)\} \times [0, 1 - i/k]) \cup ([r(j, i), v(j, i)] \times$$

$$\{1 - i/k\}) \cup (\{v(j, i)\} \times [1 - i/k, 1])$$

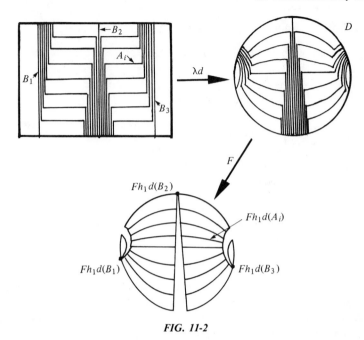

FIG. 11-2

in case j is even. Ignore the cases $j = 0$ and $j = k - 1$ for the moment. Elsewhere, in most instances such a set C is bounded by arcs A_i and A_{i+1}, and then application of (6) and Lemma 3 indicates $F\lambda d(C)$ lies in the γ-neighborhood of $\lambda d([r(j, i), v(j, i)] \times [i/k, (i + 1)/k])$, which, by (4), lies in the 2γ-neighborhood of the point $\lambda d(\{s_{2j+1}\} \times \{(2i + 1)/2k\})$; in the exceptional instances $F\lambda d(C)$ also lies in the 2γ-neighborhood of some point. In case $j = 0$ or $j = k - 1$, (7) holds because the entire image $F\lambda d([s_{2j}, s_{2j+2}] \times I)$ is small, by (4), (5), and the original choice of s.

The argument is completed by naming a homeomorphism h_1 of B^2 onto D such that

(8) h_1 agrees with λ on $p^{-1}([-1, 1] - (-s, s))$,

(9) $\rho(h_1, \lambda) < \varepsilon/2$,

(10) $h_1 d \,|\, B_j = \lambda d \,|\, B_j$, and

(11) $h_1 d(\{r(j, i)\} \times I) = \lambda d(A_i)$ for all j and for $i = 1, ..., k$.

Here h_1 is determined by a horizontal shear on $[s_{2j}, s_{2j+2}] \times I$, producing minor change in the second factor, composed with λd. That h_1 satisfies conclusion (b) follows from (1) and (8), and that it satisfies conclusion (c), from (2) and (9). The all-important conclusion (d) holds because by (10) and (11) each arc $Fh_1(g)$, unless it is near $h(S^0)$, is contained in some $F\lambda(C)$, which by (7) is sufficiently small; otherwise (1), (8), and the original choice of s certify $Fh_1(g)$ is small. Similarly, (e) holds. Finally, conclusion (f) follows quickly from (5) and the definition of F.   ■

Proof of Theorem 1.   Determine a sequence $\varepsilon_0, \varepsilon_1, \varepsilon_2, \ldots$ of positive numbers such that $\sum_j \varepsilon_j < \varepsilon$, and identify a decreasing sequence $U_0, U_1, U_2, \ldots$ of open subsets of M for which $\bigcap_j U_j = h(B^2 - S^0)$. Set $F_0 = \mathrm{Id}_M$ and $h_0 = h$. Apply uniform continuity and Lemma 3 repeatedly to obtain a sequence $\{F_j\}$ of homeomorphisms of M to itself and another sequence $\{h_j\}$ of homeomorphism of B^2 onto D such that, among other properties,

(a') $\quad F_{j+1} | M - U_j = F_j | M - U_j$,
(b') $\quad h_{j+1} | S^0 = h_j | S^0$,
(c') $\quad \rho(h_{j+1}, h_j) < \varepsilon_j$,
(d') $\quad \mathrm{diam}\, F_{j+1} h_{j+1}(g) < \varepsilon_j$, for all g in G_p,
(e') $\quad F_{j+1} h_{j+1}(g) \subset N(F_j h_j(g); \varepsilon_j)$, and
(f') $\quad$ to each $x \in M$ for which $F_{j+1}(x) \neq F(x)$ there corresponds $g_j \in G_p$
such that $\{F_{j+1}(x), F_j(x)\} \subset N(F_j(g_j); \varepsilon_j)$.

It follows from (d'), (e'), and (f') that $\{F_j\}$ is a Cauchy sequence converging to a map $f: M \to M$ which, by (a'), sends $M - D$ homeomorphically onto $M - f(D)$. Imposing additional controls in the course of the construction, we can limit the adjustment allowed in the sequence $\{h_j\}$ so that it converges to a homeomorphism (and not just a map) h' of B^2 onto D. Moreover, (d') implies that, for each $g \in G_p$, $fh'(g)$ is a point. Finally, we can wield controls similar to those on $\{h_j\}$ in combination with (e') to ensure that $fh'(g)$ and $fh'(g')$ are distinct points whenever g and g' are distinct elements of G_p. This demonstrates that D is squeezed to an arc along h', as required.   ■

Notes.   In 3-manifolds, all 2-cells and 3-cells can be squeezed to 1-cells and 2-cells (Daverman–Eaton [1]). However, in n-manifolds, $n > 3$, there exist k-cells ($k > 2$) that cannot be squeezed to lower-dimensional cells (Daverman [3]). Conditions under which a given k-cell in an n-manifold M can be squeezed are developed in Bass [1].

EXERCISES

1. Find an embedding h of B^2 into E^3 for which $h(B^2)$ cannot be squeezed to an arc along h (Hint: see Example 9.5).
2. Suppose K is a k-cell ($k > 2$) in E^n such that, for each $(k - 1)$-cell R standardly embedded in K with $R \cap \partial K = \partial R$, there exist arbitrarily small homeomorphisms λ of K to itself, fixed except near R, such that $\lambda(R)$ is flat in E^n. Show that K can be squeezed to a $(k - 1)$-cell.

12. THE DOUBLE SUSPENSION OF A CERTAIN
HOMOLOGY SPHERE

In 1963 J. Milnor (see Lashof [1]) listed what he considered to be the seven
most difficult and important problems in geometric topology. Heading that
list was the following question: if H^3 denotes a non-simply connected
homology 3-sphere, is the double suspension of H^3 homeomorphic to S^5?
(By a homology n-sphere we mean a closed n-manifold having the same
homology as S^n.) Slightly more specific and almost equally important is the
question addressed here:

The Double Suspension Problem. *Does there exist a non-simply connected
PL homology n-sphere H^n whose double suspension $\Sigma^2 H^n$ is S^{n+2}?*

Its significance lodges in the subject of manifold structures; combinatorial,
or PL, triangulations are abundant, advantageous, desirable. Anyone
attempting to understand whether an arbitrary simplicial triangulation of a
manifold is necessarily PL immediately confronts the double suspension
problem. If the answer to it is affirmative, then the simplicial triangulation
of S^{n+2} arising by double suspending the one given on H^n cannot be PL, since
the link of any 1-simplex in the suspension circle will be a copy of H^n, which
is not even topologically equivalent to S^{n+2}, let alone being PL equivalent
to it.
     The first affirmative solution, stemming from methods highly reminiscent
of the visual 3-dimensional shrinking techniques of Bing, was discovered in
1975 by R. D. Edwards [3]. After several stages of development, led by
Edwards and also J. W. Cannon [5], the problem has been completely settled.

Theorem (double suspension). *The double suspension $\Sigma^2 H^n$ of every
homology n-sphere H^n is homeomorphic to S^{n+2}.*

     Closer study of this rich theorem will be made in Section 24, after
Edwards's cell-like approximation theorem has been established, and the
unrestricted version will be derived in Section 40, after some resolution
results have been laid out. At this point the goal is a more limited one, that
of setting forth an affirmative solution to the double suspension problem.
The argument, quite different from Edwards's, is due to C. H. Giffen [1],
who in the early 1960s had made a clever observation, still a virtual secret
at the time of Edwards's work, that would have combined with known
decomposition results to solve the existence problem much sooner. Based on
what Giffen aptly calls a shift-spinning technique, his observation leads to
a short, elegant, and, for us, elementary proof of the momentous fact that
noncombinatorial triangulations do exist.

The double suspension problem is linked to decomposition theory in a straightforward way. For any particular homology sphere H^n, a metamorphosis of the double suspension problem to a decomposition problem can be bolstered by (1) finding a compact $(n + 1)$-manifold W^{n+1} bounded by H^n and (2) identifying a pseudo-spine X of W^{n+1}, i.e. a compact subset X of Int W^{n+1} such that $W^{n+1} - X$ is homeomorphic to $\partial W^{n+1} \times [0, 1) = H^n \times [0, 1)$. Of interest then is the decomposition G_X of W^{n+1}, where G_X has X as its only nondegenerate element, for clearly W^{n+1}/G_X is equivalent to cH^n, the cone on H^n. When H^n is not simply connected, there is no chance for cH^n to be a manifold, but the same cannot be said of

$$(cH^n) \times E^1 = (W^{n+1}/G_X) \times E^1.$$

The fundamental question in the double suspension problem is whether $\Sigma^2 H^n$ is a manifold. The potential nonmanifold set there lies in the suspension circle, along which symmetry properties reveal $\Sigma^2 H^n$ to be locally homeomorphic to (Int cH^n) $\times E^1$, where Int cH^n denotes the open cone. Consequently, to determine whether $\Sigma^2 H^n$ is an $(n + 2)$-manifold, one must decide whether $(W^{n+1}/G_X) \times E^1$ is an $(n + 2)$-manifold. This was precisely Edwards's original strategy: working with a certain W^4 bounded by H^3, he went to great lengths to show $G_X \times E^1$ was a shrinkable decomposition of $W^4 \times E^1$ when X was Zeeman's dunce hat, a compact contractible 2-complex. Giffen's observation provides an even simpler pseudo-spine, a 2-cell D, in the same manifold W^4. Then, according to Theorem 11.1 and Theorem 10.7, $(W^4/G_D) \times E^1$ is a 5-manifold.

It should be added that in what follows Giffen's work has been narrowed to fulfill the aim here of describing a single, specific example.

Proposition 1 (Giffen). *Let Γ denote the simple closed curve in $S^1 \times B^2$ shown in Fig. 12-1. Consider the first coordinate-preserving inclusion*

$$S^1 \times B^2 \subset S^1 \times S^2 = S^1 \times \partial B^3 \subset S^1 \times B^3.$$

Form a 4-manifold W^4 by attaching a 2-handle $h = B^2 \times B^2$ to $S^1 \times B^3$ along a regular neighborhood of Γ in $S^1 \times B^2$. Then there exists a 2-cell D in Int W^4 such that D is a locally polyhedral subset of W^4 at each point of Int D and D is a pseudo-spine for W^4.

A few comments about the manifold W^4 and its boundary H^3 are in order. Historically they were conceived for a different purpose by B. Mazur [1], who introduced them to construct an involution on S^4 having the nonsimply connected homology 3-sphere H^3 as its fixed point set. Previously M. H. A. Newman [1] had constructed similar involutions on S^n for $n > 4$; Newman readily described how such homology $(n - 1)$-spheres arise in S^n, just by taking regular neighborhoods of PL embedded, homologically trivial,

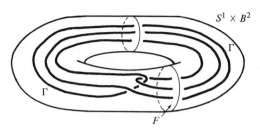

FIG. 12-1

nonsimply connected 2-complexes. Mazur built his objects in limbo and showed they could be embedded in S^4. No matter how the 2-handle h is attached to Γ in $S^1 \times B^3$, the resulting 4-manifold W^4 is contractible, essentially because the core Γ of the attaching region winds once homotopically (though not geometrically) around $S^1 \times B^2 \subset S^1 \times \partial B^3$. As a result, a homological computation shows or duality (Spanier [1, p. 305]) implies that H^3 is a homology 3-sphere. For those who know what the dunce hat is, it should not be hard to spot a dunce hat Z to which W^4 collapses. That observation led to Mazur's involution on S^4. Because $Z \times I$ is collapsible (Zeeman [1]), so is $W^4 \times I$, since

$$W^4 \times I \searrow Z \times I \searrow \text{point.}$$

Therefore, $W^4 \times 1$ is a 5-cell and its boundary S^4 is represented as the double of W^4 along $H^3 = \partial W^4$. Finally, and of utmost importance, if the 2-handle h is attached to $S^1 \times B^3$ with no unexpected twisting of the B^2-factor, one can obtain a presentation for the fundamental group G of H^3 as

$$G = \langle a, b : a^7 = b^5, b^4 = a^2 b a^2 \rangle,$$

which Mazur showed to be nontrivial.

Before looking at the proof of Proposition 1, let us see how it disposes of the double suspension problem. Almost all the machinery is already in place, with the disk-squeezing result of Section 11 and with Theorem 10.7 concerning products of arc decompositions with a line. The only other device needed is the following flatness result, to ensure that the disk-squeezing theorem can be applied.

Proposition 2. *Let D be a disk in E^n, $n > 3$, such that D is locally polyhedral modulo its boundary, and let A be an arc in D such that $A \cap \partial D \subset \partial A$. Then A is flat.*

Proof. We shall assume, without much loss of generality, that A is also locally polyhedral modulo ∂A. By rigidly rotating the axes so that the x_n-axis is parallel to no line in a 3-dimensional hyperplane determined by any four of either the vertices of A or points of ∂A, we can reposition the embedding so that vertical projection p of E^n to the horizontal plane

$$P = \{\langle x_1, \ldots, x_n \rangle \in E^n \,|\, x_n = 0\}$$

is 1–1 on A. As in the proof of Proposition 10.4, A and $p(A)$ are equivalently embedded in E^n, and the result follows from Corollary 10.4A. ∎

Theorem 3. *The double suspension $\Sigma^2 H^3$ of H^3, the Mazur 3-manifold, is a 5-manifold (and, hence, is homeomorphic to S^5).*

Proof. By Proposition 1, H^3 bounds the 4-manifold W^4 having a nice disk D as pseudo-spine. Better yet, by Proposition 2 and Theorem 11.1, W^4 has an arc α as pseudo-spine. According to Theorem 10.7 (see Exercise 10.1), (Int $W^4/G_\alpha) \times E^1$ is a 5-manifold, which implies that $\Sigma^2 H^3$ is a manifold at each point of the suspension circle, the only place where the matter was in doubt. (By Exercise 6.9, $\Sigma^2 H^3$ then is S^5.) ∎

Proof of Proposition 1. Cut the solid torus $S^1 \times B^2$ of Mazur's example apart along the disk F shown in Fig. 12-1 to obtain $[0, 1] \times B^2$, with L_1 representing the finite collection of arcs corresponding to Γ after the cutting.

Let C^3 denote the one-point compactification of $\bigcup_i [i - 1, i] \times B^2$, L_i the translate of L_1 in $[i - 1, i] \times B^2$, and $L = (\bigcup_i L_i) \cup \{\infty\}$, shown in Fig. 12-2. Obviously C^3 is a 3-cell and L is the union of two arcs, one of which, considered as a subset of E^3 under the natural inclusion there, is wild. Let $s: C^3 \to C^3$ be the shift homeomorphism translating everything (except ∞) to the right by one unit, i.e., $s(\langle t, b \rangle) = \langle t + 1, b \rangle$ for $\langle t, b \rangle \in \bigcup [i - 1, i] \times B^2$ and $s(\infty) = \infty$. Now, for the spin, construct the mapping torus T of s and in T find the mapping torus Ω of $s \,|\, L: L \to L$. (Recall that by definition T is the space resulting from $C^3 \times I$ upon identification of $\langle c, 1 \rangle$ with $\langle s(c), 0 \rangle$.)

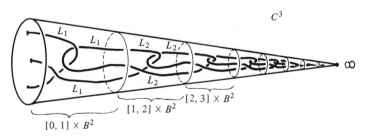

FIG. 12-2

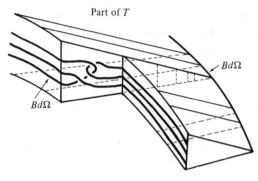

FIG. 12-3

It should be transparent that Ω is an annulus and that T is homeomorphic to $S^1 \times B^3$. Under the natural embedding of Ω in T, with $\Omega \cap \partial T = \partial \Omega$, the interesting component of $\partial \Omega$ corresponds to Γ, as suggested in Fig. 12-3. Thicken T to a new copy of $S^1 \times B^3$ in which the interesting component of $\partial \Omega$ still corresponds to Γ and the uninteresting one to $S^1 \times \{0\}$. In $W^4 = (S^1 \times B^3) \cup h$, where $h \cap (S^1 \times B^3)$ is a regular neighborhood of Γ in $S^1 \times \partial B^3$ and where $\Gamma = \partial B^2 \times \{0\} \subset B^2 \times B^2 = h$, define D as

$$D = \Omega \cup B^2 \times \{0\} \subset \text{Int } W^4,$$

which is a 2-cell, as promised.

In order to check that D is a pseudo-spine for W^4, it is mainly necessary to observe that T collapses to Ω, under an infinite sequence of collapses. Starting from the free side of $[0, 1] \times B^2 \times \{0\} \subset C^3 \times \{0\}$, these collapses deform around the S^1-factor of the mapping torus, initially freeing the first translate under s of $[0, 1] \times B^2 \times \{0\}$, thereby allowing iteration of the process to proceed. To be specific, the initial step coincides with the image in T of the collapse

$$([0, 1] \times B^2 \times I) \searrow (L_1 \times I) \cup ([0, 1] \times B^2 \times \{1\}) \cup (\{1\} \times B^2 \times I).$$

Successive repetition yields

$$W^4 \searrow (S^1 \times B^3) \cup D \searrow T \cup D \searrow \Omega \cup D = D.$$

Although properly the last of these is an infinite collapse, this possible problem causes no major complication, for the regular neighborhood theorem attests that, at any stage, the closed complement in W^4 of a small regular neighborhood of the uncollapsed part is naturally homeomorphic to $\partial W^4 \times [0, 1] = H^3 \times [0, 1]$, and so, by straightforward construction, $W^4 - D$ is homeomorphic to $H^3 \times [0, 1)$. Hence, D is a pseudo-spine for W^4. ∎

Concluding Remark. The intermediate step taken here, proving W^4 has an arc as pseudo-spine, is not the only way to complete the proof of Theorem 3, nor is it the most familiar. Another route, of comparable length and difficulty, involves a result of Bryant [1] indicating directly that $(W^4/G_D) \times E^1$ is homeomorphic to $W^4 \times E^1$. A thorough tracing of the latter path also requires addressing the flatness problem, similar to the one encountered in Proposition 2, of showing $D \times \{0\}$ to be a locally flat subset of $W^4 \times E^1$.

EXERCISES

1. Let T be a PL solid torus in $S^1 \times B^2$ having the curve Γ of Fig. 12-1 as centerline. Form a 4-manifold with boundary N^4 from $S^1 \times B^2 \times I$ by homeomorphically identifying $S^1 \times B^2 \times \{0\}$ with $T \times \{1\}$. Show that there exists a topological embedding $H: S^1 \times S^1 \to N^4$ which is a homotopy equivalence.
2. Show that the manifold N^4 of Exercise 1 can be embedded in E^4.

Remark. Surprisingly, as pointed out in Eaton–Pixley–Venema [1], the embedding $H: S^1 \times S^1 \to N^4$ cannot be approximated by a PL embedding. Matsumoto [1] has shown that when N^4 is formed without twisting the matched solid tori, no inclusion of a PL torus $S^1 \times S^1$ in N^4 is a homotopy equivalence. Hence, there exists a copy of $S^1 \times S^1$ embedded in E^4 that cannot be approximated by PL embedded tori.

13. APPLICATIONS OF THE LOCAL CONTRACTABILITY OF MANIFOLD HOMEOMORPHISM GROUPS

Now that the shrinkability criterion has been exercised to develop useful results about manifolds, it is an appropriate time to organize the data concerning various notions of shrinkability. Shrinkable decompositions of manifolds are both strongly shrinkable and ideally shrinkable; furthermore, shrinkable decompositions of manifolds can be realized by pseudo-isotopies. Establishing these statements is the main purpose of this section. The techniques marshaled for the task also reveal why certain subdecompositions of shrinkable decompositions are themselves shrinkable.

The results derived here all are based on the important fact, proved by R. D. Edwards and R. C. Kirby [1] and independently by A. V. Černavskiĭ [1, 2], that the group of self-homeomorphisms defined on a manifold is locally contractible. To prove it, Edwards and Kirby state a preliminary technical result that readily applies to situations arising in the course of this section.

Lemma (partial straightening, Edwards and Kirby [1, p. 70]). *Let $\tilde{M}$ be an n-manifold, C a compact subset of $\tilde{M}$, V a compact neighborhood of C, and $\varepsilon > 0$. Let E denote the space of all embeddings of V in $\tilde{M}$ (with the sup-norm metric). Then there exists $\delta > 0$ such that, for the δ-neighborhood P in E of the inclusion $\eta: V \to \tilde{M}$, there is a homotopy $\phi_t: P \to E$ and there is a compact neighborhood W of C in V such that, for all $\alpha \in P$,*

(a) $\phi_0(\alpha) = \alpha$,
(b) $\phi_t(\eta) = \eta$,
(c) $\phi_1(\alpha)\,|\,C = \eta\,|\,C$,
(d) $\rho(\phi_t(\alpha), \alpha) < \varepsilon$, and
(e) $\phi_t(\alpha)\,|\,V - W = \alpha\,|\,V - W$.

One should observe that for every $\alpha \in P$ the homotopy ϕ_t in E engenders an isotopy $q_t\colon \tilde{M} \to \tilde{M}$ moving no point as much as ε, leaving points outside $\alpha(W)$ fixed, starting at the identity, and, when restricted to $\alpha(C)$, ending at α^{-1}. [Here $q_t(y) = \phi_t(\alpha)(\alpha^{-1}(y))$ for $y \in \alpha(V)$.] Particularly useful in the first two results here is $\psi = q_1^{-1}$, which is an ε-homeomorphism of $\tilde{M}$ such that $\psi\,|\,\tilde{M} - \alpha(V) = \mathrm{Id}\,|\,\tilde{M} - \alpha(V)$ and $\psi\,|\,C = \alpha\,|\,C$. The existence of the isotopy q_t does not exploit the full strength of the local contractibility theorem, just something like the consequence that the group of homeomorphisms is locally pathwise-connected.

Theorem 1.   *Let G be a usc decomposition of an n-manifold M. Then G is shrinkable if and only if it is strongly shrinkable.*

Proof.   Our argument deals only with the compact case of the nontrivial implication.

Consider a G-saturated open cover $\mathcal{U}$ of M, an open set W containing N_G, and $\varepsilon > 0$. Set $X = M - W$ and find an open set V_1 in M/G such that $\pi(X) \subset V_1$ and $\pi(g) \in V_1$ implies diam $g < \varepsilon/3$. Making a preliminary refinement, if necessary, assume that diam $U < \varepsilon/3$ whenever $U \in \mathcal{U}$ and $\pi(U) \cap V_1 \neq \varnothing$. Determine another G-saturated open cover $\tilde{\mathcal{U}}$ of M that triple star-refines $\mathcal{U}$. Name $\varepsilon^* > 0$ small enough that every ε^*-subset of M/G lies in some set $\pi(\tilde{U})$, $\tilde{U} \in \tilde{\mathcal{U}}$.

Apply normality to obtain open sets V_2, V_3, and V_4 in M/G for which

$$\pi(X) \subset V_4 \subset \mathrm{Cl}\,V_4 \subset V_3 \subset \mathrm{Cl}\,V_3 \subset V_2 \subset \mathrm{Cl}\,V_2 \subset V_1.$$

Since G is shrinkable, M/G is a manifold; thus, in that manifold the partial straightening lemma yields $\delta^* > 0$ corresponding to $\mathrm{Cl}\,V_3 = V$, $\mathrm{Cl}\,V_4 = C$, and ε^*. Without loss of generality, we may assume

$$\delta^* < \min\{\varepsilon^*, \rho(\pi(X), (M/G) - V_4), \rho(\mathrm{Cl}\,V_{i+1}, (M/G) - V_i)$$
$$\text{for } i = 1, 2, 3\}.$$

The shrinkability of G gives a homeomorphism $F\colon M \to M/G$ such that $\rho(\pi, F) < \delta^*/2$.

Find $\alpha > 0$ so that diam $F^{-1}(A) < \varepsilon$ when $A \subset M/G$ and diam $A < \alpha$. Again there exists a homeomorphism $f\colon M \to M/G$ such that $\rho(\pi, f) < \min\{\alpha/2, \delta^*/2\}$. Note that $\rho(Ff^{-1}, \mathrm{Id}) < \delta^*$ and that diam $f(g) < \alpha$ for each $g \in G$.

By the partial straightening lemma, there exists an ε^*-homeomorphism $\psi: M/G \to M/G$ such that $\psi \,|\, \text{Cl } V_4 = Ff^{-1} \,|\, \text{Cl } V_4$ and $\psi \,|\, (M/G) - Ff^{-1}(V_3) = \text{Id}$. As a result, $\psi \,|\, (M/G) - V_2 = \text{Id}$.

Define $h: M \to M$ as $h = F^{-1}\psi f$. Triple star-refinement features of $\mathfrak{U}$, combined with the choices of ε^* and δ^*, imply that h is $\mathfrak{U}$-close to Id_M. Moreover, for $x \in X = M - W$, $f(x) \in V_4$, implying that

$$h(x) = F^{-1}\psi f(x) = F^{-1}(Ff^{-1})f(x) = x,$$

as required.

Finally, it must be checked that h shrinks elements of G to small size. For $g \in G$, there are two cases:

Case 1. $\pi(g) \cap V_1 = \varnothing$. Then $f(g)$ misses V_2, so $\psi \,|\, f(g) = \text{Id}$. Hence, diam $\psi f(g) < \alpha$, and diam $h(g) = \text{diam } F^{-1}\psi f(g) < \varepsilon$.

Case 2. $\pi(g) \in V_1$. In this case, $h(g) \subset \text{St}(g, \mathfrak{U})$, and any $U \in \mathfrak{U}$ meeting g has diameter $< \varepsilon/3$, so diam $h(g) < \varepsilon$. ∎

Reversing the roles of $C = \text{Cl } V_4$ and V_1 in the preceding argument leads to a valuable result about the shrinkability of subdecompositions.

Theorem 2. *Suppose G is a shrinkable decomposition of an n-manifold M, C is a closed subset of M/G, and $G(C)$ is the decomposition of M induced over C (that is,*

$$G(C) = \{\pi^{-1}(c) \,|\, c \in C\} \cup \{\{x\} \,|\, x \in M - \pi^{-1}(C)\}).$$

Then $G(C)$ is shrinkable.

Proof. Again we shall assume M to be compact. Consider a $G(C)$-saturated open cover $\mathfrak{U}_C$ and $\varepsilon > 0$. After refining $\mathfrak{U}_C$, if necessary, we can suppose that each $U \in \mathfrak{U}_C$ intersecting $\pi^{-1}(C)$ is G-saturated, since any $U \in \mathfrak{U}_C$ can be expressed as the union of $U - \pi^{-1}(C)$ and its maximal G-saturated subset. Thus, we can produce a G-saturated open cover $\mathfrak{U}$ of M such that every $U \in \mathfrak{U}$ intersecting $\pi^{-1}(C)$ is also an element of $\mathfrak{U}_C$. Let $\tilde{\mathfrak{U}}$ be another G-saturated open cover of M that double star-refines $\mathfrak{U}$.

Find $\varepsilon^* > 0$ small enough that every ε^*-subset of M/G lies in some $\pi(\tilde{U})$, $\tilde{U} \in \tilde{\mathfrak{U}}$, and let $V_1 = \text{St}(C, \pi(\tilde{\mathfrak{U}}))$. Apply normality to obtain open sets V_2, V_3, and V_4 in M/G such that

$$C \subset V_4 \subset \text{Cl } V_4 \subset V_3 \subset \text{Cl } V_3 \subset V_2 \subset \text{Cl } V_2 \subset V_1.$$

Use the partial straightening lemma to obtain $\delta^* > 0$ corresponding to the open set $(M/G) - \text{Cl } V_3$, closed set $(M/G) - V_2$, and positive number $\varepsilon^*/3$. Without loss,

$$\delta^* < \min\{\varepsilon^*/3, \rho(C, (M/G) - V_4), \rho(V_{i+1}, (M/G) - V_i) \,(i = 1, 2, 3)\}.$$

Since G is shrinkable, there exists a homeomorphism $F: M \to M/G$ such that $\rho(\pi, F) < \delta*/2 < \varepsilon*/3$ and there exists another homeomorphism $f: M \to M/G$ such that $\rho(\pi, f) < \min\{\alpha/2, \delta*/2\}$, where, as before, $\alpha > 0$ is chosen so $A \subset M/G$ and diam $A < \alpha$ implies diam $F^{-1}(A) < \varepsilon$.

By the partial straightening lemma, there exists a homeomorphism $\psi: M/G \to M/G$ such that $\psi \mid (M/G) - V_2 = Ff^{-1} \mid (M/G) - V_2$ and $\psi \mid \text{Cl } V_4 = \text{Id}$. Define $h: M \to M$ as $h = F^{-1}\psi f$.

To check that h shrinks elements of $G(C)$ to small size, consider $g = \pi^{-1}(c)$, $c \in C$. Then $f\pi^{-1}(c) \subset V_4$ and diam $f\pi^{-1}(c) < \alpha$. As a result, $h(g) = F^{-1}f(g)$, and diam $h(g) < \varepsilon$.

To check that h is $\mathfrak{U}_C$-close to Id_M, there are two cases: $x \in M - \pi^{-1}(V_1)$ and $x \in \pi^{-1}(V_1)$. In the first case, $h(x) = x$. In the second case, one computes first that $\rho(\pi, \pi h) < \varepsilon*$, which implies that h is $\tilde{\mathfrak{U}}$-close to Id_M. Thus, there exists $\tilde{U} \in \tilde{\mathfrak{U}}$ containing x and $h(x)$; furthermore, there exists $\tilde{U}_x \in \tilde{\mathfrak{U}}$ such that $x \in \tilde{U}_x$ and $\tilde{U}_x \cap \pi^{-1}(C) = \varnothing$. Since there exists $U \in \mathfrak{U}$ for which

$$\tilde{U}_x \cup \tilde{U} \subset \text{St}(\tilde{U}_x, \tilde{\mathfrak{U}}) \subset U$$

we see that $U \in \mathfrak{U}_C$ and that h is $\mathfrak{U}_C$-close to Id_M.

Corollary 2A. *If G is a usc decomposition of E^n such that $G \times E^1$ is a shrinkable decomposition of E^{n+1}, then the decomposition G_0 whose nondegenerate elements are the sets $g \times \{0\}$, $g \in H_G$, is also a shrinkable decomposition of E^{n+1}.*

Corollary 2B. *Suppose G is a usc decomposition of a manifold M, C_1 and C_2 are closed subsets of M/G such that the decomposition $G(C_2)$ induced over C_2 is shrinkable, and $\theta: M \to M$ is a proper onto map realizing $G(C_1)$. Then $\theta(G(C_2))$ is shrinkable.*

Proof. Assume M to be compact. Consider a $\theta(G(C_2))$-saturated open cover $\mathfrak{U}$ of M and $\varepsilon > 0$.

Determine $\delta > 0$ small enough that the θ-image of every δ-set has diameter less than ε.

Find a compact subset Z of $C_2 - C_1$ such that diam $\theta\pi^{-1}(c) < \varepsilon/3$ for each $c \in C_2 - Z$. After expanding Z and refining $\mathfrak{U}$, if necessary, we can suppose that any $U \in \mathfrak{U}$ meeting $\theta\pi^{-1}(C_2 - Z)$ has diameter $< \varepsilon/3$.

According to Theorem 2, the decomposition $G(Z)$ induced by $\pi: M \to M/G$ over Z is shrinkable, because it coincides with the one induced over $\pi_2\pi^{-1}(Z)$ by $\pi_2: M \to M/G(C_2)$. By Theorem 1, $G(Z)$ is strongly shrinkable. Hence, there exist a homeomorphism $h*: M \to M$ moving no points of $\pi^{-1}(C_1)$, shrinking elements of $G(Z)$ to δ-size, and staying $\theta^{-1}(\mathfrak{U})$-close to Id_M. The homeomorphism $h = \theta h*\theta^{-1}$ fulfills the requirements of this corollary. ∎

Example. The result expressed in Theorem 2 is valid only in spaces like manifolds, in which the partial straightening lemma holds. Recall the space S of Examples 7.1 and 7.2. Because the circle J_0 is locally shrinkable in S, the decomposition G^* of $S \times I$ with $H_{G^*} = \{J_0 \times \{t\} \mid t \in I\}$ is shrinkable, but decompositions induced over points of $\pi(N_{G^*})$ fail to be shrinkable and nondegenerate elements of G^* fail to be locally shrinkable in $S \times I$.

Theorem 3. *A usc decomposition G of an n-manifold M is shrinkable if and only if it is ideally shrinkable.*

Proof. Assume G to be shrinkable. Given an open set W in M containing N_G, express $\pi(W)$ as a monotone union of closed subsets $X_1, X_2, \ldots$ of M/G with $X_j \subset \text{Int } X_{j+1}$. Set $C_1 = X_1$ and $C_j = X_j - \text{Int } X_{j-1}$ for $j > 1$. By Theorems 1 and 2, the decompositions $G(C_j)$ induced over C_j are strongly shrinkable. Since no point of M/G belongs to more than two of the sets C_j, one can produce homeomorphisms h_j shrinking the individual decompositions $G(C_j)$ and also converging to a homeomorphism $h: M \to M$ that shows G to be ideally shrinkable. Details are left as an exercise. ∎

This section closes with a result displaying the relationship between the shrinkability of manifold decompositions and the presence of pseudo-isotopies. During the classical period people often went to extra efforts to substantiate the existence of pseudo-isotopies realizing various shrinkable decompositions. In 1969 T. M. Price [2] proved that shrinkable decompositions of 3-manifolds always could be realized in that manner. The result at hand does the same thing without any restriction on dimension.

Theorem 4. *If G is a shrinkable decomposition of a manifold M, then G can be realized as the end of a pseudo-isotopy. Furthermore, given any G-saturated open cover $\mathfrak{U}$ of M, there exists a pseudo-isotopy $\phi: M \times I \to M$ such that each path $\phi(\{p\} \times I)$, $p \in M$, is contained in some $U \in \mathfrak{U}$.*

Proof (sketch). Assume that M is compact. Choose $\varepsilon > 0$ small enough that each ε-subset of M/G lies in some $\pi(U)$, $U \in \mathfrak{U}$. For $\varepsilon_i = \varepsilon/2^{i+1}$ apply the partial straightening lemma (with $C = V = M/G$) to obtain $\delta_i > 0$ such that every δ_i-homeomorphism of M/G to itself is ε_i-isotopic to $\text{Id}_{M/G}$, and require $\delta_{i+1} < \delta_i$. For $i = 0, 1, \ldots$ find a homeomorphism $h_i: M \to M/G$ such that $\rho(\pi, h_i) < \delta_i/2$. Then $h_{i+1}h_i^{-1}$ is within δ_i of $\text{Id}_{M/G}$; consequently, there exists an ε_i-isotopy θ_t^i between $\text{Id}_{M/G}$ and $h_{i+1}h_i^{-1}$. Define $\Phi_t: M \to M/G$ with $\Phi_0 = h_0$ and $\Phi_1 = \pi$ by having Φ_t trace out $\theta_t^i h_i$ on the interval $[1 - 2^{-i}, 1 - 2^{-(i+1)}]$, and then define $\phi_t: M \to M$ as $\phi_t = h_0^{-1}\Phi_t$.

EXERCISES

1. Prove that the conclusion of Theorem 2 does not hold in the space $S \times I$ of the Example.

2. Let X be the decomposition space resulting from the decomposition of $S \times I$ having $J_c \times \{1\}$ as its only nondegenerate elements. Show that there is a shrinkable decomposition of X that is not strongly shrinkable.

3. Suppose G is a shrinkable decomposition of a (compact) manifold M and Z is a countable (but not necessarily closed) subset of M/G such that $G(Z)$ is usc. Show that $G(Z)$ is shrinkable. (Hint: review Section 8.)

4. Prove Theorem 3.

5. Let G be a usc decomposition of an n-manifold M such that M/G is finite-dimensional and each $\pi(g) \in M/G$ has a closed neighborhood C_g for which the decomposition $G(C_g)$ induced over C_g is shrinkable. Then G is shrinkable.

III ————————————————————————

CELL-LIKE DECOMPOSITIONS OF ABSOLUTE NEIGHBORHOOD RETRACTS

Chapter III is relatively technical in nature and investigates certain properties preserved by those usc decompositions typically considered. Up to this point cellular decompositions, which have been the most prevalent, may have seemed to be the typical ones, but, due to the phenomenon treated in Section 10, one can see how other more general decompositions could be equally workable. The appropriate concept is that of a cell-like decomposition, set forth in Section 15. Section 18 exposes the relationship between cell-likeness and cellularity for subsets of manifolds.

The optimal setting for this investigation is the family of locally compact, metric, absolute neighborhood retracts (ANRs). For cell-like usc decompositions of finite-dimensional ANRs, the decomposition spaces turn out to possess enough nice local properties to be ANRs, provided they are known to be finite-dimensional. The proviso raises the fundamental problem: could a cell-like decomposition raise dimension?

The indispensability of ANRs to the study of cell-like decompositions becomes manifest in Section 17, where the equivalence between the dimension-raising problem and an ANR-preserving problem (in the finite-dimensional realm) is demonstrated. The most profound among the sections in this chapter, Section 17 also includes a proof of the strongest current result pertaining to the fundamental problem, due to Kozlowski and Walsh, which attests that cell-like usc decompositions of 3-manifolds cannot raise dimension.

14. ABSOLUTE RETRACTS AND ABSOLUTE NEIGHBORHOOD RETRACTS

Simplicial complexes are composed from simple pieces according to a construction code whose rigidity imposes an axiomatic basis belonging more to the realm of geometry than to that of topology. Studied here is a more general class of spaces, the absolute neighborhood retracts, defined in purely topological terms. Included in this class are all finite-dimensional manifolds.

Borsuk's book [1] provides a comprehensive treatment of absolute neighborhood retracts, a topic that threads throughout the fabric of geometric topology in general and decomposition theory in particular. The topic impinges upon infinite-dimensional manifolds in beautiful and powerful ways, as displayed in Chapman's book [2].

A subset C of a space S is a *retract* (*neighborhood retract*) of S if there exists a retraction R of S to C (a retraction R' to C defined on some neighborhood U of C).

The significance attached to retract properties stems partially from their relationship to map extension properties.

Proposition 1. *A subspace C of a space S is a retract (neighborhood retract) of S if and only if each map f of C to an arbitrary space Y has a continuous extension $F: S \to Y$ (a continuous extension $F': U \to Y$ defined on some neighborhood U of C in S).*

Proof. For one direction, the extension F equals fR; for the other, choose $Y = C$ and $f = \text{Id}_C$, and then the promised extension F is a retraction. ∎

Proposition 1 justifies defining absolute retract properties in terms of map extendability. Restricting ourselves to metric spaces, we say that a metric space Y is an *absolute retract* (abbreviated as AR) if, for each closed subset A of a metric space X, every map $f: A \to Y$ has a continuous extension $F: X \to Y$; similarly, we say that Y is an *absolute neighborhood retract* (ANR) if every such map $f: A \to Y$ has a continuous extension $F': U \to Y$ defined on some neighborhood U of A in X.

Proposition 2 (invariance). *For any closed embedding e of an AR (ANR) Y in a metric space X, $e(Y)$ is a retract (neighborhood retract) of X.*

Proof. Since $e^{-1}: e(Y) \to Y$ can be extended to $F: X \to Y$, the retraction to $e(Y)$ can be defined as $R = eF$. ∎

Note that each AR Y is necessarily contractible, since the map $Y \times \{0, 1\} \to Y$ defined as $\langle y, 0 \rangle \to y$ and $\langle y, 1 \rangle \to y_0$ can be extended to $\phi: Y \times I \to Y$, which must be a contraction. Other basic properties are listed below.

Theorem 3. (a) *Every retract of an AR is an AR.*
(b) *Every retract of an ANR is an ANR.*
(c) *Every open subset of an ANR is an ANR.*
(d) *The Cartesian product of any countable collection of ARs is an AR.*
(e) *The Cartesian product of any finite collection of ANRs is an ANR.*

Corollary 3A. *The Hilbert cube I^∞ is an AR; the n-cube I^n is an AR; and the n-sphere S^n ($n = 1, 2, ...$) is an ANR.*

Proof. View S^n as a retract of the punctured $(n + 1)$-cube.

A space S is *locally contractible* at $s \in S$ if each neighborhood U of s contains a neighborhood V of s that contracts in U; S is *locally contractible* if it is locally contractible at each $s \in S$. This is just one among many properties that neighborhood retracts inherit from their overriding spaces.

Proposition 4. *Every neighborhood retract C of a locally contractible (respectively: locally compact; locally connected; locally pathwise-connected) space S is itself locally contractible (respectively: locally compact, locally connected, locally pathwise-connected).*

Proof. Name a retraction $R: W \to C$ defined on some neighborhood W of C, and consider an open subset U' of C containing some $c \in C$. Find an open subset U of S with $c \in U \subset W$ and $R(U) \subset U'$. By definition of local contractibility, U contains a neighborhood V of c in S for which there exists a contraction ϕ_t of V in U. Then $R\phi_t$ contracts $V \cap C$ in U'. ∎

Corollary 4A. *Each compact ANR Y is locally contractible, locally connected, and locally pathwise-connected.*

Proof. Apply Urysohn's metrization theorem to embed Y in I^∞, and use Proposition 2. ∎

The abbreviations CAR and CANR will be used henceforth to represent the terms "compact absolute retract" and "compact absolute neighborhood retract," respectively.

Proposition 5. *Let Y be a CANR and $\varepsilon > 0$. There exists $\delta > 0$ such that, for any two maps h_0, h_1 of an arbitrary space S to Y with $\rho(h_0, h_1) < \delta$, there is a homotopy $H: S \times I \to Y$ satisfying $H_0 = h_0$, $H_1 = h_1$, and diam $H(\{s\} \times I) < \varepsilon$ for every $s \in S$.*

Proof. Equate Y with a subset of I^∞, which is a retract of some neighborhood U under $R: U \to Y$. Determine $\delta > 0$ so that $N(C; \delta) \subset U$ and that diam $R(Z) < \varepsilon$ whenever $Z \subset N(C; \delta)$ has diameter less than δ. Given δ-close maps h_0 and h_1, one can define H_t as the image under R of the straight line homotopy connecting them. ∎

A subspace A of a space X has the *homotopy extension property* (HEP) *in X with respect to a space Y* if every map $f: (A \times I) \cup (X \times \{0\}) \to Y$ has a continuous extension $F: X \times I \to Y$; moreover, A has the *absolute homotopy extension property* (AHEP) in X if it has HEP in X with respect to every space Y. Borsuk [1] derived the key result.

Theorem 6 (homotopy extension).   *If A is a closed subset of a metric space X, then A has HEP in X with respect to every ANR Y.*

Proof.   Each map $f: (A \times I) \cup (X \times \{0\}) \to Y$ can be extended to a map $f': W \to Y$ having a larger open subset W of $X \times I$ as domain. Then A has a neighborhood V in X such that $V \times I \subset W$. Urysohn's lemma provides a map $u: X \to I$ such that $u(A) = \{1\}$ and $u(X - V) = \{0\}$. The desired extension $F: X \times I \to Y$ can be defined as $F(\langle x, t \rangle) = f'(\langle x, t \cdot u(x) \rangle)$.

Corollary 6A.   *Suppose Y is an ANR, A is a closed subset of the metric space X, and $f_0, f_1: A \to Y$ are homotopic maps. Then f_0 has a continuous extension $F_0: X \to Y$ if and only if f_1 has a continuous extension $F_1: X \to Y$. Moreover, for each $F_0: X \to Y$ extending f_0 there exists a map $F_1: X \to Y$ extending f_1 such that F_1 is homotopic to F_0.*

Corollary 6B.   *Each contractible ANR is an AR.*

Proof.   Given a map $f_0: A \to Y$ defined on a closed subset A of a metric space X, note that the contractibility hypothesis implies f_0 is homotopic to a constant mapping $f_1: A \to Y$. Since the latter obviously extends over X, so does f_0.

Corollary 6C. E^n *is an AR.*

Corollary 6D.   *If A is a closed subspace of a metric space X and $(A \times I) \cup (X \times \{0\})$ is a neighborhood retract of $X \times I$, then A has AHEP in X.*

Corollary 6D is a consequence of the proof for Theorem 6.

Corollary 6E.   *Suppose Y is a CANR and $\varepsilon > 0$. There exists $\delta > 0$ such that if f_0 is a map of a metric space X to Y, if f_A is a map of A to Y such that $\rho(f_A, f_0 \mid A) < \delta$, where A is a closed subset of X, and if U is an open subset of X containing A, then there exists a homotopy $H: X \times I \to Y$ satisfying (1) $H_0 = f_0$, (2) $H_1 \mid A = f_A$, (3) $H(\langle x, t \rangle) = f_0(x)$ for every $x \in X - U$, and (4) $\operatorname{diam} H(\{x\} \times I) < \varepsilon$ for every $x \in X$.*

The proof is left as an exercise.

Now that the basic features of ANRs have been outlined, we shall investigate some devices for unmasking them.

Theorem 7. *Suppose that the metric space Y is the union of closed sets Y_1 and Y_2 and that $Y_0 = Y_1 \cap Y_2$.*

(a) *If Y_0, Y_1, and Y_2 are ANRs (ARs), then Y is an ANR (AR).*
(b) *If Y and Y_0 are ANRs (ARs), then Y_1 and Y_2 are ANRs (ARs).*

Proof. Statement (b) is the easier to establish. In the AR case it suffices to find a retraction R of Y to, say, Y_1. Because Y_0 is an AR, there is some retraction r of Y_2 to Y_0, and r extends via the identity on Y_1 to the desired retraction R. In the ANR case it can be derived in a similar fashion.

For statement (a), consider a map $f: A \to Y$ defined on a closed subset A of a metric space X. We show how to extend f over a neighborhood of A in X. Define A_i as $f^{-1}(Y_i)$, $i = 1, 2$, and $A_0 = A_1 \cap A_2$. Furthermore, define

$$X_1 = \{x \in X \mid \rho(x, A_1) \le \rho(x, A_2)\},$$

$$X_2 = \{x \in X \mid \rho(x, A_2) \le \rho(x, A_1)\},$$

and

$$X_0 = X_1 \cap X_2.$$

(See Fig. 14-1.)

Since Y_0 is an ANR, there exists a map $f_0: W_0 \to Y_0$ defined on a closed neighborhood W_0 of A_0 in X_0 and extending $f \mid A_0$. In particular, there exists an open subset U_0 of X with $A_0 \subset U_0$ and $U_0 \cap X_0 \subset W_0$.

Since Y_i is an ANR for $i = 1, 2$, there exists a neighborhood W_i of $W_0 \cup A_i$ in X_i and there exists a map $f_i: W_i \to Y_i$ extending $f_0 \cup (f \mid A_i)$. Restrictions of f_1 and f_2 combine to give a map from $[W_1 - (X_0 - U_0)] \cup [W_2 - (X_0 - U_0)]$ to Y, and inspection reveals this domain to be an open subset containing A. ∎

Corollary 7A. *Every finite simplicial complex is an ANR.*

Corollary 7B. *If Y is an ANR and A is a closed subset of Y that is also an ANR, then A has AHEP in Y.*

Proof. With the support of Theorem 3, which indicates that $A \times I$ is an ANR, Theorem 7 shows $(Y \times \{0\}) \cup (A \times I)$ to be an ANR, and Corollary 6D does the rest.

Corollary 7C. *If K is a finite simplicial complex and L is a (finite) subcomplex, then L has AHEP in K.*

Theorem 8. *Every compact, locally contractible subset Y of E^n is an ANR.*

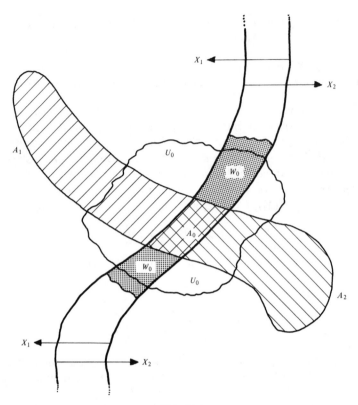

FIG. 14-1

Proof. It suffices to show that Y is a retract of some open subset of E^n. As a preliminary step, to convey the idea of the proof, we explain how this is almost the case, in that for every $\varepsilon > 0$ there exist an open set W_ε containing Y and a map $m: W_\varepsilon \to Y$ for which $\rho(y, m(y)) < \varepsilon$, $y \in Y$.

The local contractibility and compactness of Y imply that for every $\delta > 0$ there exists $\eta(\delta) > 0$ such that every $\eta(\delta)$-subset of Y is contractible in a δ-subset of Y. Start with $\delta_0 = \varepsilon/3$ and recursively define $\delta_i = \eta(\delta_{i-1})/3$ for $i \in \{1, ..., n\}$.

Triangulate E^n with simplexes of diameter less than δ_n and let W denote the union of all simplexes meeting Y. Then W contains an open set about Y.

Construct successive maps $\psi^k: W^{(k)} \to Y$, defined on the k-skeleton $W^{(k)}$ of W, as follows:

For $v \in W^{(0)}$, set $\psi^0(v) = y$, where $y \in Y$ and $\rho(y, v) < \delta_n$. This is possible because of the small mesh of the triangulation and the definition of W.

For any 1-simplex σ of W, $\psi^0 \,|\, \partial\sigma$ sends $\partial\sigma$ into a $3\delta_n = \eta(\delta_{n-1})$ subset of Y. Due to the choice of $\eta(\delta_{n-1})$, $\psi^0 \,|\, \partial\sigma$ can be extended to a map of σ into a δ_{n-1}-subset of Y. The compilation of all these extensions is ψ^1.

For the iterative step, suppose $\psi^k \colon W^{(k)} \to Y$ (where $1 \leq k < n$) has been determined so that $\psi^k(\tau)$ is contained in a δ_{n-k}-subset of Y, where τ denotes any k-simplex of W. Consider a $(k + 1)$-simplex γ of W. Specify some k-simplex τ_γ of W in $\partial\gamma$. Then every other k-simplex τ'_γ of W in $\partial\gamma$ satisfies $\psi^k(\tau'_\gamma) \subset N(\psi^k(\tau_\gamma); \delta_{n-k})$. Hence, diam $\psi^k(\partial\gamma) < 3\delta_{n-k} = \eta(\delta_{n-(k+1)})$. By choice of $\eta(\delta_{n-(k+1)})$, ψ^k extends to a map $\psi^{k+1} \colon W^{(k+1)} \to Y$ such that diam $\psi^{k+1}(\gamma) < \delta_{n-(k+1)}$, for every $(k + 1)$-simplex γ.

The desired map $m \colon W \to Y$ is ψ^n. For $w \in W$, find a simplex σ in W with $w \in \sigma$ and choose a vertex $v \in \sigma$. Then

$$\rho(w, m(w)) \leq \rho(w, v) + \rho(v, m(v)) + \rho(m(v), m(w))$$

$$< \delta_n + \delta_n + \delta_0 \leq 3\delta_0 = \varepsilon.$$

In order to retool this machinery for the theorem at hand, triangulate $W - Y$ so that diameters of simplexes go to zero near Y. Then, in the manner above, carefully build a map $m' \colon W - Y \to Y$, working up through the various skeleta. In order to extend m' via the identity to provide a retraction of W to Y, the only additional item needed is a rule demanding efficiency: if v if a vertex of $W - Y$, choose $m'(v)$ so that $\rho(v, m'(v)) = \rho(v, Y)$; if σ is a k-simplex ($k > 0$) of $W - Y$, and if d is the greatest lower bound of the diameters of all contractions of $m'(\partial\sigma)$ in Y, then choose m' so diam $m'(\sigma) \leq 2d$. ∎

Corollary 8A. *Every finite-dimensional, locally contractible, compact metric space is an ANR.*

Corollary 8B. *Every compact n-manifold is an ANR.*

Finite-dimensionality stands as an essential hypothesis in Corollary 8A. Borsuk has described an example of a compact, locally contractible subset Z of I^∞ that cannot be an ANR because, for each $n > 0$, Z retracts to some n-sphere, which is prohibited on homological grounds.

EXERCISES

1. Show that every locally compact, separable ANR can be embedded in I^∞ so that its image is a retract of some open subset of I^∞.
2. Prove Corollary 6E.
3. Show that every ANR is locally contractible.

4. Suppose Y_1 and Y_2 are closed subsets of the metric space Y such that $Y = Y_1 \cup Y_2$ and that Y_1, Y_2, and $Y_1 \cap Y_2$ are ARs. Prove that Y is an AR.
5. Suppose U_1 and U_2 are open subsets of a metric space and that U_1 and U_2 are ANRs. Show that $U_1 \cup U_2$ is an ANR.

15. CELL-LIKE SETS

A compact subset C of a space X is *cell-like in X* if, for each neighborhood U of C in X, C can be contracted to a point in U. For instance, the Whitehead continuum, Example 9.7, is cell-like in E^3. The fact that it is not cellular in E^3 separates the concepts of cell-likeness and cellularity.

Lemma 1. *Suppose the metric space X contains a compact subset C that is cell-like in X and e is an embedding of C in an ANR Y. Then $e(C)$ is cell-like in Y.*

Proof. Let U be a neighborhood of $e(C)$ in Y. Since Y is an ANR, there exist a neighborhood V of C in X and a map $\tilde{e}: V \to U$ extending e. Name a contraction ϕ_t of C in V. Then $\tilde{e}\phi_t e^{-1}$ provides a contraction of $e(C)$ in U. ∎

Thus, unlike cellularity, which depends on the embedding, the property of being cell-like is invariant under embeddings in ANRs. It makes sense to define cell-likeness in absolute terms.

A compact metric space C is *cell-like* if, for every embedding e of C in an ANR Y, $e(C)$ is cell-like in Y. This should not be confused with the term "pointlike," which pertains to embeddings. Cell-likeness is equivalent to the shape theory concept of having the shape of a point.

Proposition 2. *A compact metric space C is cell-like if and only if it can be embedded as a cell-like subset of some metric space X.*

Corollary 2A. *Each compact, contractible, metric space is cell-like.*

Corollary 2B. *Each cellular subset of a manifold is cell-like.*

Next, a homotopy-theoretic characterization.

Theorem 3. *A compact metric space C is cell-like if and only if every map f of C into an ANR Y is homotopic to a constant.*

Proof. ($\Rightarrow$) Since C is compact metric, there is an embedding e of C in the Hilbert cube I^∞, and $e(C)$ is cell-like in I^∞ by definition. Consider a map f of C into an ANR Y. There exist a neighborhood V of $e(C)$ in I^∞ and a map $F: V \to Y$ extending fe^{-1}. A contraction ϕ_t of $e(C)$ in V leads to a homotopy $F\phi_t e$ between f and the constant map $F\phi_1 e$.

($\Leftarrow$)   Suppose h is an embedding of C in an ANR Y and U is a neighborhood of $h(C)$ in Y. Since U is also an ANR, by hypothesis h is homotopic in U to a constant, which means that $h(C)$ is contractible in U.   ∎

Corollary 3A.   *A CANR is cell-like if and only if it is contractible.*

Proposition 4.   *Let C denote a compact subset of an ANR Y. Then C is cell-like in Y if and only if, for each neighborhood U of C, some neighborhood V of C in U is contractible in U.*

Proof.   Suppose C is cell-like in Y and U is a neighborhood of C in Y. Let ϕ_t denote a contraction of C in U. Define a map $f \colon A \to U$ on the closed subset $A = (C \times I) \cup (U \times \{0, 1\})$ of $U \times I$ as $f(\langle u, 0 \rangle) = u$ and $f(\langle u, 1 \rangle) = \phi_1(C) =$ point for $u \in U$ and as $f(\langle c, t \rangle) = \phi_t(c)$ for $c \in C$. Since U is also an ANR, f extends to a map $F \colon W \to U$ defined on a neighborhood W of A in $U \times I$. Then C has a neighborhood V with $V \times I \subset W$, and $F \mid V \times I$ acts as the desired contraction of V.

The other implication is obvious.   ∎

Corollary 4A.   *If C is a cell-like subset of E^n ($n > 1$) and U is a connected open set containing C, then $U - C$ is connected.*

Proof.   According to Hurewicz–Wallman [1, p. 100], a compact subset of E^n separates E^n if and only if it admits a map to S^{n-1} that is not null homotopic. Thus, $E^n - C$ is connected. Use the diagram below to verify $\tilde{H}_0(U - C) \cong 0$.

$$
\begin{array}{ccccc}
H_1(U, U - C) & \longrightarrow & \tilde{H}_0(U - C) & \longrightarrow & \tilde{H}_0(U) \cong 0. \\
\Big\downarrow \cong \text{(excision)} & & & & \\
0 \cong H_1(E^n) \longrightarrow & H_1(E^n, E^n - C) & \longrightarrow & \tilde{H}_0(E^n - C) \cong 0. & \quad \blacksquare
\end{array}
$$

Corollary 4B.   *If C is a cell-like subset of an n-manifold M ($n > 1$) and V is a connected open subset of M containing C, then $V - C$ is connected.*

Proof.   The relevant portion of the homology sequence for the pair $(V, V - C)$ is just

$$H_1(V, V - C) \to \tilde{H}_0(V - C) \to \tilde{H}_0(V) \cong 0,$$

which reveals the sufficiency of showing that $H_1(V, V - C) \cong 0$. By duality (see Spanier [1, p. 342]) $H_1(V, V - C) \cong \check{H}_c^{n-1}(C)$. It follows quickly from the definition of cell-likeness that $\check{H}_c^{n-1}(C) \cong 0$; alternately, it follows from Exercise 15.4 that C is homeomorphic to a cellular subset C^* of some Euclidean space E^m ($m \geq n$) and from duality again that

$$\check{H}_c^{n-1}(C) \cong H_{m+1-n}(E^m, E^m - C^*) \cong H_{m+1-n}(E^m, E^m - \text{origin}) \cong 0.$$

Corollary 4C.   *Each cell-like subset C of a 2-manifold M is cellular in M.*

Proof.   This argument pertains to the case $M = E^2$. The general case will follow by lifting a neighborhood V of C in M to an open subset $\tilde{V}$ in $\tilde{M}$, the universal cover of M, which must be topologically E^2 or S^2. Cellularity of the lifted copy of C in $\tilde{V}$ will imply cellularity of C in M.

Let U denote an arbitrary neighborhood of C in E^2. Find a smaller neighborhood V of C that contracts in U, and then build a compact 2-manifold with boundary H in E^2 such that $C \subset \text{Int } H \subset H \subset V$, ∂H is polygonal, and Int H is connected. If ∂H is connected, the planar Schönflies theorem will imply that H is a 2-cell, as required for cellularity. If ∂H fails to be connected, cut away at H until the boundary is connected. This is possible because, by Corollary 4A, $H - C$ is connected, so a polygonal arc can be strung through $H - C$ from one component of ∂H to another and carefully thickened to a disk D so that $H' = \text{Cl}(H - D)$ is a compact 2-manifold with boundary such that $\partial H'$ is polygonal and Int H' is connected but that $\partial H'$ has fewer components than ∂H. Eventually this yields a 2-cell B satisfying $C \subset \text{Int } B \subset B \subset U$.   ∎

The final corollary records the cohomological property used in proving Corollary 4B.

Corollary 4D.   *If C is a cell-like set, then $\tilde{H}^*(C) \cong 0$ (Čech cohomology, with any coefficient group).*

Theorem 5.   *If G is a shrinkable usc decomposition of a locally contractible metric space S, then each $g \in G$ is cell-like in S.*

Were local contractibility a topological property of certain open sets in themselves rather than of their inclusion into larger sets, this would be a direct consequence of Proposition 5.12. Nevertheless, the proof fits right into the pattern for Proposition 5.11, with an extra detail.

Proof.   Fix $g_0 \in G$ and a neighborhood W of g_0 in S. As before, there exist open subsets U_1, U_2, and U_3 of S/G such that $\pi^{-1}(U_1) \subset W$ and

$$\pi(g) \in U_{i+1} \subset \bar{U}_{i+1} \subset U_i \qquad (i = 1, 2).$$

Let $\mathcal{U}$ denote the G-saturated open cover of S specified as

$$\{S - \pi^{-1}(\bar{U}_2), \ \pi^{-1}(U_1 - \bar{U}_3), \ \pi^{-1}(U_2) - g_0, \ \pi^{-1}(U_3)\}.$$

For each $s \in S$ find a neighborhood V_s of s that is contractible in some $U \in \mathcal{U}$. Apply the shrinkability hypothesis to obtain a homeomorphism $h: S \to S$ limited by $\mathcal{U}$ and shrinking G to size $\mathcal{V} = \{V_s \mid s \in S\}$. Thus, there exist $V \in \mathcal{V}$ and $U \in \mathcal{U}$ such that $h(g_0) \subset V$ and V is contractible in U;

as a result, $g_0 \subset h^{-1}(V)$ and $h^{-1}(V)$ is contractible in $h^{-1}(U)$. All that remains is the verification, exactly like the one given to prove Proposition 5.11, that $h^{-1}(U) \subset W$. ∎

The pertinence of cell-likeness to decomposition theory is signaled by the two corollaries below.

Corollary 5A. *If G is a shrinkable usc decomposition of an ANR, then each $g \in G$ is cell-like.*

Corollary 5B. *If G is a usc decomposition of an n-manifold M such that $G \times E^k$ is a shrinkable decomposition of $M \times E^k$ for some integer k, then each $g \in G$ is cell-like.*

EXERCISES

1. Show that the product of any finite or countable collection of cell-like spaces is cell-like.
2. If X_1 and X_2 are cell-like subsets of the Hilbert cube I^∞ such that $X_1 \cap X_2$ is a point, then $X_1 \cup X_2$ is cell-like.
3. Show that the suspension of any cell-like space is cell-like.
4. Show that each cell-like subset of E^n admits a cellular embedding in some Euclidean space E^m where $m \geq n$.
5. If X is a compact, connected subset of E^2 such that $E^2 - X$ is connected, then X is cellular.
6. Let $f: W_1 \to W_2$ be a proper surjective map between two n-manifolds with boundary such that $f^{-1}(\partial W_2) = \partial W_1$ and $f \mid \text{Int } W_1$ is 1-1. Then f is cell-like.

16. *UV* PROPERTIES AND DECOMPOSITIONS

The cell-like property can be refracted prismatically to bring to light an infinite family of related properties, which were introduced and first analyzed by S. Armentrout [2, 3]. The setting requires a space X, a subset A, and a nonnegative integer n. Then A has Property $n - UV$ in X if to each neighborhood U of A in X there corresponds another neighborhood V of A in U such that every map of ∂B^{n+1} into V extends to a map of B^{n+1} into U; A has Property UV^n in X if it has Property $k - UV$ in X for $k \in \{0, 1, ..., n\}$; and A has Property UV^ω in X if it has Property $k - UV$ in X for all integers $k \geq 0$.

Proposition 1. *Every cell-like subset A of an ANR Y has Property UV^ω in Y.*

Proposition 15.4 makes this obvious. Almost as obvious, in view of methods exploited early in Sections 14 and 15, is the invariance result.

Proposition 2. *Suppose the compact subset A of an ANR Y_1 has Property n − UV in Y_1 and e is an embedding of A in another ANR Y_2. Then e(A) has Property n − UV in Y_2.*

Proof. For any neighborhood U of $e(A)$ in Y_2, the hypothesis that Y_2 is an ANR provides a neighborhood U' of A in Y_1 and a map $f: U' \to U$ extending e. Then U' contains a neighborhood V' of A fulfilling the condition expressed as Property $n − UV$. The fact that V' is also an ANR leads in turn to a neighborhood V^* of $e(A)$ in U and a map $f': V^* \to V'$ extending $e^{-1}: e(A) \to A$.

Let Z denote the closed subset $(A \times I) \cup (V^* \times \{0, 1\})$ of $X = V^* \times I$, and let $h: Z \to U$ denote the map defined as $h(\langle v, 0 \rangle) = v, h(\langle v, 1 \rangle) = ff'(v)$ for $v \in V^*$ and as $h(\langle e(a), t \rangle) = e(a)$ for $a \in A$ and $t \in I$. Then h extends to a map $h: W \to U$ defined on a neighborhood W of Z in $X = V^* \times I$. As a result, V^* contains another neighborhood V of $e(A)$ such that $V \times I \subset W$, which means that $ff' \mid V$ is homotopic in U, under $h \mid V \times I$, to the inclusion map $V \to U$.

Consider a map $\mu: \partial B^{n+1} \to V$. Because $f'\mu: \partial B^{n+1} \to V'$ extends to a map $F': B^{n+1} \to U', ff'\mu: \partial B^{n+1} \to U$ extends to $fF': B^{n+1} \to U$. Moreover, $ff'\mu$ is homotopic in U to μ, which implies μ is null homotopic in U, as required. ∎

Corollary 2A. *Suppose the compact subset A of an ANR Y_1 has Property UV^n (UV^ω) in Y_1 and e is an embedding of A in another ANR Y_2. Then e(A) has Property UV^n (UV^ω) in Y_2.*

Just as we did with cell-likeness, we reformulate these properties in absolute terms. A compact metric space A is said to have Property $n − UV$ (Property UV^n; Property UV^ω) provided $e(A)$ has the appropriate property in Y, under every embedding e of A in every ANR Y.

In reasonable spaces the simplest property, Property UV^0, is equivalent to connectedness.

Proposition 3. *Suppose X is a locally pathwise-connected Hausdorff space and A is a compact subspace. Then A has Property UV^0 in X if and only if A is connected.*

The proof is an exercise.

Corollary 3A. *Every cell-like set is connected.*

Theorem 4. *Suppose A is a compact subset of a metric space X such that A has Property UV^n in X, and suppose U is a neighborhood of A. Then there exists a neighborhood V of A in U such that, for every map $f: K \to V$ of an n-dimensional finite simplicial complex K, f is homotopic in U to a constant map.*

Proof. Because A satisfies Property UV^n in X, there exists a finite sequence $V_{n+1} \supset V_n \supset \cdots \supset V_k \supset \cdots V_0$ of neighborhoods of A, each one in U, such that every map of ∂B^{k+1} into V_k ($0 \le k \le n$) extends to a map of B^{k+1} into V_{k+1}. Set $V = V_0$.

Given a map f of a finite n-complex K into V, we shall explain how to extend f to a map F of the cone cK on K into U. If L denotes a subcomplex of K, we shall use cL to denote the cone on L, naturally embedded as a subcone of cK, and for $k \in \{0, 1, \ldots, n\}$ we shall use $K^{(k)}$ to represent the k-skeleton of K.

Let $M_0 = K \cup \{c\}$ (where K is equated with the base of cK) and $M_k = K \cup cK^{(k-1)}$ ($k > 0$). Extend $f: K \to V$ to $f_0: M_0 \to V_0$ by arbitrarily choosing $f_0(c) \in V_0$. Assuming f_k is a map of M_k into V_k, obtain a map $f_{k+1}: M_{k+1} \to V_{k+1}$ extending f_k as follows: each $\sigma \in M_{k+1}$ not in M_k is a $(k + 1)$-cell and f_k is defined on $\partial\sigma$, sending it into V_k; the extensions over the various simplexes σ, promised in the above description of the neighborhoods V_i, fit together neatly as such a map f_{k+1}. The final map f_{n+1} provides a homotopy in $V_{n+1} \subset U$ between f and the constant mapping $K \to f_0(c)$. ∎

Corollary 4A. *If the compact subset A of a simplicial n-complex has Property UV^n, then A is cell-like.*

Corollary 4B. *If the compact subset A of E^n has Property UV^n, then A is cell-like.*

Corollary 4C. *A compact subset A of E^n is cell-like if and only if A has Property UV^1 and $\check{H}_q(A; \mathbb{Z}) \cong 0$ (Čech homology) for $q > 0$.*

Proof. The elementary forward implication involves, in part, a homological version of Corollary 14.4D. The reverse implication depends on the following Hurewicz theorem from Lacher [1]: if A has Property UV^{k-1} and $\check{H}_k(A; \mathbb{Z}) \cong 0$ ($k \ge 2$), then A has Property UV^k.

Proposition 5. *A CANR Y has Property UV^n if and only if Y is n-connected.*

Proof. Let e be an embedding of Y in I^∞. Since $e(Y)$ is an ANR, there exists a retraction $r: U \to e(Y)$ defined on some neighborhood U of $e(Y)$.

Suppose Y has Property UV^n, and consider a map $f: \partial B^{k+1} \to Y$. The definition of Property UV^n implies the existence of a map $F: B^{k+1} \to U$ extending ef. Then $e^{-1}rF: B^{k+1} \to Y$ extends f.

Conversely, suppose Y is n-connected. The crux of the implication is that each neighborhood U' of $e(Y)$ contains a smaller neighborhood V' of $e(Y)$ such that the inclusion $V' \hookrightarrow U'$ and $r|V'$ are homotopic in U' (via the straight-line homotopy). One can show that a given map $f: \partial B^{k+1} \to V'$ is homotopic to a constant in U' by first running through this homotopy to rf and then, invoking the n-connectedness of Y, finding another homotopy in $e(Y)$ between rf and a constant map.   ∎

Returning to the subject of decompositions, as usual we call a usc decomposition G of a space S a UV^n *decomposition* if each $g \in G$ has Property UV^n in S. Although parallelity suggests that a cell-like decomposition G of S should be one in which each $g \in G$ is cell-like in S, that is *not* the way this fundamental concept will be treated. Instead, by a *cell-like decomposition G* of a metric space X we mean a usc decomposition G of X such that every $g \in G$ is a cell-like set. Of course, in case G is a cell-like decomposition of an ANR X, then each $g \in G$ necessarily is cell-like in X.

Armentrout and T. M. Price [1] developed virtually all the material to be presented in what remains of Section 16. Following their lead, we find it a convenient technical abbreviation to speak of a collection $\mathcal{W}$ of subsets from a space S as an *n-homotopy star-refinement of another collection* $\mathcal{U}$ of subsets if, for each $W \in \mathcal{W}$, there exists $U \in \mathcal{U}$ such that

(i) $\mathrm{St}(W, \mathcal{W}) \subset U$ and
(ii) every map $f: \partial B^{k+1} \to \mathrm{St}(W, \mathcal{W})$ extends to a map $F: B^{k+1} \to U$
      $(k = 0, 1, ..., n)$.

Standard refinement operations settle the result below.

Lemma 6.   *Suppose X is a metric space, G is a UV^n decomposition of X, Z is a subset of X/G, and $\mathcal{U}$ is a cover of Z by open subsets of X/G. Then there exists another cover $\mathcal{W}$ of Z by open subsets of X/G such that $\pi^{-1}(\mathcal{W})$ n-homotopy star-refines $\pi^{-1}(\mathcal{U})$.*

Lemma 6 aids in showing that whenever G satisfies certain UV Properties then $\pi: X \to X/G$ behaves, to a limited extent, approximately like a fibration. The proof of this, given for the ensuing theorem, represents the central technique of this section.

Theorem 7 (approximate lifting).   *Suppose X is a metric space; G is a UV^{n-1} decomposition $(n \geq 1)$ of X; K is a finite simplicial n-complex; L is a subcomplex of K; f is a map of K to X/G; F_L is a map of L to X such that $\pi F_L = f \mid L$; and $\varepsilon > 0$.*
   Then there exists a map $F: K \to X$ such that $\rho(f, \pi F) < \varepsilon$ and $F \mid L = F_L$.

Proof.   For each $z \in f(K)$ there exists a G-saturated open subset U_z of X such that $z \in \pi(U_z)$ and $\mathrm{diam}\ \pi(U_z) < \varepsilon/2$. This generates a G-saturated open cover $\mathcal{U}_n = \{U_z \mid z \in f(K)\}$ of $\pi^{-1}f(K)$. Repeated application of Lemma 6 provides successive open covers $\mathcal{U}_{n-1}, ..., \mathcal{U}_1, \mathcal{U}_0$ of $\pi^{-1}f(K)$ by G-saturated open subsets of X such that $\mathcal{U}_i$ $(n - 1)$-homotopy star-refines $\mathcal{U}_{i+1}$ $(i = 0, 1, ..., n - 1)$.

Subdivide K into simplexes so small that to each σ in the new complex, still denoted as K, there corresponds $U_\sigma^0 \in \mathcal{U}_0$ with $f(\sigma) \subset \pi(U_\sigma^0)$. We shall produce successive extensions $F_0, F_1, ..., F_n$ of F_L, where F_i is a map defined on L and the i-skeleton of K, such that πF_i is close to f, appropriately restricted.

First, extend F_L across $L \cup K^{(0)}$ to a map F_0 such that for $v \in K^{(0)}$ $\pi F_0(v) = f(v)$. Given a 1-simplex σ of K and the promised $U_\sigma^0 \in \mathcal{U}_0$ for which $f(\sigma) \subset \pi(U_\sigma^0)$, we see that $F_0(\partial \sigma) \subset U_\sigma^0$. Because $\mathcal{U}_0$ $(n-1)$-homotopy star-refines $\mathcal{U}_1$, F_0 extends to a map F_1 of $L \cup K^{(1)}$ into X such that $F_1 \mid \partial \sigma = F_0 \mid \partial \sigma$ and $F_1(\sigma)$ lies in some element of $\mathcal{U}_1$, for every 1-simplex σ in K.

Suppose inductively we have defined F_i on $L \cup K^{(i)}$ so that $F_i \mid L \cup K^{(0)} = F_0$ and that for every $\sigma \in K^{(i)}$ there exists $U_\sigma^i \in \mathcal{U}_i$ for which $F_i(\sigma) \subset U_\sigma^i$. In order to extend F_i to a map $F_{i+1}: L \cup K^{(i+1)} \to X$, consider $\tau \in K^{(i+1)} - K^{(i)}$: fix a simplex γ in $\partial \tau$, find $U_\gamma^i \in \mathcal{U}_i$ with $F_i(\gamma) \subset U_\gamma^i$, note that $F_i(\partial \tau) \subset \operatorname{St}(U_\gamma^i, \mathcal{U}_i)$, and apply the prearranged fact that $\mathcal{U}_i$ $(n-1)$-homotopy star-refines $\mathcal{U}_{i+1}$ to produce F_{i+1}, where $F_{i+1} \mid \tau$ satisfies $F_{i+1} \mid \partial \tau = F_i \mid \partial \tau$ and $F_{i+1}(\tau)$ lies in some element of $\mathcal{U}_{i+1}$.

The desired map $F: K \to X$ is the final map F_n. Yet to be shown, however, is that πF is close to f. Consider a point x in K and a simplex σ with $x \in \sigma \in K$. By construction, $F(\sigma) \subset U_\sigma^n \in \mathcal{U}_n$, which forces $\operatorname{diam} \pi F(\sigma) < \varepsilon/2$. Furthermore, there exists $U \in \mathcal{U}_n$ such that $f(\sigma) \subset \pi(U)$, so $\operatorname{diam} f(\sigma) < \varepsilon/2$ as well. Choose a vertex v of σ and recall that $\pi F(v) = f(v)$. This yields

$$\rho(f(x), \pi F(x)) \leq \rho(f(x), f(v)) + \rho(\pi F(v), \pi F(x)) < \varepsilon/2 + \varepsilon/2 = \varepsilon. \quad \blacksquare$$

A pair (A_1, A_2) of subspaces of a space S, where $A_2 \subset A_1$, is said to be *n-connected* if, for each map $f: (B^k, \partial B^k) \to (A_1, A_2)$, where $k \in \{0, 1, ..., n\}$, there is a homotopy $H_t: (B^k, \partial B^k) \to (A_1, A_2)$ such that $H_0 = f$ and $H_1(B^k) \subset A_2$.

Corollary 7A. *Suppose X is a metric space, G is a UV^n decomposition of X, and W_1, W_2 are open subsets of X/G with $W_2 \subset W_1$ and (W_1, W_2) n-connected. Then $(\pi^{-1}(W_1), \pi^{-1}(W_2))$ is n-connected.*

Given a UV^{n-1} decomposition G of X and a map f of an n-complex K into X/G, one learns from Theorem 7 how to find a map $F: K \to X$ such that πF is close to f. If G is a UV^n decomposition of X, one can find $F: K \to X$ such that not only is πF close to f but it is homotopic to f under a homotopy that stays close to f.

Theorem 8. *Suppose X is a metric space; G is a UV^n decomposition of X; K is a finite simplicial n-complex; L is a subcomplex of K; f is a map of K into X/G, F_L is a map of L to X such that $\pi F_L = f \mid L$; and $\varepsilon > 0$.*

Then there exists a map F of K to X with $F|L = F_L$ and there exists a homotopy H_t of K in X/G such that $H_0 = \pi F$, $H_1 = f$, $H_t|L = f|L$, and $\rho(f, H_t) < \varepsilon$ for all $t \in [0, 1]$.

The lemma below supplies the main ingredient for the proof.

Lemma 9.   *Under the hypotheses of Theorem 8, there exists $\delta > 0$ such that, for any two maps α_0, α_1 of K in X extending F_L with $\rho(f, \pi\alpha_i) < \delta$ for $i \in \{0, 1\}$, there is a homotopy h_t of K in X satisfying*

(a) $h_0 = \alpha_0$,
(b) $h_1 = \alpha_1$,
(c) $h_t|L = F_L$, *and*
(d) $\rho(f, \pi h_t) < \varepsilon$ *for all $t \in [0, 1]$.*

The argument establishing the lemma is based on the same techniques used to establish Theorem 7. With Lemma 9 in hand, the proof of Theorem 8 proceeds like the one showing the point $a \in A \subset S$ [with $a \in \text{Cl}(S - A)$] to be pathwise-accessible from $S - A$ if $S - A$ is locally pathwise-connected at a; this argument was outlined earlier in Theorem 13.4.

Theorem 10.   *Suppose X is a metric space; G is a UV^n decomposition of X; W is an open subset of X/G; $P: X \to X/G$ represents the projection map; and $w \in P^{-1}(W)$. Then $P_*: \pi_k(P^{-1}(W), w) \to \pi_k(W, P(w))$ is an isomorphism for $k \in \{0, 1, ..., n\}$.*

Proof.   To see that P_* is onto, name a map $\mu: \partial B^{k+1} \to W$ sending base point $s \in \partial B^{k+1}$ to $P(w)$. By Theorem 8 there exists a map $\tilde{\mu}: \partial B^{k+1} \to P^{-1}(W)$ sending s to w such that $P\tilde{\mu}$ is homotopic to μ in W. Thus, $P_*[\tilde{\mu}] = [\mu]$. To see that P_* is one-to-one, consider a map $F: (\partial B^{k+1}, s) \to (P^{-1}(W), w)$ such that $P_*[F]$ is trivial. Then PF extends to a map f of B^{k+1} to W, and by Theorem 7, f lifts (approximately) to a map F' of B^{k+1} to $P^{-1}(W)$ such that $F'|\partial B^{k+1} = F$, indicating that $[F]$ is trivial.

Corollary 10A.   *Suppose X is a metric space; G is a UV^ω decomposition of X; $P: X \to X/G$ denotes the decomposition map; W is an open subset of X/G; $w \in P^{-1}(W)$; and k is a nonnegative integer. Then $P_*: \pi_k(P^{-1}(W), w) \to \pi_k(W, P(w))$ is an isomorphism.*

Corollary 10B.   *Suppose Y is an ANR; G is a cell-like decomposition of Y; $P: Y \to Y/G$ denotes the decomposition map; W is an open subset of Y/G; $w \in P^{-1}(W)$; and k is a nonnegative integer. Then $P_*: \pi_k(P^{-1}(W), w) \to \pi_k(W, P(w))$ is an isomorphism.*

The UV properties offer a convenient framework for describing certain other local properties. We say that a space S is *locally n-connected,* abbreviated as LC^n, if each point $s \in S$ has Property UV^n in S; we also write that S is $n - LC$ or LC^ω if the analogous UV Property holds for each $s \in S$.

Theorem 11. *If G is a UV^n decomposition of a metric space X, then X/G is LC^n.*

This is an easy consequence of Theorem 8.

Corollary 11A. *If G is a UV^ω decomposition of a metric space X, then X/G is LC^ω.*

Corollary 11B. *If G is a cell-like decomposition of an ANR Y, then Y/G is LC^ω.*

A glance back to the proof of Theorem 14.8 will reveal that it actually confirms the stronger statement below.

Theorem 12. *If A is a compact LC^{n-1} subset of E^n or of any simplicial n-complex, then A is an ANR.*

Corollary 12A. *If G is a UV^ω decomposition of a CANR Y such that Y/G is finite-dimensional, then Y/G is an ANR.*

Corollary 12B. *If G is a cell-like decomposition of a (locally) compact ANR Y such that Y/G is finite-dimensional, then Y/G is an ANR.*

EXERCISES

1. Prove Proposition 3.
2. Prove Theorem 8.
3. Suppose K is a finite simplicial n-complex with the fixed point property and G is a UV^{n-1} decomposition of K. Show that K/G also has the fixed point property.
4. Show that every map of a simplicial n-complex K to an n-connected space X is null homotopic.
5. Suppose G is a UV^ω decomposition of a CAR Y such that Y/G is finite-dimensional. Show that Y/G is a CAR.

17. CELL-LIKE DECOMPOSITIONS AND DIMENSION

Consider a usc decomposition G of a locally compact ANR Y. Under what conditions will Y/G also be an ANR? First principles (Section 5) indicate it will be if G is shrinkable, but that is a rather unsatisfactory answer, for shrinkability is an exceedingly strong property. On one hand it is very difficult to detect and on the other, as we shall see, it is unduly restrictive. A better answer is derived in this section: Y/G is an ANR in case G is cell-like and Y/G is finite-dimensional. This brings up another issue: if Y is finite-dimensional, when will Y/G also be finite-dimensional?

Our investigation into these questions begins with a look at the setting where both Y and Y/G are ANRs and involves a useful notion of controlled homotopy equivalence. A decomposition map $\pi\colon X \to X/G$ is said to be a *fine homotopy equivalence* if, for each open cover $\mathcal{W}$ of X/G, there exists a map $F\colon X/G \to X$ such that πF is $\mathcal{W}$-homotopic to the identity [which means there exists a homotopy h_t between πF and $\mathrm{Id}_{X/G}$ such that, for each $z \in X/G$, some $W_z \in \mathcal{W}$ contains $h_t(z)$ for all t] and that $F\pi$ is $\pi^{-1}(\mathcal{W})$-homotopic to Id_X. In case X is compact and metric, this translates into the assertion that for every $\varepsilon > 0$ there is a map $F\colon X/G \to X$ such that πF is ε-homotopic to $\mathrm{Id}_{X/G}$ and $F\pi$ is connected to Id_X via a homotopy H_t for which πH_t is an ε-homotopy.

The tie with cell-like decompositions is exposed in the following result, originally developed by G. Kozlowski [1] and W. E. Haver [1].

Theorem 1. *Suppose Y and Y/G are locally compact ANRs, where G is a usc decomposition of Y. Then G is cell-like if and only if the decomposition map $\pi\colon Y \to Y/G$ is a fine homotopy equivalence.*

Proof. Suppose π is a fine homotopy equivalence. Fix $g \in G$ and an open subset U of Y containing g. Without loss of generality, U is G-saturated. Using the open cover $\mathcal{W} = \{\pi(U), Y/G - \pi(g)\}$ of Y/G, apply the hypothesis to determine a map $F\colon Y/G \to Y$ such that $F\pi$ is $\pi^{-1}(\mathcal{W})$-homotopic to Id_Y via a homotopy $h_t\colon Y \to Y$ satisfying $h_0 = \mathrm{Id}_Y$ and $h_1 = F\pi$. Then $h_t \mid g$ acts as a contraction of $g = h_0(g)$ to the point $F\pi(g)$ and $h_t(g) \subset U$ for each $t \in I$, which shows g to be cell-like in Y.

Next suppose that G is cell-like. We consider only the case in which Y is compact. The central idea in the proof can be isolated by regarding both Y and Y/G as simplicial complexes, for then Theorem 16.7 quickly provides a lift $F\colon Y/G \to Y$, with πF close to $\mathrm{Id}_{Y/G}$. When it is sufficiently close, πF is ε-homotopic to $\mathrm{Id}_{Y/G}$, forcing $\pi F\pi$ to be ε-homotopic to π as well, and then their two lifts $F\pi$ and Id_Y have projections so close to π that, according to Lemma 16.9, they are connected in Y by a homotopy whose image under π moves points less than ε.

Growing out of this idea, the argument for arbitrary CANRs imposes, in addition, a kind of double-entry bookkeeping to account for the way a CANR can be replaced by, or treated as, a simplicial complex.

Lemma 2 (tube lemma). *Suppose S is a compact subset of I^∞ and $\eta > 0$. Then there exist an integer $k > 0$ and a finite simplicial complex P in $[-1, 1]^k$ such that*

$$S \subset P \times Q_k \subset N(S; \eta),$$

where Q_k denotes the product of all but the first k $[-1, 1]$-factors in I^∞.

Proof. After covering S by basic open subsets of I^∞ in $N(S; \eta)$, one invokes compactness to transform the problem at hand into an analogous problem for $S \subset [-1, 1]^k$. ∎

Continuing with the proof that π is a fine homotopy equivalence, we arrange Y and Y/G as embedded subsets of I^∞. Being ANRs, they have closed neighborhoods U and U' there, endowed with retractions $r: U \to Y$ and $r': U' \to Y/G$.

Let $\varepsilon > 0$. Find $\gamma > 0$ such that diam $\pi(A) < \varepsilon/8$ whenever $A \subset Y$ and diam $A < \gamma$, and then find $\eta > 0$ such that diam $r(B) < \gamma$ whenever $B \subset U$ and diam $B < \eta$. Apply Lemma 2 to locate a finite complex P in $[-1, 1]^k$ with $Y \subset P \times Q_k \subset N(Y; \eta) \cap U$. Increase k, if necessary, so that diam$(\{z\} \times Q_k) < \eta$ for all $z \in [-1, 1]^k$ (this entails stripping some $[-1, 1]$-factors from Q_k and adjoining them to P). Name the projection map $p: P \times Q_k \to P$ and the inclusions $e: P \to P \times \{0\} \subset P \times Q_k$ and $i: Y \to P \times Q_k$. At this point we have maps

$$P \underset{e}{\overset{p}{\rightleftarrows}} P \times Q_k \underset{r}{\overset{i}{\leftrightarrows}} Y,$$

and there exists a natural η-homotopy ψ_t between $ep: P \times Q_k \to P \times \{0\}$ and $\mathrm{Id}_{P \times Q_k}$. As a result,

(1) $r\psi_t i$ is a γ-homotopy between $repi$ and $ri = \mathrm{Id}_Y$, and
(2) $\pi r\psi_t i$ is an $(\varepsilon/8)$-homotopy in Y/G between $\pi repi$ and π.

The map $\pi re: P \to Y/G$ comes equipped with a lift $\alpha_0 = re$ to Y. By Lemma 16.9, there exists $\delta > 0$ such that if $\alpha_1: P \to Y$ is an approximate lift with $\rho(\pi\alpha_1, \pi re) < \delta$, then $\alpha_0 = re$ and α_1 are related by a homotopy $\theta_t: P \to Y$ with $\theta_0 = \alpha_0$, $\theta_1 = \alpha_1$, and $\rho(\pi\theta_t(q), \pi re(q)) < \varepsilon/8$ for all $q \in P$ and $t \in I$.

Proposition 14.5 gives $\beta > 0$ such that any two β-close maps from a space S into Y/G are $(\varepsilon/8)$-homotopic in Y/G. There also exists $\eta' > 0$ such that diam $r'(B') < \min\{\delta/2, \beta\}$ whenever $B' \subset U'$ and diam $B' < \eta'$. Apply Lemma 2 again to obtain a finite complex P' in $[-1, 1]^m$ with $Y/G \subset P' \times Q_m \subset N(Y/G; \eta') \cap U'$, and also with each set $\{z\} \times Q_m$ of size less than η'. As before, name the projection $p': P' \times Q_m \to P'$ and name the inclusions $e': P' \to P' \times \{0\} \subset P' \times Q_m$ and $i': Y/G \to P' \times Q_m$. Thus, we have maps

$$P' \underset{e'}{\overset{p'}{\rightleftarrows}} P' \times Q_m \underset{r'}{\overset{i'}{\leftrightarrows}} Y/G,$$

plus a natural η'-homotopy ψ'_t in $P' \times Q_m$ between $e'p'$ and $\mathrm{Id}_{P' \times Q_m}$, which yields

(3) $r'\psi_t'i'$ is a homotopy in Y/G between $r'e'p'i'$ and $r'i' = \mathrm{Id}_{Y/G}$ moving points less than $\min\{\delta/2, \beta\}$.

Now we have established the existence of all the spaces and maps diagrammed below. According to Theorem 16.7, there exists a map $f: P' \to Y$ satisfying

$$\rho(\pi f, r'e') < \min\{\delta/2, \beta\}.$$

Define the desired map $F: Y/G \to Y$ as $F = fp'i'$.

$$P \underset{e}{\overset{p}{\rightleftarrows}} P \times Q_k \underset{r}{\overset{i}{\leftrightharpoons}} Y$$
$$\Big\downarrow \pi$$
$$P' \underset{e'}{\overset{p'}{\rightleftarrows}} P' \times Q_m \underset{r'}{\overset{i'}{\leftrightharpoons}} Y/G$$

Having prescribed the map, we must still show that it is a fine homotopy inverse to π. To see that πF is ε-homotopic to $\mathrm{Id}_{Y/G}$, note that, by choice of β, πf is $(\varepsilon/8)$-homotopic to $r'e'$ via a homotopy θ_t', and then $\theta_t'p'i'$ serves as a homotopy between $\pi fp'i' = \pi F$ and $r'e'p'i'$. According to (3), $r'e'p'i'$ is $(\varepsilon/8)$-homotopic to $\mathrm{Id}_{Y/G}$. Strung together, the latter two homotopies provide an $(\varepsilon/4)$-homotopy between πF and $\mathrm{Id}_{Y/G}$.

To demonstrate that $F\pi$ is $\pi^{-1}(\varepsilon)$-homotopic to Id_Y, we shall produce three homotopies, one between Id_Y and $repi$, another between $repi$ and $F\pi repi$, and a third between $F\pi repi$ and $F\pi$. Consider the lift $\alpha_1: P \to Y$ given by $\alpha_1 = F\pi re$. Then

$$(4) \quad \rho(\pi\alpha_1, \pi re) = \rho(\pi fp'i'\pi re, \pi re)$$
$$\leq \rho(\pi fp'i', \mathrm{Id}_{Y/G})$$
$$\leq \rho(\pi fp'i', r'e'p'i') + \rho(r'e'p'i', \mathrm{Id}_{Y/G})$$
$$\leq \rho(\pi f, r'e') + \rho(r'e'p'i', \mathrm{Id}_{Y/G})$$
$$< \delta/2 + \delta/2 = \delta,$$

where the final inequality follows from approximation properties of f and from (3). By our choice of δ, there exists a homotopy $\theta_t: P \to Y$ such that $\theta_0 = re$, $\theta_1 = F\pi re$, and diam $\pi\theta(\{q\} \times I) < \varepsilon/8$ for every $q \in P$. The homotopy $r\psi_t i$ of (1) between Id_Y and $repi$ and the homotopy $\theta_t pi$ between $repi$ and $F\pi repi$ combine to give a homotopy between Id_Y and $F\pi repi$ whose image under π moves points less than $\varepsilon/4$. Finally, as in (4) above,

$$(5) \quad \rho(\pi F\pi, \pi) \leq \rho(\pi F, \mathrm{Id}_{Y/G})$$
$$= \rho(\pi fp'i', \mathrm{Id}_{Y/G})$$
$$\leq \rho(\pi f, r'e') + \rho(r'e'p'i', \mathrm{Id}_{Y/G})$$
$$< 2\beta < \varepsilon/4.$$

Applying (2) and (5), we see that $F\pi r\psi_t i$ is a homotopy between $F\pi repi$ and $F\pi$ whose image under π moves points along tracks of diameter less than $3\varepsilon/4$. Altogether, these yield a homotopy between Id_Y and $F\pi$ whose π-image has tracks of diameter less than ε, as required. ∎

Corollary 1A. *If G is a usc decomposition of a metric space X and $\pi: X \to X/G$ is a fine homotopy equivalence, then each $g \in G$ is cell-like in X.*

The corollary holds because the proof given for the comparable implication in Theorem 1 used neither compactness properties nor properties of ANRs.

Theorem 1 paves the way for a quick proof of the statement that the uniform limit of cell-like mappings between ANRs is cell-like. A (proper) map $f: X \to X'$ between locally compact metric spaces is said to be *cell-like* if each point-inverse $f^{-1}(x')$, $x' \in X'$, is a cell-like set.

Before addressing the limit of cell-like mappings, we need a technical result.

Lemma 3. *Suppose G is a usc decomposition of a metric space X such that $\pi: X \to X/G$ is a fine homotopy equivalence. Suppose V is an open subset of X/G and C is a compact subset of V with C being contractible in V. Then $\pi^{-1}(C)$ is contractible in $\pi^{-1}(V)$.*

Proof. Name a contraction $\theta_t: C \to V$. Use compactness to find a set V' such that $\mathcal{V} = \{V, V'\}$ is an open cover for X/G and $\theta_t(C) \cap V' = \varnothing$ for all t. Then obtain a map $F: X/G \to X$ for which $F\pi$ is $\pi^{-1}(\mathcal{V})$-homotopic to Id_X.

The inclusion $\pi^{-1}(C) \to \pi^{-1}(V)$ is homotopic to $F\pi \mid \pi^{-1}(C)$. The latter, in turn, is homotopic to the constant map $F\theta_1\pi \mid \pi^{-1}C$, via the homotopy $F\theta_t\pi$. It is easily seen that both homotopies operate in $\pi^{-1}(V)$, as required. ∎

Theorem 4. *Suppose Y and Y′ are ANRs and suppose $\{f_i \mid i = 1, 2, \ldots\}$ is a sequence of proper cell-like maps of Y onto Y′ converging uniformly to $f: Y \to Y'$. Then f is a proper cell-like map.*

Proof. Uniform convergence ensures f is continuous and onto. The proof that it is also proper is left as an exercise. What remains is a verification of its cell-likeness, taken from Lacher's survey of cell-like mappings (Lacher [2]).

For $y' \in Y'$ and a neighborhood O of $f^{-1}(y')$, find neighborhoods U, V, and W of y' in Y' such that

$$\bar{W} \subset V \subset \bar{V} \subset U \subset \bar{U},$$

where $\bar{U}$ is compact, $f^{-1}(\bar{U}) \subset O$, and $\bar{W}$ is contractible in V.

Claim 1.   *There exists an integer $N > 0$ such that $f_n^{-1}(V) \subset f^{-1}(U)$ for all $n \geq N$.*

Proof.   Otherwise, there would be points $x_n \in f_n^{-1}(V) - f^{-1}(U)$ for infinitely many n. In the case where Y is compact we can assume, without loss of generality, $\{x_n\}$ converges to some $x_0 \in Y$. Then, of course, $\{f(x_n)\}$ converges to $f(x_0)$, implying $f(x_0) \notin U$ because no $f(x_n) \in U$. Due to the uniform convergence of $\{f_n\}$, $\{f_n(x_n)\}$ also converges to $f(x_0)$. Since $f_n(x_n) \in V$, it follows that $f(x_0) \in \bar{V} \subset U$, a contradiction.

In the noncompact case, the same argument works after passage to one-point compactifications, which is permissible by properness (Proposition 3.2) and uniform convergence.

Claim 2.   *There exists an integer $M > 0$ such that $f^{-1}(y') \subset f_n^{-1}(W)$ for all $n \geq M$.*

Verification, which is similar to that for Claim 1, is left as an exercise.

Choose $n = N + M$. Then $f^{-1}(y') \subset f_n^{-1}(W) \subset f_n^{-1}(V) \subset O$. Using the result of Theorem 1 that f_n is a fine homotopy equivalence [according to Theorem 3.5, f_n essentially functions as the decomposition map for the cell-like, usc decomposition $\{f_n^{-1}(z) \mid z \in Y'\}$], we invoke Lemma 3 to conclude that $f_n^{-1}(\bar{W})$, and thus $f^{-1}(y')$, is contractible in $f_n^{-1}(V) \subset O$.   ∎

Remark.   The convergence of $f_n \to f$ required in Theorem 3 need not be uniform throughout Y but only over compact subsets of Y'.

Mentioned next is a well-known fact from dimension theory. In this context, an *ε-map* of a metric space S to another space X means a map $f: S \to X$ such that $\operatorname{diam} f^{-1}(x) < \varepsilon$ for each $x \in X$.

Proposition 5.   *Suppose S is a compact metric space and $n \geq 0$ is an integer such that for each $\varepsilon > 0$ there exists an ε-map f_ε of S into an n-dimensional metric space X_ε. Then $\dim S \leq n$.*

Proof.   This hinges on the existence of $\delta > 0$ such that

$$\operatorname{diam} f_\varepsilon^{-1}(A) < \varepsilon \qquad \text{whenever } A \subset X_\varepsilon \text{ and } \operatorname{diam} A < \delta.$$

Since $f_\varepsilon(S)$ is n-dimensional, there is a δ-map g of $f_\varepsilon(S)$ to some finite n-complex P; then $gf_\varepsilon: S \to P$ is an ε-map, and S must be n-dimensional (Hurewicz–Wallman [1, p. 72]).   ∎

Proposition 6.   *If G is a usc decomposition of a locally compact metric space X such that $\pi: X \to X/G$ is a fine homotopy equivalence, then $\dim(X/G) \leq \dim X$.*

Proof. Obviously the result holds in case X is infinite-dimensional. Otherwise, for each $\varepsilon > 0$, the definition of fine homotopy equivalence promises a map $F_\varepsilon \colon X/G \to X$ such that πF_ε is $(\varepsilon/2)$-close to $\mathrm{Id}_{X/G}$, causing F_ε to be an ε-map. Proposition 5 certifies that $\dim(X/G) \le \dim X$.   ∎

Corollary 6A.   *Suppose Y is a locally compact ANR, G is a cell-like decomposition of Y, and Y/G is an ANR. Then $\dim(Y/G) \le \dim Y$.*

Corollary 6B.   *Cell-like maps of one locally compact ANR onto another do not raise dimension.*

A fundamental example due to J. L. Taylor [1] describes a cell-like decomposition G of I^∞ for which I^∞/G is not a CANR. No such example has been found to date on a finite-dimensional CANR. The remaining question stands as the most significant unresolved problem about cell-like decompositions. Specifically:

Question.   If G is a cell-like decomposition of a finite-dimensional CANR Y, is Y/G a CANR?

Equivalent Question.   If G is a cell-like decomposition of a finite-dimensional compact metric space X, is X/G finite-dimensional?

In compact metric spaces the analogue of Corollary 6B is nearly valid, except that the conclusion propounds a drastic either/or—either a given cell-like map does not increase dimension or it raises the dimension to infinity.

Theorem 7.   *Suppose X is a compact metric space and G is a cell-like decomposition of X. Then either $\dim X/G \le \dim X$ or $\dim X/G = \infty$.*

Proof.   It suffices to show that $\dim X/G \le \dim X$ in case both X and X/G are finite-dimensional. Letting $n = \dim X$, we shall produce, for each $\varepsilon > 0$, an ε-map of X/G to an n-dimensional simplicial complex.

Think of X as a subset of a finite-dimensional CANR Y (for instance, $Y = I^{2n+1}$) and extend G trivially to the decomposition G^{T} of Y. Then $\pi(Y) = Y/G^{\mathrm{T}}$ is composed of $\pi(X) \approx X/G$ and $\pi(Y - X) \approx Y - X$ (because $\pi \,|\, Y - X$ is a homeomorphism of $Y - X$ onto its image). As a result, $\pi(Y)$, expressed as a union of two finite-dimensional subspaces, is itself finite-dimensional (Hurewicz–Wallman [1]) and Corollary 16.12B indicates that Y/G^{T} is a CANR.

Name $\varepsilon > 0$, and then compute $\delta > 0$ so that $\mathrm{diam}\,\pi(A) < \varepsilon/3$ whenever $A \subset Y$ has diameter less than δ. Since $\dim X = n$, there exists a δ-map μ of X to a finite n-complex P. Because P is an ANR, μ extends to another δ-map $\bar{\mu} \colon U \to P$ defined on some neighborhood U of X in Y.

According to Theorem 1, π is a fine homotopy equivalence. In particular, this gives a map $f: Y/G^T \to Y$ such that $f\pi(X) \subset U$ and πf is $(\varepsilon/3)$-homotopic to the identity. Then $\bar{\mu}f \mid \pi(X): \pi(X) \approx X/G \to P$ is necessarily an ε-map, for $f^{-1}\bar{\mu}^{-1}(p) \subset N(\pi(\bar{\mu}^{-1}(p)); \varepsilon/3)$. ■

Corollary 7A. *Cell-like maps between finite-dimensional compact metric spaces do not raise dimension.*

In view of these results, it is appropriate to seek conditions on a finite-dimensional space X implying $\dim X/G \leq \dim X$. Very little is known; the theorems, for the most part, pertain to surprisingly low-dimensional spaces. We shall start with an easily grasped result about cell-like decompositions of 1-dimensional compacta. Its proof forms the design of the one with which we end, dealing with cell-like decompositions of 3-manifolds.

Theorem 8. *If G is a cell-like decomposition of a 1-dimensional compact metric space X, then* $\dim(X/G) \leq 1$.

This will be proved by producing an ε-map of X/G to a 1-complex P. To accomplish this, several facts about 1-complexes will be employed. They include:

(1) If $\tilde{P}$ is a connected 1-complex such that $\pi_1(\tilde{P})$ is trivial and if K is a connected subcomplex of $\tilde{P}$, then $\pi_1(K)$ is trivial.

(2) A (compact) connected 1-complex K is contractible iff $\pi_1(K)$ is trivial.

(3) If $\tilde{P}$ is a simply connected 1-complex and X is a compact, connected subset of $\tilde{P}$, then X is contractible. This follows because, after subdivision of $\tilde{P}$, X underlies a subcomplex.

(4) If f is a map of ∂B^{k+1} ($k \geq 2$) to a 1-complex P, then f can be extended to a map F of B^{k+1} to $f(\partial B^{k+1})$.

Proof of (4). Consider the universal covering $p: \tilde{P} \to P$. Since ∂B^{k+1} is simply connected, f lifts to $\tilde{f}: \partial B^{k+1} \to \tilde{P}$ with $p\tilde{f} = f$. By (3), $\tilde{f}(\partial B^{k+1})$ is contractible, so $\tilde{f}$ extends to $\tilde{F}: B^{k+1} \to \tilde{f}(\partial B^{k+1})$. Define F as $p\tilde{F}$. ■

(5) If K is a simplicial complex and f is a map of its 2-skeleton $K^{(2)}$ to a 1-complex P, then f extends to a map F of K to P such that $F(\sigma) = f(\sigma \cap K^{(2)})$ for every simplex σ of K.

This fact, the key one, follows directly by extending f over successive skeleta, with (4) being used to obtain extensions on the top-dimensional simplexes of those skeleta.

Proof of Theorem 8. Embed X in a CANR Y and extend G trivially to the decomposition G^T of Y, with decomposition map $\pi: Y \to Y/G^T$. Regard Y/G^T as a subset of I^∞.

Given $\varepsilon > 0$, find $\delta > 0$ such that diam $\pi(A) < \varepsilon$ whenever $A \subset Y$ and diam $A < \delta$. The 1-dimensionality of X implies the existence of a δ-map μ of X to a 1-complex P, which then extends to another δ-map $\bar{\mu} \colon O \to P$ defined on some neighborhood O of X in Y.

Construct a G^{T}-saturated cover $\mathfrak{U}_2$ of X by open subsets of O such that diam $\pi(U_2) < \varepsilon$ for each $U_2 \in \mathfrak{U}_2$. Apply Lemma 16.6 to obtain G-saturated covers $\mathfrak{U}_1$ and $\mathfrak{U}_0$ by open subsets of O, where $\mathfrak{U}_1$ 1-homotopy star-refines $\mathfrak{U}_2$ and $\mathfrak{U}_0$ 0-homotopy star-refines $\mathfrak{U}_1$.

Determine $\beta \in (0, \varepsilon)$ such that 3β-subsets of $\pi(X)$ lie in some $\pi(U_0)$, $U_0 \in \mathfrak{U}_0$.

In the next step $\pi(X)$ is replaced by a finite complex. Lemma 2 provides a finite complex K in $[-1, 1]^k$ for which

$$\pi(X) \subset K \times Q_k \subset N(\pi(X); \beta).$$

Without loss of generality, k is large enough that the projection map $R \colon K \times Q_k \to K$, where K is equated with $K \times \{0\}$, moves points less than ε. Subdivide $K = K \times \{0\}$ so that mesh $K < \beta$, and toss out any unnecessary simplexes of K, so that $R\pi(X)$ meets each simplex of K whose interior is an open subset of K.

The aim of what follows is to complete the diagram below, by producing appropriate maps f and F.

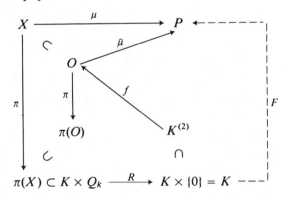

For each vertex $v \in K$ there exists some $x_v \in \pi(X)$ such that $\rho(v, x_v) < \beta$. Choose $f(v) \in \pi^{-1}(x_v)$.

As a result, for each 1-simplex $\sigma \in K$, $\pi f(\partial \sigma) \subset N(\sigma; \beta)$, which is a set of diameter less than 3β. The choice of β ensures that $f(\partial \sigma)$ lies in some $U_0 \in \mathfrak{U}_0$, and the determination of $\mathfrak{U}_0$ as a 0-homotopy refinement of $\mathfrak{U}_1$ guarantees that f, as defined on the vertex set $K^{(0)}$ of K, extends to a map on the 1-skeleton $K^{(1)}$ so that $f(\sigma) \subset U_\sigma \in \mathfrak{U}_1$ for all $\sigma \in K^{(1)}$.

Focus on a 2-simplex τ of K and 1-simplex σ in $\partial\tau$. As in the proof of Theorem 16.7, $f(\partial\tau) \subset \mathrm{St}(U_\sigma, \mathfrak{U}_1)$. Since $\mathfrak{U}_1$ 1-homotopy star-refines $\mathfrak{U}_2$,

$f \mid \partial\tau$ extends to a map $f: \tau \to P$ with $f(\tau) \subset U_\tau \in \mathcal{U}_2$. Doing this throughout gives the required map $f: K^{(2)} \to O$ such that $f(\tau) \subset U_\tau \in \mathcal{U}_2$ for all $\tau \in K^{(2)}$.

By fact (5) mentioned prior to this proof, $\bar\mu f$ extends to the required map $F: K \to P$ with $F(\gamma) = \bar\mu f(\gamma \cap K^{(2)})$ for all $\gamma \in K$.

This finishes the constructions; the argument concludes with the verification that $FR \mid \pi(X)$ is an 11ε-map. A primary ingredient is that $\rho(\pi f, \text{incl.}) < 3\varepsilon$: each $z \in K^{(2)}$ lies in some 2-simplex $\tau \in K^{(2)}$ with vertex v, and

$$\rho(\pi f(z), z) \leq \rho(\pi f(z), \pi f(v)) + \rho(\pi f(v), v) + \rho(v, z)$$

$$\leq \operatorname{diam} \pi f(\tau) + \beta + \operatorname{diam} \sigma < 3\varepsilon.$$

Consequently, for $p \in P$ and $z \in (\bar\mu f)^{-1}(p)$, $\pi f(z) \in \pi\bar\mu^{-1}(p)$, which implies $z \in N(\pi\bar\mu^{-1}(p); 3\varepsilon)$. Since diam $\pi\bar\mu^{-1}(p) < \varepsilon$, due to controls imposed on $\bar\mu$, it follows that $\bar\mu f$ is a 7ε-map. Turning to F, we recall that to each $z' \in K \cap F^{-1}(p)$ there corresponds $z \in K^{(2)}$ in the same simplex γ as z', with $F(z') = \bar\mu f(z) = p$, which shows $F^{-1}(p) \subset N(f^{-1}\bar\mu^{-1}(p); \varepsilon)$; in other words, F is a 9ε-map. Finally, because R moves points less than $\beta < \varepsilon$, we see in similar fashion that FR is an 11ε-map. ∎

Monotone decompositions of 1-manifolds are shrinkable, but monotone decompositions of 1-dimensional compacta need not be. In contrast to what was just established, monotone decompositions of 1-dimensional compacta can raise dimension: for any compact metric space S Proposition 4.4 promises a monotone usc decomposition G of E^3 such that S embeds in E^3/G, and a reinspection of that construction reveals N_G to be a 1-dimensional compactum and $\pi(N_G)$ to be topologically equivalent to S.

Theorem 9. *If G is a cell-like decomposition of a 2-manifold M, then* $\dim M/G \leq 2$.

Proof. Focus on one point $\pi(g) \in M/G$, and let B_ε denote the frontier in M/G of the ε-neighborhood of $\pi(g)$. In order to prove that the (inductive) dimension of M/G at $\pi(g)$ is ≤ 2, it is enough to show $\dim B_\varepsilon \leq 1$ for all but a countable collection of the numbers ε.

The collection $\{\pi^{-1}(B_\varepsilon)\}$ consists of pairwise disjoint closed subsets of M. If $\dim \pi^{-1}(B_\delta) = 2$, then $\pi^{-1}(B_\delta)$ must contain a nonempty open subset of M (Hurewicz–Wallman [1, p. 46]), and the separability of M prevents an uncountable collection of the sets $\pi^{-1}(B_\varepsilon)$ from having nonvoid interior. As a result, $\dim \pi^{-1}(B_\varepsilon) \leq 1$ for all but a countable collection of the sets $\pi^{-1}(B_\varepsilon)$, and for those B_ε such that $\dim \pi^{-1}(B_\varepsilon) \leq 1$, Theorem 8 establishes $\dim B_\varepsilon \leq 1$. ∎

Corollary 9A. *If G is a cell-like decomposition of a 2-dimensional CANR Y, then* $\dim Y/G \leq 2$.

Proof. K. Sieklucki [1] has shown that no n-dimensional CANR contains uncountably many pairwise disjoint n-dimensional closed subsets. The proof of Theorem 9 settles the corollary. ∎

Logically this proof pattern establishes the hypothetical statement: if cell-like decompositions of k-dimensional compacta do not raise dimension, then cell-like decompositions of $(k + 1)$-manifolds do not raise dimension either. This statement has never been put to wider-ranging use because it is not known whether or not cell-like decompositions of the next type, those defined on 2-dimensional compacta, can raise dimension. It is known, however, that cell-like decompositions of 3-manifolds do not. The original argument is due to G. Kozlowski and J. J. Walsh [1]; the version to be presented here is based on notes of F. D. Ancel. Before embarking on the proof, we mention the salient features of 3-manifolds.

A space S is said to be *aspherical* if, for each $k > 1$, every map of the k-sphere ∂B^{k+1} into S is null homotopic. When M is a connected (triangulated) manifold and $\tilde{M}$ represents its universal cover, then M is aspherical iff $\tilde{M}$ is contractible. Moreover, when M is a 3-manifold and $\tilde{M}$ is noncompact, then M is aspherical iff $H_2(\tilde{M}; \mathbb{Z})$ is trivial, which holds, in turn, iff $\pi_2(M)$ is trivial. The chief support for this conclusion is the Hurewicz isomorphism theorem (Spanier [1, p. 394]).

The fundamental 3-dimensional result is the sphere theorem of C. D. Papakyriakopolous [1]: if M is an orientable 3-manifold with $\pi_2(M)$ nontrivial, then there is a bicollared 2-sphere in M that is not contractible in M.

It helps to be able to discern the noncontractible 2-spheres on geometric grounds. A bicollared 2-sphere S in a 3-manifold M is contractible there iff S bounds a compact, contractible 3-manifold-with-boundary C (a homotopy 3-cell) in M. The derivation requires showing that some component of $M - S$ has compact closure C, passing to the universal cover of M to verify that $\pi_1(C)$ is trivial (for some such C), and invoking duality theory to conclude that $H_*(C; \mathbb{Z})$ is trivial, which then implies the contractibility of C.

With an argument founded upon the foregoing discussion, we derive the lemma below.

Lemma 10. *Let G denote a cell-like decomposition of a 3-manifold M and $\mathcal{U}$ a G-saturated open cover of M. Then there exists an open cover $\mathcal{W} = \{W_g \mid g \in G\}$ of M refining $\mathcal{U}$, where $g \subset W_g$ for all $g \in G$ and where every finite intersection of elements of $\mathcal{W}$ is aspherical.*

Proof. Without loss of generality M is connected. Initially we refine $\mathcal{U}$ so no pair of its elements covers M.

Fix $g \in G$. Choose some $U_g \in \mathfrak{U}$ containing g and find another neighborhood V_g of g that contracts in U_g. There exists a compact, connected 3-manifold-with-boundary Q_g such that

$$g \subset \text{Int } Q_g \subset Q_g \subset V_g \subset U_g.$$

If ∂Q_g is not connected, one can join up the various components of ∂Q_g by arcs in $Q_g - g$ (see Corollary 15.4B) and can delete tubular neighborhoods of these arcs to produce Q_g as above having connected boundary. Let W_g denote the interior of Q_g.

We claim that W_g is aspherical. If not, $\pi_2(W_g) \neq 0$ and W_g contains a bicollared 2-sphere S that is not homotopically trivial in W_g. However, S is contractible in U_g, so it bounds a homotopy 3-cell C there. Since C cannot be confined to W_g, C must contain ∂Q_g (due to its connectedness), and $W_g \cup C = Q_g \cup C$ is a compact manifold (without boundary). This is impossible, for it gives

$$M = W_g \cup C \subset U_g \neq M.$$

We complete the argument by showing that if W_1, $W_2 \in \mathfrak{W}$, then $W_1 \cap W_2$ is aspherical (the same fact for finite intersections follows inductively). If not, as in the preceding paragraph $W_1 \cap W_2$ contains a bicollared 2-sphere S not contractible in $W_1 \cap W_2$. In this situation S bounds homotopy 3-cells C_1 and C_2 in W_1 and W_2, respectively. Either $C_1 = C_2$, in which case an obvious contradiction is at hand, or $C_1 \cap C_2 = S$. The latter is impossible since then $C_1 \cup C_2$ would be a compact 3-manifold, giving

$$M = C_1 \cup C_2 \subset W_1 \cup W_2 \neq M. \quad \blacksquare$$

Lemma 10 will be employed in conjunction with the following.

Lemma 11. *Suppose K is a finite complex; $p \geq 0$ is an integer; $f: K^{(p)} \to M$ is a map; and $\mathfrak{W}$ is an open cover of M satisfying*

(a) *if $W_1, \ldots, W_m \in \mathfrak{W}$ then every map of the r-sphere S^r (for all $r \geq p$) into $\bigcap_i W_i$ is homotopically trivial;*
(b) *for each $\sigma \in K$, some $W(\sigma) \in \mathfrak{W}$ contains $f(\sigma \cap K^{(p)})$.*

Then f extends to a map $F: K \to M$ such that $F(\sigma) \subset W(\sigma)$ for each $\sigma \in K$.

Proof. Set $W^*(\sigma) = \bigcap\{W(\tau) \mid \sigma \subset \tau \in K\}$ and note that $f(\sigma \cap K^{(p)}) \subset W^*(\sigma)$ for all $\sigma \in K$.

Let $F^{(p)} = f$. For $k > p$ we construct inductively a map $F^{(k)}: K^{(k)} \to X$ satisfying $F^{(k)} \mid K^{(k-1)} = F^{(k-1)}$ and $F^{(k)}(\sigma \cap K^{(k)}) \subset W^*(\sigma)$ for all $\sigma \in K$. Assuming $F^{(k-1)}$ has been obtained, one defines $F^{(k)}$ on an arbitrary k-simplex $\gamma \in K$ using (a) above to extend $F^{(k-1)} \mid \partial \gamma$, which by hypothesis sends $\partial \gamma$ into $W^*(\gamma)$, to a map of γ into $W^*(\gamma)$.

To verify the inductive hypothesis that $F^{(k)}(\sigma \cap K^{(k)}) \subset W^*(\sigma)$, it must be shown that whenever σ', σ, $\tau \in K$ where

$$\sigma' \subset \sigma \cap K^{(k)} \subset \sigma \subset \tau,$$

then $F^{(k)}(\sigma') \subset W(\tau)$. It suffices to show $F^{(k)}(\sigma') \subset W^*(\sigma')$. If dim $\sigma' = k$, this occurs by construction; if dim $\sigma' < k$, then $F^{(k)} | \sigma' = F^{(k-1)} | \sigma'$, and this occurs for inductive reasons.   ∎

Theorem 12. *If G is a cell-like decomposition of a 3-manifold M, then* $\dim(M/G) \le 3$.

Proof.   The global strategy, suggested by Proposition 5, involves finding an approximate right inverse of the decomposition map π—namely a map $\psi \colon M/G \to M$ for which $\pi\psi$ is close to $\mathrm{Id}_{M/G}$. To a large extent the specifics follow the outline of Theorem 8, with Lemma 11 serving in place of the facts about 1-complexes.

Look at a compact subset X of M/G and positive number ε, and regard M/G as a subset of I^∞. We shall prove that there exists a map $\psi \colon \pi(X) \to M$ such that $\pi\psi \,|\, \pi(X)$ is ε-close to the inclusion. Proposition 5 then will certify that M/G is (locally) 3-dimensional.

Construct a G-saturated open cover $\mathfrak{U}_3$ of M such that diam $\pi(U_3) < \varepsilon$ for all $U_3 \in \mathfrak{U}_3$. Identify an open cover $\mathcal{W} = \{W_g \,|\, g \in G\}$ refining $\mathfrak{U}_3$ and satisfying the conclusion of Lemma 10. Although $\mathcal{W}$ might not be G-saturated, it has a G-saturated open refining cover $\mathcal{W}'$. Apply Lemma 16.6 to obtain G-saturated open covers $\mathfrak{U}_2$, $\mathfrak{U}_1$, and $\mathfrak{U}_0$ of X, such that $\mathfrak{U}_2$ star-refines $\mathcal{W}'$, $\mathfrak{U}_1$ 1-homotopy star-refines $\mathfrak{U}_2$, and $\mathfrak{U}_0$ 0-homotopy star-refines $\mathfrak{U}_1$. Determine $\beta \in (0, \varepsilon)$ for which every 3β-subset of $\pi(X)$ lies in some $\pi(U_0)$, $U_0 \in \mathfrak{U}_0$.

Exactly as in the proof of Theorem 8, replace $\pi(X)$ by a finite complex; find a complex K so that

$$\pi(X) \subset K \times Q_k \subset N(\pi(X); \beta)$$

and $K \times Q_k$ projects to $K = K \times \{0\}$ under a map R moving points less than β. Then build a map $f \colon K^{(2)} \to M$ in precisely the same fashion as done previously, so that πf is ε-close to the inclusion of $K^{(2)} = K^{(2)} \times \{0\}$ in I^∞.

We argue that for each k-simplex $\sigma \in K$, where $k \ge 2$, some $W(\sigma) \in \mathcal{W}$ contains $f(\sigma \cap K^{(2)})$. Choose a 2-simplex $\gamma \in K$ with $\gamma \subset \sigma$, and name $U_\gamma \in \mathfrak{U}_2$ with $f(\gamma) \subset U_\gamma$. It follows routinely that $f(\sigma \cap K^{(2)}) \subset \mathrm{St}(U_\gamma, \mathfrak{U}_2)$, which lies in some $W(\sigma) \in \mathcal{W}$ because of star-refinement properties.

The combination of Lemma 10 and Lemma 11, applied with $p = 2$, ensures the existence of a map $F \colon K \to M$ extending f such that $F(\sigma) \subset W(\sigma)$ for each $\sigma \in K$. Then πF is 3ε-close to the inclusion, since given $z \in K$ one can find $z' \in K^{(2)}$ in a common simplex σ and can check that

$$\rho(\pi F(z), z) \leq \rho(\pi F(z), \pi F(z')) + \rho(\pi f(z'), z') + \rho(z', z)$$
$$\leq \operatorname{diam} \pi(W(\sigma)) + \varepsilon + \beta < 3\varepsilon.$$

Because R moves points less than ε, $\pi FR \mid \pi(X)$ is 4ε-close to the inclusion, and $FR \mid \pi(X): \pi(X) \to M$ is an 8ε-map. ∎

Remark. Kozlowski and Walsh establish the stronger result in [1] that if G is a cell-like decomposition of a space Z that embeds in some 3-manifold, then $\dim(Z/G) \leq 3$. Walsh [2] has discovered another, more algebraic proof of Theorem 12, in which he substantiates the 3-dimensionality of all locally compact, metric, homology 3-manifolds. [These are locally homologically trivial but possibly not locally 1-connected spaces X having finite homological dimension and for which $\check{H}_*(X, X - \{x\}; \mathbb{Z}) \cong H_*(E^3, E^3 - \{0\}; \mathbb{Z})$.] Whenever G is a cell-like decomposition of a 3-manifold, M/G is a homology 3-manifold.

The route retraced here in proving Theorem 12 has an additional payoff. It can be adapted for the following result about dimension-preserving features held by certain geometrically nice cell-like decompositions. Filling in the details constitutes an extremely valuable review of the earlier techniques, and the reader is strongly urged to develop a proof.

Theorem 13. *If G is a usc decomposition of E^n into convex sets, then $\dim E^n/G \leq n$.*

EXERCISES

1. Suppose $f: Y_1 \to Y_2$ is a cell-like surjective map between CANRs. Show that B is a cell-like subset of Y_2 iff $f^{-1}(B)$ is cell-like in Y_1.
2. If $f: Y_1 \to Y_2$ and $f': Y_2 \to Y_3$ are cell-like surjective maps between ANRs, prove that $f'f: Y_1 \to Y_3$ is cell-like.
3. Suppose $f: X_1 \to X_2$ is a cell-like surjective map between finite-dimensional metric spaces, where X_1 is a compact subset of E^n. Show that B is a cell-like subset of X_2 iff $f^{-1}(B)$ is cell-like in E^n.
4. If $f: X_1 \to X_2$ and $f': X_2 \to X_3$ are cell-like surjective maps between finite-dimensional metric spaces, where X_1 is a compact subset of E^n, prove that $f'f: X_1 \to X_3$ is cell-like.
5. Show that the uniform limit of proper, surjective maps $f_n: X \to X'$ between locally compact metric spaces is a proper, surjective map.
6. Prove Claim 2 made in the argument given for Theorem 4.
7. A map $m: X \to Z$ between compact metric spaces is said to be *approximately right invertible* if, for each $\varepsilon > 0$, there exists a map $\mu: Z \to X$ such that $\operatorname{dist}(z, m\mu(z)) < \varepsilon$ for all $z \in Z$. Show that if Z is compact and $m: X \to Z$ is approximately right invertible, then $\dim Z \leq \dim X$.

8. Suppose Y is a CANR having the fixed point property and G is a cell-like decomposition of Y such that Y/G is an ANR. Prove that Y/G has the fixed point property.

9. Prove Theorem 13.

18. THE CELLULARITY CRITERION AND DECOMPOSITIONS

Peripherally related to the UV properties of Section 16 is a property introduced by D. R. McMillan, Jr. [1], referred to as the (or as McMillan's) cellularity criterion, measuring features pertaining to the embedding of subspaces instead of simply measuring absolute properties of those subspaces. A subset D of a space S is said to *satisfy the cellularity criterion* (in S) if each neighborhood U of D in S contains another neighborhood V of D such that every map of ∂B^2 into $V - D$ can be extended to a map of B^2 into $U - D$.

The naturality of the cellularity criterion, as well as a partial justification for the name, is manifested in the elementary observation below.

Proposition 1. *Every cellular subset of an n-manifold ($n > 2$) satisfies the cellularity criterion.*

Proposition 2. *Suppose the T_2-space S contains a closed subset D satisfying the cellularity criterion and suppose $f \colon B^2 \to S$ is a map for which $f(\partial B^2) \to S - D$. Then $f \mid \partial B^2$ extends to a map $F \colon B^2 \to S - D$; moreover, for each neighborhood U of D in $S - f(\partial B^2)$, there exists a map $F_U \colon B^2 \to S - D$ such that $F_U(f^{-1}(U)) \subset U$ and $F_U \mid X = f \mid X$, where X denotes the component of $B^2 - f^{-1}(U)$ containing ∂B^2.*

Proof. Let U denote some neighborhood of D, with $f(\partial B^2) \cap U = \varnothing$. Since D satisfies the cellularity criterion, U contains a neighborhood V of D such that every loop in $V - D$ is null homotopic in $U - D$. Construct a compact 2-manifold-with-boundary H in B^2 with

$$f^{-1}(D) \subset \operatorname{Int} H \subset H \subset f^{-1}(V) \subset \operatorname{Int} B^2.$$

Enumerate the components $J_1, ..., J_k$ of $\bar{K} \cap H$, where K represents the component of $B^2 - H$ containing ∂B^2. (Then K contains the component X named in the statement of this proposition. See Fig. 18-1.)

Let B_i denote the 2-cell in B^2 bounded by J_i. The construction ensures that the B_i's are pairwise disjoint. Since $J_i \subset \partial H \subset f^{-1}(V - D)$, $f \mid J_i$ extends to a map $F_i \colon B_i \to U - D$. Extend $f \mid \bar{K}$ to $F \colon B^2 \to S - D$ by setting $F \mid B_i = F_i$ for $i \in \{1, ..., k\}$. ∎

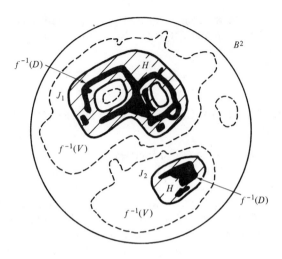

FIG. 18-1

Proposition 3. *Suppose S is a locally pathwise-connected T_2-space; D is a closed subset of S such that* (a) *D has Property UV^1 in S,* (b) *D satisfies the cellularity criterion (in S), and* (c) *for each connected open neighborhood W of D, $W - D$ is connected; and $s_0 \in S - D$. Then the inclusion-induced homomorphism $\pi_1(S - D, s_0) \to \pi_1(S, s_0)$ is an isomorphism.*

Proof. Proposition 2 indicates that the homomorphism is one-to-one, so the only issue is surjectivity. Consider a map $\lambda: I \to S$ with $\lambda(0) = \lambda(1) = s_0$. Specify a connected neighborhood V of D in S such that each loop in V is contractible in S, and then determine a finite number of pairwise disjoint subarcs $A_1, \ldots, A_k$ of Int I in $\lambda^{-1}(V)$ whose interiors cover $\lambda^{-1}(D)$. Apply (c) above to connect each pair of points $\lambda(\partial A_i)$ by a path in $V - D$, and construct a new map $\bar\lambda: I \to S - D$ such that $\bar\lambda(A_i) \subset V - D$ and that $\bar\lambda$ agrees with λ off $\bigcup A_i$, using these paths as guides. Because the loops produced by the two maps $\bar\lambda \mid A_i$ and $\lambda \mid A_i$ on the various A_i's are all contractible in S, $\bar\lambda$ and λ are homotopic rel ∂I.　■

Corollary 3A. *If C is a cell-like subset of an n-manifold M ($n \geq 3$) that satisfies the cellularity criterion in M and if $s_0 \in M - C$, then $\pi_1(M - C, s_0) \to \pi_1(M, x_0)$ is an isomorphism.*

See also Proposition 16.1 and Corollary 15.4B.

For those decompositions having a relatively mild UV property, the decomposition map faithfully preserves the matter of satisfying the cellularity criterion, at least for saturated closed subsets of the domain.

Proposition 4. *Suppose G is a UV^1 decomposition of a metric space X and D is a closed subset of X/G. Then D satisfies the cellularity criterion in X/G if and only if $\pi^{-1}(D)$ satisfies the cellularity criterion in X.*

Proof. Details demand another formal drill on the techniques from Section 16. We will run through the forward implication.

Assuming D satisfies the cellularity criterion in X/G, consider a G-saturated open subset U of X containing $\pi^{-1}(D)$. Then $\pi(U)$ contains a neighborhood V of D such that every loop in $V - D$ is null homotopic in $\pi(U) - D$. Given a map $F: \partial B^2 \to \pi^{-1}(V) - \pi^{-1}(D)$, one can extend πF to a map $f: B^2 \to \pi(U) - D$, which lifts (approximately) to a map $F': B^2 \to U - \pi^{-1}(D)$ with $F' \mid \partial B^2 = F$, by Theorem 16.7. ∎

Corollary 4A. *Suppose G is a cell-like decomposition of an ANR Y and D is a closed subset of Y/G. Then D satisfies the cellularity criterion in Y/G if and only if $\pi^{-1}(D)$ satisfies the cellularity criterion in Y.*

The cellularity criterion is noteworthy because it is a sufficient as well as a necessary condition for a cell-like subset of an n-manifold ($n > 3$) to be cellular. This pivotal fact, an important result from the topology of manifolds but one which requires the development of too much background data to be included here, was proved largely by McMillan [1]. For $n \geq 5$ his argument is based on engulfing methods and for $n = 3$, insofar as it is known to apply, on special 3-dimensional results such as Dehn's lemma. In case $n = 4$ the result recently was established by M. H. Freedman [1], as an outgrowth of his solution to the 4-dimensional Poincaré conjecture.

Theorem 5. *A cell-like subset C of an n-manifold M ($n \geq 4$) is cellular in M if and only if C satisfies the cellularity criterion in M; similarly, a cell-like subset C of E^3 is cellular there if and only if it satisfies the cellularity criterion in E^3.*

In the latter case it is essential that C lie in E^3, where every bicollared 2-sphere bounds a 3-spell. Should the 3-dimensional Poincaré conjecture turn out to be false, there will be a compact, contractible 3-manifold-with-boundary F (necessarily bounded by a 2-sphere) in a closed 3-manifold Σ^3 homotopy equivalent but not homeomorphic to S^3, with $\Sigma^3 - F$ homeomorphic to E^3, and F will satisfy the cellularity criterion but it will not be cellular in Σ^3, for otherwise Proposition 6.7 would show Σ^3 to be the 3-sphere.

Corollary 5A. *Let M be an n-manifold ($n \geq 3$) and C a cell-like subset of M. Then $C \times \{0\}$ is cellular in $M \times E^1$.*

Proof. It suffices to show that $C \times \{0\}$ satisfies the cellularity criterion in $M \times E^1$. Given a neighborhood U of $C \times \{0\}$ there, find a neighborhood W of C in M and $\delta > 0$ such that $W \times (-\delta, \delta) \subset U$. Since C is cell-like, it has a connected open neighborhood V in W such that each loop in V is contractible in W.

Let Z_+ denote $V \times (-\delta, \delta) - (C \times (-\delta, 0])$ and let Z_- denote $(V \times (-\delta, \delta)) - (C \times [0, \delta))$. By Corollary 15.4B, $Z_+ \cap Z_- = (V - C) \times (-\delta, \delta)$ is connected. Clearly Z_+ deformation retracts to $V \times \{\delta/2\}$, and each loop there is contractible in $W \times \{\delta/2\}$. The Siefert–van Kampen theorem attests that the inclusion-induced homomorphism

$$\pi_1(V \times (-\delta, \delta) - (C \times \{0\})) = \pi_1(Z_+ \cup Z_-) \to \pi_1(U - (C \times \{0\}))$$

is trivial. ∎

As a result, the problem of whether there exists a cell-like decomposition G of a compactum X such that $\dim(X/G) > \dim X$ is equivalent to the problem of whether there exists a cellular decomposition G^* of some Euclidean space E^n such that $\dim(E^n/G^*) > n$.

Corollary 5B. *Suppose X is a finite-dimensional, compact metric space and G is a cell-like decomposition of X such that $\dim(X/G) > \dim X$. Then there exist an integer n and a cellular decomposition G^* of E^n such that $\dim(E^n/G^*) = \infty$.*

Proof. Let $k = \dim X$ and $n = 2k + 2$. By Theorem 17.7, $k \geq 2$, so $n \geq 6$. By Hurewicz–Wallman [1, p. 56], X can be embedded in E^{2k+1}, thought of as $E^{2k+1} \times \{0\} \subset E^n$. The desired decomposition G^* is the trivial extension of the image of G under this embedding. ∎

Let A denote a subset of a space S and $a \in A \cap \mathrm{Cl}(S - A)$. We say that $S - A$ is *locally k-connected at a*, or in abbreviated form that $S - A$ is *k-LC at a*, provided that each neighborhood U of a contains another neighborhood V such that each map of ∂B^{k+1} into $V - A$ can be extended to a map of B^{k+1} into $U - A$. We also say that A is *locally k-co-connected* (*k-LCC*) provided $S - A$ is *k-LC* at each $a \in A \cap \mathrm{Cl}(S - A)$.

Corollary 5C. *Let G denote a cell-like decomposition of an n-manifold M ($n \geq 4$). Then $g \in G$ is cellular in M if and only if $\pi(M - g)$ is $1 - LC$ at the point $\pi(g)$.*

See Proposition 4 as well as Theorem 5. Of course, the analogue of Corollary 5C in which the n-manifold M is replaced by E^3 also is valid.

Corollary 5C reveals that whether a cell-like decomposition of a manifold is a cellular decomposition can be determined by examining local properties of the decomposition space.

Corollary 5D. *Suppose G is a cellular decomposition of an n-manifold M* *(n ≥ 4) and G' is a cell-like decomposition of M such that M/G' is homeo-* *morphic to M/G. Then G' is cellular.*

Corollary 5E. *If G is a cell-like decomposition of an n-manifold M (n ≥ 4)* *such that M/G is an n-manifold, then G is cellular.*

EXERCISES

1. Suppose X is a nowhere dense, compact subset of a space S such that X has Property $1 - UV$ in S and X is k-LCC, $k \in \{0, 1\}$. Show that X satisfies the cellularity criterion in S.

2. Let K denote a finite simplicial complex, L a subcomplex of K to which K collapses, and e an embedding of K in an n-manifold M ($n > 2$) such that $e(K)$ satisfies the cellularity criterion in M. Show that $e(L)$ also satisfies the cellularity criterion.

3. For $i \in \{1, 2\}$ let C_i denote a cell-like subset of an n_i-manifold, where $n_i > 1$. Show that $C_1 \times C_2$ satisfies the cellularity criterion in $M_1 \times M_2$.

4. Suppose G is a cellular decomposition of S^n, where $n \geq 3$, and f is a map of S^n onto S^n/G such that the decomposition G_f induced by f is cell-like. Show that G_f is cellular.

5. Show that the truth of Corollary 5E for $n = 3$ implies the truth of the 3-dimensional Poincaré conjecture.

6. Let X denote a cell-like subset of S^n, $n \geq 2$, and ΣX its natural suspension in $\Sigma S^n \approx S^{n+1}$. Show that ΣX is cellular in S^{n+1} iff X has simply connected complement in S^n.

IV _____

THE CELL-LIKE APPROXIMATION THEOREM

Chapter IV is directed toward a single goal—the proof of the fundamental result of manifold decomposition theory, Edwards's cell-like approximation theorem. Several ingredients must be prepared. Included among them are: a filtration device, applicable to finite-dimensional decompositions, for refracting the aggregate shrinking problem into a sequence of 0-dimensional problems; an amalgamation device for transforming 0-dimensional decomposition problems into null sequence problems; Štan'ko's concept of embedding dimension, which governs the successful solution of these null sequence problems; and the disjoint disks property, a minimal kind of general position property measured in decomposition spaces. When present, this disjoint disks property permits restructuring of the aggregate decompositions, back in the source manifold, so the nondegeneracy sets have the appropriate embedding dimension; the aggregate decompositions are shrinkable because all the related null sequence decompositions, created by refracting and amalgamating, are shrinkable.

Because the disjoint disks property is a property of decomposition spaces, it persists when a given decomposition is restructured, under approximations to the natural decomposition map with other cell-like maps. In a sense, its chief disadvantage is its applicability just to (cell-like) decompositions of manifolds having dimension at least 5. One could compose a strong brief to support the contention that cell-like decompositions of such manifolds are better understood than those of dimensions 3 and 4 precisely because the disjoint disks property holds only in the former.

The concept of embedding dimension delineates a notion of tameness for arbitrary compacta embedded in n-manifolds. Given a null sequence decomposition consisting of cell-like compacta having embedding dimension no more than $n - 3$, one can perform shrinking using methods inspired by those of Section 8. The taming theory undergirding these operations demands greater familiarity with results and techniques from PL topology and embedding theory than needed in the preceding chapters. Section 21 provides an outline of what is needed, plus an explication of the functional properties of embedding dimension.

19. CHARACTERIZING SHRINKABLE DECOMPOSITIONS OF MANIFOLDS—THE SIMPLE TEST

This section represents a prelude to more powerful subsequent sections that disclose how a certain disjoint disks property characterizes shrinkable decompositions of n-manifolds ($n \geq 5$). The results derived, valid without restriction on n, provide an effective means for analyzing decompositions of 3-manifolds, somewhat similar to those stemming from the disjoint disks property. The arguments, originally given by J. W. Cannon [4], revive the spirit of Bing's early work on shrinkability.

Let G be a usc decomposition of an n-manifold M, and let A and B be disjoint closed subsets of M. We say that G *inessentially spans A and B* if for each G-saturated open cover $\mathcal{U}$ of M there exists a homeomorphism $h: M \to M$ such that h is $\mathcal{U}$-close to Id_M and $\pi h(A) \cap \pi h(B) = \varnothing$ [equivalently, no element of G meets both $h(A)$ and $h(B)$].

The first item below should be self-evident; it is a rephrasing of what was used in Section 9 to demonstrate the nonshrinkability of the examples described there. The second, a strong converse to the first, is nearly the intended characterization.

Proposition 1. *If G is a shrinkable decomposition of an n-manifold M, then G inessentially spans each pair of disjoint closed subsets of M.*

Theorem 2. *Suppose G is a monotone usc decomposition of an n-manifold M. Then G is shrinkable if and only if G inessentially spans each pair of disjoint, bicollared $(n - 1)$-spheres in M.*

Proof. Assume M to be compact, and consider $\varepsilon > 0$. The all-important initial step consists of locating finitely many pairs $(A_1, B_1), ..., (A_k, B_k)$ of bicollared $(n - 1)$-spheres in M, each pair being disjoint, such that every connected subspace of M having diameter at least ε intersects both A_i and B_i

for some i. To do this, for each $x \in M$ choose a pair of n-cells C_x and C_x' with bicollared boundaries and satisfying

$$x \in \operatorname{Int} C_x \subset C_x \subset \operatorname{Int} C_x' \qquad \text{and} \qquad \operatorname{diam} C_x' < \varepsilon;$$

use compactness to extract a finite subcollection $C_1, \ldots, C_k$ of the smaller cells whose interiors cover M, and then set $A_i = \partial C_i$ and $B_i = \partial C_i'$ for $i \in \{1, \ldots, k\}$.

We shall produce a homeomorphism $H: M \to M$ shrinking each $g \in G$ to ε-size and satisfying $\rho(\pi, \pi H) < \varepsilon$. By hypothesis there exists a self-homeomorphism h_1 of M satisfying

$$\rho(\pi, \pi h_1) < \varepsilon/k \qquad \text{and} \qquad \pi h_1(A_1) \cap \pi h_1(B_1) = \varnothing.$$

Since $h_1(A_2)$ and $h_1(B_2)$ are disjoint bicollared $(n-1)$-spheres, there exists another self-homeomorphism h_2 of M such that

$$\pi h_2 h_1(A_2) \cap \pi h_2 h_1(B_2) = \varnothing \qquad \text{and} \qquad \rho(\pi, \pi h_2) < \varepsilon/k;$$

we require, in addition, that

$$\rho(\pi, \pi h_1) < (\tfrac{1}{4}) \cdot \rho(\pi h_1(A_1), \pi h_1(B_1))$$

to ensure

$$\pi h_2 h_1(A_1) \cap \pi h_2 h_1(B_1) = \varnothing.$$

Continuing in this fashion, we determine homeomorphisms $h_1, h_2, \ldots, h_k$ of M to itself such that, for $i \in \{1, \ldots, k\}$,

$$\rho(\pi, \pi h_i) < \varepsilon/k \qquad \text{and} \qquad \pi h_k \cdots h_1(A_i) \cap \pi h_k \cdots h_1(B_i) = \varnothing.$$

Set $H = (h_k \cdots h_1)^{-1}$. According to the above, no $g \in G$ intersects both $H^{-1}(A_i)$ and $H^{-1}(B_i)$ for any index i; equivalently, no set $H(g)$ meets both A_i and B_i. The initial choice of pairs (A_i, B_i) thereby implies that each $H(g)$ has diameter less than ε, and it is easy to see that $\rho(\pi, \pi H) < \varepsilon$. ■

With a slightly different construction at the outset, the second half of the preceding argument can be recirculated to establish the desired characterization, which involves cells instead of spheres.

Theorem 3. *Suppose G is a monotone usc decomposition of an n-manifold M. Then G is shrinkable if and only if G inessentially spans every pair of disjoint, locally flat $(n-1)$-cells in M.*

Remark. M. Starbird [1] has found some improvements to Theorem 3 for the case $n = 3$. In particular, for any cell-like decomposition G of a 3-manifold M such that $\pi(N_G)$ is 0-dimensional, he has shown G to be shrinkable if it inessentially spans every pair of locally flat 2-cells in M whose boundaries miss N_G.

EXERCISES

1. Prove Theorem 2 for noncompact manifolds M.
2. Prove Theorem 3.

20. AMALGAMATING DECOMPOSITIONS

Let G denote a usc decomposition of a (compact) metric space S. Another usc decomposition G^* of S is called an ε-*amalgamation of* G, where $\varepsilon > 0$, provided

(i)   for each $g \in G$ there exists $g^* \in G^*$ such that $g \subset g^*$ and
(ii)   for each $g^* \in G^*$ there exists $g \in G$ such that $\pi(g^*) \subset N(\pi(g); \varepsilon)$.

In almost equivalent words, G^* is a G-saturated usc decomposition and diam $\pi(g^*) < 2\varepsilon$ for each $g^* \in G^*$. When one inspects the diagram,

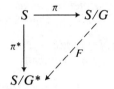

one sees that the implicitly defined rule $F = \pi^*\pi^{-1}$ is a (continuous) function and that, for each point $s^* \in S/G^*$, $F^{-1}(s^*)$ has diameter less than 2ε.

The purpose behind this concept is unveiled in what otherwise may seem to be a totally innocuous result.

Theorem 1. *Suppose G is a usc decomposition of a compact metric space S such that, for each $\varepsilon > 0$, there exists a shrinkable ε-amalgamation G^* of G. Then G itself is shrinkable.*

Proof. Fix $\varepsilon > 0$. The goal here is the usual one of producing a homeomorphism $h: S \to S$ shrinking elements of G to ε-size while satisfying $\rho(\pi, \pi h) < \varepsilon$.

By hypothesis there exists a shrinkable $(\varepsilon/2)$-amalgamation G^* of G. According to (ii) of the definition above, each $g^* \in G^*$ is contained in some set $\pi^{-1}(N(\pi(g); \varepsilon/2))$, where $g \in G$; thus, the open cover $\{\pi^{-1}(N(\pi(g); \varepsilon/2)) \mid g \in G\}$ has a G^*-saturated open refinement $\mathfrak{U}$. The shrinkability of G^* gives rise to a homeomorphism $h: S \to S$ that is $\mathfrak{U}$-close to Id_S and that shrinks each $g^* \in G^*$ to size less than ε. Certainly then h shrinks each element of G, necessarily found in some element of G^*, to size ε, and h being $\mathfrak{U}$-close to Id_S implies that $\rho(\pi(s), \pi h(s)) < \varepsilon$ for each $s \in S$. ∎

Most of the amalgamations to be used reassemble elements from a decomposition G for which $\pi(N_G)$ has dimension 0. To help cull out such decompositions now and to provide a measure of nondegeneracy sets later on, in Section 36, we say that a decomposition G of a metric space is *k-dimensional* if $\dim \pi(N_G) \leq k$ and that G is *closed-k-dimensional* if $\dim \text{Cl } \pi(N_G) \leq k$.

Proposition 2. *Let S denote a separable metric space. Then every finite usc decomposition of S is closed-0-dimensional and every countable usc decomposition of S is 0-dimensional.*

One might note that each of the decompositions described in Section 9 is closed-0-dimensional. Generally, one can produce countable, cellular decompositions of S^n $(n > 0)$ that are not even closed-$(n - 1)$-dimensional, by just inserting a null sequence of cellular (nondegenerate) sets densely throughout S^n.

Proposition 3. *If G is a k-dimensional usc decomposition of a finite-dimensional separable metric space S, then S/G is finite-dimensional.*

Proof. View S/G as $\pi(N_G) \cup \pi(S - N_G)$. Since π is a closed map, $\pi \mid S - N_G$ sends $S - N_G$ homeomorphically onto $\pi(S - N_G)$, implying

$$\dim \pi(S - N_G) = \dim(S - N_G) \leq \dim S < \infty.$$

Consequently, S/G, the union of two finite-dimensional spaces, is itself finite-dimensional (Hurewicz–Wallman [1, p. 32]). ∎

Corollary 3A. *If G is a cell-like, k-dimensional usc decomposition of a finite-dimensional, locally compact ANR Y, then Y/G is an ANR.*

Proof. See also Corollary 16.12B.

Next, a method for amalgamating certain closed-0-dimensional decompositions of manifolds. The techniques can be traced back to the 1960s, suggested in the work of Andrews and Rubin [1] and given more explicitly by R. D. Edwards and R. T. Miller [1] and by C. P. Pixley and W. T. Eaton [1].

Proposition 4. *If G is a closed-0-dimensional, cell-like usc decomposition of S^n (or any compact n-manifold), $n \geq 3$, and if $\varepsilon > 0$, then there exists a finite, cell-like, ε-amalgamation G^* of G.*

Proof. Let C denote $\text{Cl } \pi(N_G)$ in S^n/G. Since C is compact and 0-dimensional, the open cover $\{N(c; \varepsilon) \mid c \in C\}$ of C has a finite subcover, which, in turn, admits a refinement $\mathcal{W} = \{W_i \mid i = 1, ..., m\}$ consisting of pairwise disjoint, connected open subsets of S^n/G and covering C.

For $i \in \{1, ..., m\}$ let $C_i = C \cap W_i$. Then $\pi^{-1}(C_i)$ can be written as the intersection of a nested sequence $\{M_j\}$ of compact n-manifolds-with-boundary in $\pi^{-1}(W_i)$, where each component of M_{j+1} is contractible in M_j (because it lies very close to some $\pi^{-1}(c)$, $c \in C_i$). By the same reasoning used to prove Corollary 15.4B, $\pi^{-1}(C_i)$ separates no component of any M_j; therefore, in $M_j - \pi^{-1}(C_i)$ tubes can be threaded through the interior of each part of M_j to join up its various boundary components, thereby ensuring that every component of M_j has connected boundary.

Run arcs in $\pi^{-1}(W_i)$ from a preselected component T_1 of M_1 to each of the others, so that no arc meets M_1 other than in its endpoints and that no two such arcs intersect. Let A_1 denote the union of these arcs, and let $S_1 = A_1 \cup M_1$. Then S_1 is a connected subset of $\pi^{-1}(W_i)$ and, of course, M_2 contracts in S_1. (See Fig. 20-1.)

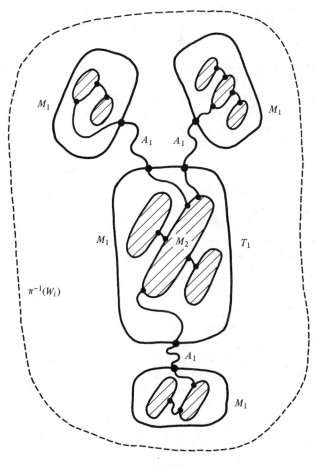

FIG. 20-1

To simplify the next step, focus on one component P_1 of M_1 and select a component T_2 of $P_1 \cap M_2$. Run arcs in P_1 from T_2 to each of the other components of $P_1 \cap M_2$ and, in addition, run an arc (arcs) in P_1 from T_2 to the point(s) of $P_1 \cap A_1$. Do this, as above, so that no arc meets M_2 elsewhere besides its endpoints and no two such arcs intersect. Let A_2 denote the union of all these arcs, obtained similarly for every component P of M_1, and let S_2 denote $A_1 \cup A_2 \cup M_2$. Again, S_2 is a connected subset of S_1 and M_3 contracts in S_2, but, more important, S_2 contracts in S_1 (see Exercise 5).

Iterating this procedure, we form compact, connected sets $\{S_j \mid j = 1, 2, \ldots\}$ such that $M_{j+1} \subset S_{j+1} \subset S_j$ and S_{j+1} contracts in S_j.

Now G^* is defined to have exactly m nondegenerate elements, one in each $\pi^{-1}(W_i)$, and the nondegenerate element in $\pi^{-1}(W_i)$ is $\bigcap S_j$, which is cell-like because for any neighborhood U some stage S_{j+1} is contractible in $S_j \subset U$. It should be obvious that G^* is a finite cell-like decomposition and that it is an ε-amalgamation of G.   ∎

The amalgamation process acts on certain nonshrinkable decompositions, like those of Section 9, to reinforce our belief in the existence of cell-like but noncellular sets. Given a closed-0-dimensional decomposition G of S^n that is cellular but nonshrinkable, we can see that when $\alpha > 0$ is sufficiently small, no finite α-amalgamation G^* of G can be cellular, since the resultant shrinkability of G^* would conflict with Theorem 1.

Corollary 4A.   *The following statements are equivalent*:

(a)   *For each closed-0-dimensional cell-like decomposition G of S^n, $G \times S^1$ is a shrinkable decomposition of $S^n \times S^1$*;

(b)   *For each finite cell-like decomposition G^* of S^n, $G^* \times S^1$ is a shrinkable decomposition of $S^n \times S^1$*;

(c)   *For each cell-like subset A of S^n, $G_A \times S^1$ is a shrinkable decomposition of $S^n \times S^1$*.

Results comparable to the above hold with E^n replacing S^n and/or E^1 replacing S^1, but the compactness of the manifolds mentioned in Corollary 4A averts more cumbersome statements.

In a sense, 0-dimensional decompositions of a compact manifold can be viewed as the countable union of closed-0-dimensional ones. Since the latter, cell-like or not, can be amalgamated as finite decompositions, one suspects that 0-dimensional decompositions should admit amalgamations as countable decompositions. The fundamental amalgamation result, due to R. D. Edwards [4], attests that each cell-like 0-dimensional decomposition can be amalgamated not merely as a countable but, even better, into a null sequence, cell-like decomposition. The amalgamation procedure also accommodates some built-in peripheral controls of subsequent value.

Theorem 5. *Suppose G is a 0-dimensional, cell-like decomposition of a compact n-manifold M, where $n \geq 3$; $p \leq n - 2$ is a nonnegative integer; $\{T_i \mid i = 1, 2, \ldots\}$ is a sequence of (curvilinear) triangulations of M whose p-skeleta $T_i^{(p)}$ all miss N_G; U is an open subset of M containing N_G; and $\varepsilon > 0$.*

Then there exists a cell-like decomposition K of M such that K is an ε-amalgamation of G, H_K forms a null sequence, and $N_G \subset N_K \subset U - \bigcup_i T_i^{(p)}$.

Proof. Since $\pi(N_G) \subset \pi(U)$ and $\pi(N_G)$ is 0-dimensional, U can be trimmed back so that $\pi(V)$ has diameter less than ε for every component V of the trimmed U. Then any monotone amalgamation K of G satisfying $N_K \subset U$ will be, of necessity, an ε-amalgamation.

For $j = 1, 2, \ldots$ let Q_j denote the union of those $g \in G$ having diameter at least $1/j$.

Part I. Architecture. The desired amalgamation K will appear after an infinite sequence of constructions. The first stage, somewhat like what was done in Proposition 4, merely requires expressing Q_1 as the union of finitely many pairwise disjoint (G-saturated) compacta $X_1^1, X_2^1, \ldots, X_{m(1)}^1$ that can be thickened to pairwise disjoint, G-saturated, connected open sets V_i^1 ($i \in \{1, \ldots, m(1)\}$) for which

$$X_i^1 \subset V_i^1 \subset \mathrm{Cl}\, V_i^1 \subset U - T_1^{(p)}$$

and V_i^1 is contractible in U.

The second stage construction is typical. Express $Q_1 \cup Q_2$ as the union of pairwise disjoint compacta $X_1^2, \ldots, X_{m(2)}^2$, where $m(2) \geq m(1)$, with

(a₂) $X_i^1 \subset X_i^2 \subset V_i^1$ for $1 \leq i \leq m(1)$,

such that these X_i^2's can be thickened to pairwise disjoint, G-saturated, connected open sets V_i^2 ($i \in \{1, \ldots, m(2)\}$) satisfying

(b₂) $X_i^2 \subset V_i^2 \subset \mathrm{Cl}\, V_i^2 \subset U - (T_1^{(p)} \cup T_2^{(p)})$,
(c₂)   for $m(1) < i \leq m(2)$, diam $V_i^2 < 1$, and
(d₂)   for $1 \leq i \leq m(1)$, $\mathrm{Cl}\, V_i^2$ is included in V_i^1 null homotopically.

The initial part of this, the determination of the X_i^2's, incorporates a conceptually significant step. Generally the set $Q_2 - Q_1$ is noncompact, which prohibits the covering of $Q_2 - Q_1$ with a finite number of compact sets from $U - Q_1$. By allowing the first stage sets X_i^1 to expand, in order to absorb some of Q_2 in $\bigcup V_i^1$, one can cover the rest of Q_2 by a finite number of compacta. In practice, this can be accomplished by forming sets Z_i ($i \in \{1, \ldots, m(1)\}$) that are both open and closed in the 0-dimensional set $\pi(Q_1 \cup Q_2)$ and that satisfy $\pi(X_i^1) \subset Z_i \subset \pi(V_i^1)$ and then by setting

$X_i^2 \subset \pi^{-1}(Z_i)$. Eventually the compact set $Q_2 - \bigcup_{i=1}^{m(1)} X_i^2$ will be fragmented into compacta X_i^2 ($i \in \{m(1) + 1, ..., m(2)\}$), and these will be thickened, as before, to open sets V_i^2 satisfying conditions (b₂) and (c₂). Logically, however, this fragmentation will not occur until after the sets X_i^2, $i \in \{1, ..., m(1)\}$, have been thickened to V_i^2's, for it is necessary to have the leftover fragments placed individually in pathwise connected open subsets of $M - \bigcup_{i=1}^{m(1)} \text{Cl } V_i^2$.

The remaining part, the determination of the other V_i^2's satisfying conditions (d₂) and (b₂), demands a bit of extra wrangling. We shall describe how to produce a connected open subset W_1 of M/G such that

$$\pi(X_1^2) \subset W_1 \subset \text{Cl } W_1 \subset \pi(V_1^1) - \pi(T_2^{(p)})$$

and $\text{Cl } W_1$ is contractible in $\pi(V_1^1)$, for then, as an elementary consequence of Theorem 17.1, the closure of $V_1^2 = \pi^{-1}(W_1)$ will be contractible in V_1^1. Since each point of $\pi(X_1^2)$ has a connected neighborhood in the ANR M/G that contracts in $\pi(V_1^1 - T_2^{(p)})$, one can extract a finite cover by neighborhoods of this type and refine to obtain another finite cover of $\pi(X_1^2)$ by pairwise disjoint, compact, connected sets $R_1, ..., R_r$ in $\pi(V_1^1 - T_2^{(p)})$. The restriction $p \leq n - 2$ ensures that $V_1^1 - T_2^{(p)}$ is arcwise connected, and the hypothesis $T_2^{(p)} \cap N_G = \varnothing$ ensures that $\pi(V_1^1) - \pi(T_2^{(p)})$ is arcwise connected as well. Hence, there exists an arc α_1 in $\pi(V_1^1 - T_2^{(p)})$ from R_1 to one of the other R_i's, say R_2, meeting $\bigcup R_i$ only in its endpoints. Similarly, there exists an arc α_2 in $\pi(V_1^1 - T_2^{(p)})$ joining $R_1 \cup \alpha_1 \cup R_2$ to one of the other R_i's and intersecting $\alpha_1 \cup (\bigcup R_i)$ only in its endpoints. Continuing in this fashion, eventually we will have identified arcs $\alpha_1, ..., \alpha_{r-1}$ such that, after a reordering of the R_i's, α_{j+1} joins $R_{j+1} \cup [\bigcup_{k=1}^{j}(\alpha_k \cup R_k)]$ to R_{j+2} while intersecting the union of the various R_i's and of the lower-indexed α_k's only in its own endpoints. As in the proof of Proposition 4, $(\bigcup R_i) \cup (\bigcup \alpha_i)$ is contractible in $\pi(V_1^1 - T_2^{(p)})$, and certainly then the former has a neighborhood W_1 whose closure also is contractible in $\pi(V_1^1 - T_2^{(p)})$.

The jth stage ($j > 1$) construction proceeds exactly like the second stage, having $Q_1 \cup \cdots \cup Q_j$ expressed as the union of pairwise disjoint (G-saturated) compacta $X_1^j, ..., X_{m(j)}^j$, with

(a$_j$) $X_i^{j-1} \subset X_i^j \subset V_i^{j-1}$ for $1 \leq i \leq m(j - 1)$,

such that there exist pairwise disjoint, G-saturated, connected open sets V_i^j ($i \in \{1, ..., m(j)\}$) satisfying

(b$_j$) $X_i^j \subset V_i^j \subset \text{Cl } V_i^j \subset U - (T_1^{(p)} \cup \cdots \cup T_j^{(p)})$,
(c$_j$) for $m(j - 1) < i \leq m(j)$ diam $V_i^j < 1/(j - 1)$,
(d$_j$) for $1 \leq i \leq m(j - 1)$ $\text{Cl } V_i^j$ is included in V_i^{j-1} null homotopically.

Part II. Amalgamation. Define a decomposition K by specifying its nondegeneracy set as follows: $X \in H_K$ iff there exist positive integers j_0 and t, with $t \leq m(j_0)$, such that

$$X = \bigcap_{j \geq j_0} \mathrm{Cl}\ V_t^j.$$

According to conditions (d_j), the sets $\{\mathrm{Cl}\ V_t^j\}_j$ form a decreasing nest; thus, given a neighborhood W of X, one can obtain an integer $s \geq j_0$ for which $V_t^s \subset W$, and condition (d_{s+1}) implies that $\mathrm{Cl}\ V_t^{s+1}$, as well as X, is contractible in $V_t^s \subset W$. Consequently, X is cell-like.

Clearly each $g \in H_G$ lies in some $X \in H_K$. Moreover, conditions (c_j) imply that H_K forms a null sequence, and conditions (b_j) imply that $N_K \subset U - \bigcup_i T_i^{(p)}$. Hence, K is an ε-amalgamation of G, as required. ∎

A weak application of Theorem 5, which hints at its greater potential, indicates how the peripheral controls provided via certain skeleta can be regulated to force cellularity of the amalgamated decomposition. (Such controls simultaneously demand cellularity of the original decomposition.)

Proposition 6. *Suppose G is a 0-dimensional, cell-like decomposition of an n-manifold M, where $n \geq 4$; $\{T_i\}$ is a sequence of simplicial triangulations of M such that* $\mathrm{mesh}\ T_i = \max\{\mathrm{diam}\ \sigma \mid \sigma \in T_i\} \to 0$ *as $i \to \infty$ and such that each 2-skeleton $T_i^{(2)}$ misses N_G; and $\varepsilon > 0$. Then there exists a cellular usc decomposition K of M such that K is an ε-amalgamation of G, H_K forms a null sequence, and $N_K \subset M - \bigcup T_i^{(2)}$.*

Proposition 6 is an immediate result of Theorem 5 and the ensuing lemma.

Lemma 7. *Suppose X is a cell-like subset of an n-manifold M ($n \geq 4$) and $\{T_i\}$ is a sequence of simplicial triangulations of M such that* $\mathrm{mesh}\ T_i \to 0$ *as $i \to \infty$ and that $T_i^{(2)} \cap X = \emptyset$ for all i. Then X is cellular in M.*

Proof. This will follow from Theorem 18.5, once it is established that X satisfies the cellularity criterion. One way to do this is to apply Exercise 18.1. A slightly more direct way is to start with an open neighborhood U of X, to name a compact neighborhood W of X in U, to find another neighborhood V of X that contracts in W, and to consider a loop L in $V - X$. Choose an integer k so large that L is homotopic in $V - X$ to a loop L' in $T_k^{(2)}$ and that each $\sigma \in T_k$ meeting W lies in U. The key fact to be used is that the 2-skeleton of a complex carries the fundamental group; consequently, for the union P of all simplexes from T_k that meet W, the natural homomorphism $\pi_1(P \cap T_k^{(2)}) \to \pi_1(P)$ is an isomorphism. Since L' contracts in $W \subset P$, L' must contract in $P \cap T_k^{(2)} \subset U - X$, and thus L contracts in $U - X$ as well.

EXERCISES

1. Suppose G is a usc decomposition of a CANR Y such that, for each $\varepsilon > 0$, G has a cell-like ε-amalgamation. Show that G is cell-like.
2. Suppose G is a usc decomposition of an n-manifold M such that, for each $\varepsilon > 0$, G has a cellular ε-amalgamation. Show that G is cellular.
3. Suppose G is a usc decomposition of a compact metric space X such that, for each $\varepsilon > 0$, G has an ε-amalgamation K where H_K is a null sequence. Show that G is 0-dimensional.
4. Let G denote a usc decomposition of an n-manifold M such that M/G is also an n-manifold and $\pi(N_G)$ is contained in a tame Cantor set in M/G. Use amalgamations to prove that G is shrinkable.
5. Suppose D_0 and D_1 are disjoint continua in a metric space S such that both D_0 and D_1 are contractible in S, and suppose $\alpha: [0, 1] \to S$ is an embedding such that $\alpha(0) \in D_0$, $\alpha(1) \in D_1$, and $\alpha(I) \cap (D_0 \cup D_1) = \alpha(\partial I)$. Show that $D_0 \cup \alpha(I) \cup D_1$ contracts in S.

21. THE CONCEPT OF EMBEDDING DIMENSION

Up to this point we have been relatively carefree in our treatment of simplicial complexes, allowing context to make plain whether the emphasis should fall on the collection of simplexes or on the underlying point set. Throughout this section we shall exert more precision, beginning with a brief review of terminology from PL topology.

A *simplicial complex* K is a locally finite collection of simplexes in some Euclidean space such that (1) each face of any $\sigma \in K$ is also an element of K and (2) if $\sigma_1, \sigma_2 \in K$ and $\sigma_1 \cap \sigma_2 \neq \varnothing$, then $\sigma_1 \cap \sigma_2$ is a face of both σ_1 and σ_2. (A *face* of a simplex σ is another simplex determined by some subset of the vertices of σ.) The *underlying point set of K*, the set of all points in simplexes of K, will be denoted as $|K|$.

By a (simplicial) triangulation of a space M we mean a pair (K, ψ), where K is a simplicial complex and ψ is a homeomorphism of $|K|$ onto M. If (K, ψ) is an arbitrary triangulation of E^n, then it is probably curvilinear, in the sense that $\psi(\sigma)$, $\sigma \in K$, represents something homeomorphic to a simplex, not necessarily possessing any convexity or linearity compatible with the linear structure of E^n. A *rectangular triangulation* of, say, E^n can be described simply as a simplicial complex K for which $|K| = E^n$, for then each $\sigma \in K$ must be a simplex in E^n.

Let K and K' denote simplicial complexes. Then K' is a *subdivision* of K if $|K| = |K'|$ and if each $\sigma' \in K'$ is contained in some $\sigma \in K$.

A simplicial complex K is a PL n-ball (or a PL n-cell) if there exist subdivisions K' of K and T' of T, the complex consisting of the standard

n-simplex Δ^n and its faces, such that K' is simplicially isomorphic to T'. A PL *triangulation* of an n-manifold M is a triangulation (K, ψ) of M such that, for each vertex $v \in K$, the star of v in K, written as $\mathrm{St}(v; K)$, is a PL n-ball. Here $\mathrm{St}(v; K)$ means the complex consisting of all $\sigma \in K$ for which $v \in \sigma$, plus their faces.

An n-manifold M is called a PL n-*manifold* if it has a specified PL triangulation. Then M can be covered by compatible coordinate charts such that the simplicial structure transferred from Δ^n to M by these charts matches the structure imposed by the triangulation. Accordingly, it makes sense to define a *rectilinear triangulation* of a PL manifold; however, without reference to the specified triangulation, there would be no means for distinguishing rectilinear from curvilinear.

A map f from a simplicial complex P into a PL n-manifold M is a *PL map* if there exist subdivisions P' of P and K' of K, where (K, ψ) denotes the specified triangulation of M, such that $\psi^{-1}f$ maps simplexes of P' linearly onto simplexes of K'; a *PL embedding*, of course, is a PL map which also happens to be an embedding.

The celebrated work of R. C. Kirby and L. C. Siebenmann [1] attests that not every n-manifold admits a PL triangulation. Whether it must admit some (simplicial) triangulation is an important open question. In case $n = 3$, E. E. Moise [1] and later R. H. Bing [4] proved that each 3-manifold has a triangulation, and it is not terribly difficult to prove that each such triangulation is a PL triangulation.

Two triangulations (K_1, ψ_1) and (K_2, ψ_2) of a manifold M are *equivalent* if there exist subdivisions K_1' and K_2' of K_1 and K_2, respectively, as well as a simplicial isomorphism $K_1' \to K_2'$. Standard results of PL topology establish that, when (K_1, ψ_1) and (K_2, ψ_2) are equivalent, then (K_1, ψ_1) is a PL triangulation iff (K_2, ψ_2) is. Thus, in speaking of a PL manifold, one presumes the existence of a specific PL triangulation (K_1, ψ_1) but deals willingly with arbitrary triangulations from the same equivalence class. In particular, this means that, for an arbitrary PL triangulation (K_1, ψ_1) and any subdivision K_1' of K_1, the triangulation (K_1', ψ_1) represents something in the same class.

A subset X of an n-manifold M endowed with a triangulation (K, ψ) is a *subpolyhedron* (with respect to said triangulation) if there exists a subdivision K' of K such that $\psi^{-1}(X)$ underlies a subcomplex of K'. More generally, X is a *tamely embedded polyhedron* if there exists a homeomorphism h of M onto itself such that $h(X)$ is a subpolyhedron; equivalently, X is a tamely embedded polyhedron iff there exists a triangulation (K_X, ψ_X) equivalent to (K, ψ) for which X is a subpolyhedron relative to (K_X, ψ_X). It is worth noting that every subpolyhedron and, thus, every tamely embedded polyhedron must be a closed subset of M.

These concepts should be contrasted with that of an (abstract) *polyhedron*, namely a space admitting a triangulation. The earlier terms require compatibility between the structure on the polyhedron and the triangulation of the overriding manifold.

With this as background we are now prepared to explore the concept of embedding dimension, or demension, which was formalized as a comprehensive theory by M. A. Štan'ko [1, 2]. A valuable source is a survey by R. D. Edwards [1], who outlined Štan'ko's work, stressed its usefulness, and expanded it. That source could be particularly beneficial since the treatment given the topic here is somewhat cursory, intended merely as background to make plausible the applications that follow.

The focus turns from abstract dimension theory to the geometry of E^n or, more generally, of any PL n-manifold M. Then $X \subset M$ has *embedding dimension* $\leq k$, written as dem $X \leq k$, if for each $(n - k - 1)$-dimensional tamely embedded polyhedron P in M and each open cover $\mathcal{V}$ of M, there exists a homeomorphism $h: M \to M$ such that $h(P) \cap X = \varnothing$ and h is $\mathcal{V}$-close to Id_M. As usual, one says that X has *embedding dimension* k provided dem $X \leq k$ but not dem $X \leq k - 1$. Loosely put, dem $X = k$ iff X behaves rather like a k-dimensional subpolyhedron of M.

The extreme cases are easily understood. Clearly, when $X \subset M$ a PL n-manifold, dim $X = n$ iff dem $X = n$, for each is equivalent to X containing a nonvoid open subset of M; hence, dim $X \leq n - 1$ iff dem $X \leq n - 1$. On the other hand, when X denotes a 0-dimensional abstract polyhedron, dem $X = 0$ for trivial reasons. However, the existence of Cantor sets X in S^n ($n \geq 3$) for which $\pi_1(S^n - X)$ is nontrivial indicates that dim $X = 0$ does not always imply dem $X = 0$.

Two features are obvious from the definition: embedding dimension k is invariant under homeomorphisms of the ambient manifold, and embedding dimension $\leq k$ is hereditary.

Proposition 1.   *If $X \subset M$ with dem $X = k$ and if h is a homeomorphism of M onto itself, then dem $h(X) = k$.*

Proposition 2.   *If $Y \subset X \subset M$ and dem $X \leq k$, then dem $Y \leq k$.*

When the manifold M has a PL triangulation, the delicate general position adjustments ordinarily done in E^n can be performed equally well in M by transferring from M back to E^n in chart-by-chart fashion. This yields:

Proposition 3.   *If X is a tamely embedded subpolyhedron of M, then dem $X = \dim X$.*

The philosophical attitude that dem $X = k$ essentially means X behaves like a k-dimensional subpolyhedron is most valid in case X is compact.

In many senses, however, embedding dimension functions just as well with σ-compact subsets of the manifolds as with subcompacta, due to the following:

Theorem 4. *Suppose* $X_1, X_2, \ldots$ *are compact subsets of* M *such that* dem $X_i \leq k$ *for all* i. *Then* dem$(\bigcup X_i) \leq k$.

The proof is fairly routine, primarily requiring controls governing a Cauchy sequence of self-homeomorphisms on M sufficient to guarantee the limit is a homeomorphism. Besides that, say for the case of compact M, given a tamely embedded polyhedron P and $\varepsilon > 0$, one first procures an $(\varepsilon/2)$-homeomorphism h_1 such that $h_1(P) \cap X_1 = \varnothing$. Then one adjusts the tame $h_1(P)$ off X_2 via a homeomorphism h_2, moving points less than $\varepsilon/4$ and less than $\frac{1}{4}\rho(h_1(P), X_1)$ as well, so that $h_2 h_1(P) \cap (X_1 \cup X_2) = \varnothing$. This is continued with care to maintain the desired properties in the limit. ■

A similar argument reveals that dem X dominates dim X.

Theorem 5. *For any* σ-*compact subset* X *of a PL manifold* M, dem $X \geq$ dim X.

Proof. We suppose $M = E^n$ and dem $X = k$. Since the case $k = n$ is trivial, we assume $k < n$. By Theorem 4 it suffices to study the case in which X is compact.

Consider the Nöbeling space $\mathfrak{N}_n^k$, described in Hurewicz–Wallman [1, p. 29], consisting of all points $\langle x_1, \ldots, x_n \rangle$ in E^n at most k of whose coordinates are rational. Its complement, the subset $\mathfrak{L}_n^{k+1}$ consisting of all points with at least $k + 1$ rational coordinates, can be naturally viewed as a countable union $(n - k - 1)$-dimensional affine hyperplanes P_j, individually determined by assigning a rational number in each of $(k + 1)$ coordinates while allowing complete freedom in the remainder. The hypothesis dem $X \leq k$ implies that X can be pushed off each hyperplane P_j. With meticulous care, like that suggested for the proof of Theorem 4, we can produce a limiting homeomorphism h of E^n onto itself such that $h(X) \cap \mathfrak{L}_n^{k+1} = \varnothing$, forcing $h(X) \subset \mathfrak{N}_n^k$. Since dim $\mathfrak{N}_n^k = k$, this establishes that dim $X \leq k$, as required. ■

Another mechanism at work here is a basic geometric duality. Given an n-simplex σ, identify its k-skeleton K and the $(n - k - 1)$-complex P dual to K in R, where R denotes the (complex associated with the) first barycentric subdivision of σ; namely

$$P = \{\tau \in R \mid \tau \cap |K| = \varnothing\}.$$

Then $\sigma - |P|$ deformation retracts to K. Moreover, each simplex of R is uniquely expressible as the geometric join of a simplex from K and a simplex from P (one possibly empty); hence, if X is a compact subset of $\sigma - |P|$ and if W is an open subset of σ containing $|K|$, there exists a homeomorphism of σ onto itself compressing X in W while fixing $|K \cup P|$. Expanding this, when X is a subset of a PL n-manifold M and dem $X \le k$, then for each PL triangulation T of M we see that X can be pushed off the $(n - k - 1)$-complex P dual to $T^{(k)}$ in T', the first barycentric subdivision of T, causing X to lie in an open regular neighborhood of some k-spine, a portion of $T^{(k)}$ near X, which neighborhood deformation retracts to the spine via a homotopy moving points less than mesh T. Furthermore, if X is compact it can be compressed arbitrarily near that spine by a homeomorphism of M moving points less than mesh T. Such neighborhoods are characteristic of the sets $X \subset M$ satisfying dem $X \le k$.

Theorem 6. *Let X be a subset of a compact PL n-manifold M. Then dem $X \le k$ if and only if, for each $\varepsilon > 0$, there exists a k-dimensional polyhedron K tamely embedded in M and there exists an open regular neighborhood V of K that ε-deformation retracts to K along the "lines" of the regular neighborhood, such that $X \subset V \subset N(X; \varepsilon)$.*

Two important results from PL topology will be applied frequently; their proofs fall outside the scope of this text. The first is due to R. H. Bing and J. M. Kister [1], promising an ambient isotopy "covering" a homotopy between PL embeddings in the trivial range.

Theorem 7. *Let M denote a PL n-manifold; $\mathfrak{U} = \{U_\alpha\}$ a collection of open subsets of M; K a finite simplicial k-complex, with $2k + 2 \le n$; f_0 and f_1 PL embeddings of K in M; and $F: K \times I \to M$ a map such that $F_0 = f_0$, $F_1 = f_1$, and, for every $x \in |K|$ on which F_t is nonconstant, some $U_x \in \mathfrak{U}$ contains $F(\{x\} \times I)$.*
Then there exists a PL isotopy θ_t of M to itself, with $\theta_0 = \mathrm{Id}$, such that $\theta_1 f_0 = f_1$; moreover, for every $p \in M$ on which θ_t is nonconstant, there exist $U_\alpha, U_\beta \in \mathfrak{U}$ such that $\theta_t(p) \subset U_\alpha \cup U_\beta$ (for all $t \in I$).

Corollary 7A. *If X is a σ-compact subset of a PL n-manifold M such that $\dim X \le n - 2$ and $n \ge 4$, then dem $X \le n - 2$.*

Proof. Consider the situation where X is compact. For each 1-complex K in M, there is a short homotopy between the inclusion and a (PL) embedding of K in $M - X$.   ∎

The second important result stems from the homotopy-implies-isotopy results developed with engulfing techniques. This was independently derived by J. L. Bryant [2] and M. A. Štan'ko [2].

Theorem 8. *Suppose X is a compact subset of a PL n-manifold M, $n \geq 5$, such that X is 1-LCC in M and $\dim X \leq n - 3$. Then $\dim X = \operatorname{dem} X$.*

According to a result of T. Homma [1] and R. H. Bing [6], Theorem 8 also holds when $n = 3$. R. D. Edwards recently announced that it holds when $n = 4$ as well.

Theorem 8 affords a quick method for dispensing with most cases in another key result. It should be mentioned that Theorem 8 is not a necessary tool—the result could be established from first principles; however, doing that would require further coping with the technical intricacies of PL topology.

Theorem 9. *Let X denote a σ-compact subset of a PL n-manifold M. Then $\operatorname{dem} X \leq k$ if and only if M has a sequence of PL triangulations $\{T_i\}$, each equivalent to the specified one, such that mesh $T_i \to 0$ as $i \to \infty$ and $X \cap T_i^{(n-k-1)} = \phi$.*

Proof. The forward implication is straightforward in all dimensions; we consider the reverse implication only for $n \geq 5$ and for compact X.

The hypothesis about the triangulations T_i implies, among other things, that $\dim X \leq k$, because of the nice maps from X (indeed, from $M - T_i^{(n-k-1)}$) to the k-complex dual to $T_i^{(n-k-1)}$ in the first barycentric subdivision of T_i. When $k = n$, $n - 1$, or $n - 2$, $\operatorname{dem} X$ is known to be bounded by k, so that the only case of interest is $k \leq n - 3$. But then X is 1-LCC (any small loop in $M - X$ can be adjusted to lie in some $T_i^{(2)}$, and contractibility in a small subset of M implies contractibility in a small subset of $T_i^{(2)} \subset M - X$), and Theorem 8 settles the matter. ∎

Corollary 9A. *If X is a cell-like subset of a PL n-manifold M and $\operatorname{dem} X \leq n - 3$, then X is cellular.*

Proof. See also Lemma 20.7.

Summary. For a σ-compact subset X of M^n (PL) where $n \geq 4$:

(1) $\operatorname{dem} X \geq \dim X$;
(2) $\operatorname{dem} X \geq \operatorname{dem} Y$ for all $Y \subset X$;
(3) $\operatorname{dem} X = \max\{\operatorname{dem} C \mid C$ is a compact subset of $X\}$;
(4) $\operatorname{dem} X = \dim X$ unless $\operatorname{dem} X = n - 2$ and $\dim X < n - 2$.

Anomalies compared with Property (4) above appear in case $n = 3$. The tangled continua X of H. G. Bothe [1] and of D. R. McMillan, Jr., and H. Row [1] satisfy $\dim X = 1 = n - 2$ and $\operatorname{dem} X = 2$. Such anomalies arise because nice 1-dimensional objects in E^3 can be knotted.

Next, material reinforcement for the viewpoint that objects of embedding dimension k behave like k-dimensional tame polyhedra.

Proposition 10.   *If X and X' are compact subsets of a PL n-manifold M such that dem X + dem $X' < n$, then for each $\varepsilon > 0$ there exists an ε-homeomorphism h of M onto itself such that $h(X) \cap X' = \varnothing$.*

Sketch of proof.   Apply Theorem 9 to obtain a triangulation T with small mesh such that $X \cap T^{(n-k-1)} = \varnothing$, where $k = \text{dem } X$, and let P denote the k-complex dual to $T^{(n-k-1)}$ in T', the first barycentric subdivision of T. Because dem X + dem $X' < n$, there exists a small homeomorphism g of M^n for which $g(P) \cap X' = \varnothing$. The linear structure in T' between $T^{(n-k-1)}$ and P facilitates the naming of another small (PL) homeomorphism ψ compressing X so close to P that $g\psi(X) \cap X' = \varnothing$.   ∎

The following generalization of Proposition 10 is needed for Section 22.

Proposition 11.   *Let X and X' denote compact subsets of a PL n-manifold M, and let $\varepsilon > 0$. Then there exists an ε-homeomorphism h of M onto itself such that*

$$\text{dem}[h(X) \cap X'] \le \text{dem } X + \text{dem } X' - n.$$

Sketch of proof.   This is accomplished by producing $(\varepsilon/2)$-homeomorphisms ψ and ψ' of M such that

$$\text{dem}[\psi(X) \cap \psi'(X')] \le \text{dem } X + \text{dem } X' - n$$

and by setting $h = (\psi')^{-1}\psi$. These are obtained as limits, bolstered at the various stages by a triangulation T_i of small mesh and by small homeomorphisms θ_i, θ_i' of M such that $\theta_i(X) \cap \theta_i'(X')$ lies in a thin regular neighborhood (in the sense of Theorem 6) of some m-complex, where $m = \text{dem } X + \text{dem } X' - n$. Let $k = \text{dem } X$ and $k' = \text{dem } X'$. Typical general position modifications give rise to a very small (PL) homeomorphism f of M to itself for which $f(T_i^{(k)}) \cap T_i^{(k')}$ is an m-complex P. Consequently, there exist small (limited by mesh T_i) homeomorphisms θ_i moving X (more precisely, its image after previous adjustments) so close to $T_i^{(k)}$ and θ_i' moving X' so close to $T_i^{(k')}$ that $f\theta_i(X) \cap \theta_i'(X')$ is confined to a thin regular neighborhood of P. Subsequent motion must be controlled to preserve this property in the limit.   ∎

Finally, this section closes by presenting two technical results with important forthcoming roles to play. They promise the capability of a kind of shadow-building and give the expected bounds on embedding dimension of the shadows. A related kind of polyhedral shadow-building forms the backbone of most engulfing arguments.

Proposition 12.   *For any compact subset Y of Int B^n, there exists a starlike-equivalent compact subset X of Int B^n such that $X \supset Y$ and dem $X \le$ dem $Y + 1$.*

Proof. Adjust Y so that it avoids the origin 0. Choose a sequence of triangulations $\{T_i\}$ of ∂B^n with mesh going to zero. Let $k = \text{dem } Y$, and let P_i denote the geometric cone in B^n from 0 over the $(n - k - 2)$-skeleton of T_i.

We want Y to miss each P_i, which can be achieved by a small perturbation, exactly as in the proof of Theorem 5. As a result, we have a homeomorphism f of Int B^n onto itself for which $f(Y) \cap P_i = \varnothing$ for all i.

Name the radial retraction r from $B^n - 0$ to ∂B^n, and let $Z = rf(Y)$. Note that by construction Z has embedding dimension $\leq k$ in ∂B^n. Determine a compact subcone X^* of the cone from 0 over Z such that $f(Y) \subset X^* \subset \text{Int } B^n$, and define X as $f^{-1}(X^*)$. According to Proposition 1, dem $X = \text{dem } X^*$, so it suffices to show that dem $X^* \leq k + 1$.

One way to do this is to construct "triangulations" of B^n by coning from 0 over the triangulations T_i of ∂B^n and then truncating the result by cutting along finite families of $(n - 1)$-spheres concentric with ∂B^n; this provides "triangulations" R_i of B^n consisting of convex linear cells (when B^n is regarded as I^n), with mesh R_i tending toward zero and with the $(n - k - 2)$-skeleton of each R_i missing X^*. These function enough like ordinary PL triangulations that the argument of Theorem 8 applies.

Another method is to observe that dim $X^* \leq \text{dim } Z + 1 \leq k + 1$ because dim $Z \leq \text{dem } Z \leq k$. Specializing to $n \geq 5$, we know dem $X^* = \text{dim } X^*$ unless (possibly) $k + 1 < n - 2$. In the remaining cases, Theorem 7 attests that dem $X^* = \text{dim } X^* \leq k + 1$ provided X^* is 1-LCC, which follows quickly (an exercise) from the fact that Z is 1-LCC in ∂B^n. ∎

The last result serves as a relative version of Proposition 12.

Proposition 13. *Suppose Y is a compact subset of Int $B^n \subset B^n \subset E^n$, A is a closed subset of E^n, and $\varepsilon > 0$. Then there exist ε-homeomorphisms f_Y and f_A of E^n, supported in compact sets, and there exists a pair (X, X_A) of compact subsets of B^n, each starlike with respect to the origin 0, satisfying*

(a) *$f_Y(Y) \subset X$ and dem $X = \text{dem } Y + 1$,*
(b) *$X \cap f_A(A) \subset X_A$ and dem $X_A \leq \text{dem } X + \text{dem } A + 1 - n$.*

Proof. In the manner described before, without regard to A, construct a homeomorphism f_Y of E^n to itself, fixed outside B^n, adjusting Y so as to place it in a cone X, where ∂B^n contains a compact set Z with embedding dimension (relative to ∂B^n) bounded by dem Y, and where X corresponds to the cone from 0 over Z. Then dem $X \leq \text{dem } Y + 1$. Attaching another cone from 0, if necessary, we can assume dem $X = \text{dem } Y + 1$.

In a similar manner, construct another small homeomorphism f_A' of E^n to itself, fixed outside a neighborhood of B^n (not just outside B^n, because $A \cap \partial B^n$ could be a snarl), adjusting $A \cap B^n$ so as to place it in a cone C_A, where this time ∂B^n contains a compactum Z_A with embedding dimension (relative to ∂B^n) bounded by dem A and where the cone C_A from 0 over Z_A contains $f_A'(A) \cap B^n$.

The decisive measure is initiated by a minor perturbation of ∂B^n minimizing the intersection of Z and Z_A. By Proposition 11 there exists a small homeomorphism h_∂ of ∂B^n such that

$$\text{dem}[Z \cap h_\partial(Z_A)] \le \text{dem } Z + \text{dem } Z_A - (n - 1)$$

$$\le \text{dem } Y + \text{dem } A + 1 - n.$$

Now h_∂ extends over B^n by coning from 0, which further extends to a homeomorphism h of E^n, by tapering h_∂ off the identity on a collar in $E^n - \text{Int } B^n$. Then h simplifies intersections between X and C_A, with bounds given by:

$$\text{dem}[X \cap h(C_A)] = 1 + \text{dem}[Z \cap h_\partial(Z_A)]$$

$$\le 1 + (\text{dem } Y + \text{dem } A + 1 - n)$$

$$\le \text{dem } X + \text{dem } A + 1 - n.$$

Set $f_A = hf_A'$ and $X_A = X \cap h(C_A)$ to complete the proof. ∎

EXERCISES

1. If X is a compact subset of a PL n-manifold M and dem $X \le k$, then X is i-LCC for $i \in \{0, 1, ..., n - k - 1\}$.

2. If the compact subset X of ∂B^n has dem $X < n - 3$, then the cone over X from the origin is 1-LCC in B^n.

3. Let G denote a cell-like decomposition of E^n such that H_G forms a null sequence of (contractible) tamely embedded polyhedra, each of dim p, where $2p + 4 \le n$. Show that G is shrinkable by using the Bing–Kister theorem to find ε-amalgamations $\tilde{G}$ of G such that $H_{\tilde{G}}$ forms a null sequence of starlike-equivalent compact polyhedra.

22. SHRINKING SPECIAL 0-DIMENSIONAL DECOMPOSITIONS

The heart of this section is the proof of the following result.

Proposition 1. *Let G be a cell-like decomposition of a (PL) n-manifold M such that H_G forms a null sequence and dem $g \le n - 3$ for each $g \in G$. Then G is shrinkable.*

The argument, which is lengthy, incorporates one of the last bare-handed shrinking techniques to be described here. Before pursuing it, however, we look at its amalgamated consequence.

Theorem 2. *Let G be a cell-like, 0-dimensional decomposition of a (PL) n-manifold M such that* dem $N_G \le n - 3$. *Then G is shrinkable.*

Proof. According to Theorem 21.9, there exists a sequence $\{T_i\}$ of triangulations of M for which $N_G \cap T_i^{(2)} = \varnothing$. For any $\varepsilon > 0$, Theorem 20.5 provides a cell-like decomposition K of M such that K is an ε-amalgamation of G, H_K forms a null sequence, and, significantly, $N_K \cap (\bigcup_i T_i^{(2)}) = \varnothing$. Therefore, Theorem 21.9 again reveals that dem $N_K \le n - 3$, and Proposition 1 attests that K is shrinkable.

Armed with Proposition 1, for each $\varepsilon > 0$ we have produced a shrinkable ε-amalgamation of G. According to Theorem 20.1, this indicates that G itself is shrinkable. ∎

Proof of Proposition 1. Enumerate the elements $g_1, g_2, \ldots$ of H_G.

Special Case. Each $g_i \in H_G$ is a subpolyhedron of M and $2 \dim g_i + 2 \le n$.

This special case is a worthy guide for the general case. It traces out the same proof pattern and identifies the appropriate shrinking technique, but in a setting permitting use of the more familiar PL, as opposed to embedding dimension, methodology.

Given $g_0 \in H_G$, a neighborhood U of g_0, and $\delta > 0$, we will find a homeomorphism F of M onto itself satisfying (1) $F \mid M - U = \text{Id}$, (2) diam $F(g_0) < \delta$, and (3) for each $g \in G$ either $F(g) = g$ or diam $F(g) < \delta$. The shrinkability of G will be implied then by Theorem 7.5.

To do this, first restrict U so that every $g \in G$, other than g_0, touching U has diameter less than $\delta/3$. Since g_0 is cellular by Corollary 21.9A, there exists a PL n-cell B in U with $g_0 \subset \text{Int } B$. (Alternatively, one can obtain such a B by applying the Bing–Kister result.) Fix a point $c \in \text{Int } B - g_0$ and regard B as the PL cone from c over ∂B. Let C_0 denote the subcone consisting of all points on any line segment from c to ∂B that meets g_0 (i.e., C_0 equals the cone from c over the natural projection of g_0 to ∂B). With B regarded in this fashion as a cone, C_0 is a starlike subset of B, but with B considered *in situ*, C_0 represents a starlike-equivalent subset of U. Furthermore, C_0 is a $(\dim g_0 + 1)$-subpolyhedron of M in U.

Triangulate $U - g_0$ so $C_0 - g_0$ is a subcomplex and adjust M via a (PL) homeomorphism ψ fixing g_0, supported in U, and placing the vertices of $C_0 - g_0$ in general position with respect to those of each g_i. Because

$$\dim C_0 + \dim g_i = \dim g_0 + 1 + \dim g_i < n,$$

$\psi(C_0) \cap N_G = g_0$. (One can attain the same ends from demension-theoretic methods, using the facts that dem $C_0 = \dim C_0$ and dem $g_i = \dim g_i$.)

As in the proof of Lemma 8.5 or by Exercise 8.2, the starlike-equivalent set $\psi(C_0)$ can be shrunk to size δ without causing the sizes of other g_i's touching U to swell to size δ, under a homeomorphism F supported in U. (Exercise 1 suggests another method for concluding the proof in this case.)

General Case. This requires an iterated coning and general positioning, similar to procedures employed for codimension ≤ 3 engulfing. The inductive device is set forth as the lemma below.

Lemma 3 (*a*-demensional shrinking). *Under the hypothesis of Proposition 1, suppose A is a closed subset of M such that $a = \dim A \leq n - 2$, $g_m \in H_G$, and $\eta > 0$. Then there exists a neighborhood W of g_m in $N(g_m; \eta)$ and there exists a homeomorphism ψ of M onto itself, supported in W, such that $\dim \psi(g_i) < \eta$ for each $g_i \in G$ satisfying $\psi(g_i) \cap W \cap A \neq \varnothing$.*

We shall postpone the proof of this lemma until after exploiting it to complete the proof of Proposition 1.

Exactly as in the proof of the special case, given $g_0 \in H_G$, a neighborhood U of g_0, and $\delta > 0$, we intend to describe a homeomorphism F of M onto itself supported in U such that $\dim F(g_0) < \delta$ and, for the other $g \in G$, either $F(g) = g$ or $\dim F(g) < \delta$.

Start as before by trimming back the neighborhood U of g_0 so that each g_i ($\neq g_0$) meeting U has diameter less than $\delta/3$ and, using the cellularity of g_0 (see Corollary 21.9A), by locating an n-cell B in U for which $\text{Int } B \supset g_0$. Fix a point $c \in \text{Int } B - g_0$ and think of B as the geometric cone from c over ∂B. Apply Proposition 21.12 to obtain a starlike-equivalent subset X of $\text{Int } B$ such that $g_0 \subset X$ and $\dim X \leq \dim g_0 + 1 \leq n - 2$. We want to stress that X is typical of the objects A to which Lemma 3 applies. For technical simplicity we suppress the homeomorphism from B to the starlike cell as well as the homeomorphism of B to itself carrying X to a starlike set, and treat X as a starlike subset of $\text{Int } B$.

At this spot one must recall the method used in Lemma 8.5 to shrink a starlike object X while regulating the sizes of nearby elements from a specified null sequence. The shrinking was achieved as a composition of a sequence $f_1, f_2, \ldots, f_{k-1}$ of short radial moves, each disturbing points less than $\delta/3$, with f_j sending $f_{j-1} \cdots f_1(X)$ into a starlike subset of itself and with f_j restricted so as to move no point outside an arbitrary small neighborhood V_j of $f_{j-1} \cdots f_1(X)$, determined after $f_1, \ldots, f_{j-1}$ had been identified.

To begin that process, we find f_1 as above supported in $V_1 = U$. Note that $\dim f_1(g) < \delta$ for each $g \in H_G$ in U except g_0.

Consider any one, say g_m, of the finite collection of elements of G ($\neq g_0$) for which $\dim f_1(g_m) \geq \delta/3$ and $f_1(g_m) \cap f_1(X) \neq \varnothing$. Construct an $f_1(G)$-saturated neighborhood Q_m of $f_1(g_m)$ such that $f_1(g_0) \cap Q_m = \varnothing$, $\dim Q_m < \delta$, and no two such neighborhoods intersect. Choose $\eta > 0$ sufficiently small that $\eta < \delta/3$ and $N(f_1(g_m); \eta) \subset Q_m$. (From here on we treat g_m as the only one of these sets; when others are present, we perform similar adjustments near each.) By Lemma 3 there exists a neighborhood W_m of $f_1(g_m)$ in $N(f_1(g_m); \eta)$, as shown in Fig. 22-1, and there exists a homeomorphism ψ_1 supported in W_m such that diam $\psi_1(f_1(g_i)) < \eta$ whenever $\psi_1 f_1(g_i) \cap W_m \cap f_1(X) \neq \varnothing$. As a result, for every $g \in G$ except $g = g_0$, $\psi_1 f_1(g) \cap f_1(X) \neq \varnothing$ implies diam $\psi_1 f_1(g) < \delta/3$. More generally, for all $g \in G$ in U except $g = g_0$, diam $\psi_1 f_1(g) < \delta$. Finally, note that $\psi_1 f_1(g_0) = f_1(g_0) \subset f_1(X)$.

Determine a neighborhood V_2 of $f_1(X)$ in V_1 so close to $f_1(X)$ that each set $\psi_1 f_1(g)$ meeting V_2 (except for $g = g_0$) has diameter less than $\delta/3$. Then, as in Lemma 8.5, find a $(\delta/3)$-homeomorphism f_2 supported in V_2 pulling the starlike set $f_1(X)$ inward another notch. Apply Lemma 3, as in the preceding paragraph, to obtain a homeomorphism ψ_2 supported in $V_2 - f_2 f_1(g_0)$ and rectifying sizes of the elements from $f_2 \psi_1 f_1(G)$ that meet $f_2 f_1(X)$, so that, for each $g \in G$ in U other than g_0,

$$\text{diam } \psi_2 f_2 \psi_1 f_1(g) < \delta$$

and, if $\psi_2 f_2 \psi_1 f_1(g) \cap f_2 f_1(X) \neq \varnothing$,

$$\text{diam } \psi_2 f_2 \psi_1 f_1(g) < \delta/3.$$

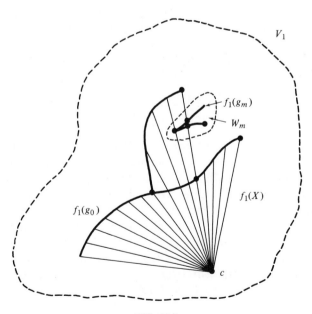

FIG. 22-1

Repeat this procedure, producing in order homeomorphisms $f_1, \psi_1, f_2,$ $\psi_2, f_3, \ldots, \psi_{k-2}, f_{k-1}$. The composition $F = f_{k-1}\psi_{k-2}f_{k-2} \cdots \psi_1 f_1$ has the desired effect. Not only does it shrink g_0 to δ-size [in analyzing this, a crucial aspect is the part of the construction ensuring that $\psi_j f_j \cdots \psi_1 f_1(g_0) = f_j \cdots f_1(g_0)$], but F does so with control: if $f_j \psi_{j-1} \cdots f_1(g)$ becomes dangerously large (diam $\geq \delta/3$), either ψ_j rectifies the problem or ψ_j pushes $f_j \psi_{j-1} \cdots f_1(g)$ off $f_j \cdots f_1(X)$, which allows the later radial compressions to sweep past without affecting the new image. This completes the proof of Proposition 1. ∎

Proof of Lemma 3. The argument proceeds by induction on $a = \operatorname{dem} A$, beginning with the vacuous case $a = -1$. To be sure, general position (embedding dimension) methods quickly dispose of the cases $a = 0, 1, 2$ as well; see Proposition 21.10.

Consider a closed subset A of M with $\operatorname{dem} A = a\,(\leq n - 2)$, and suppose inductively that the $(a - 1)$-demensional version of Lemma 3 is true. The scheme to be used is a variation on the one just described—$g_m \in H_G$ will be "shrunk," in successive stages, by small adjustments, until it misses A, and none of the elements nearby will be stretched to large size.

Choose a neighborhood W of g_m in $N(g_m; \eta)$ so that any $g\,(\neq g_m)$ from G intersecting W has diameter less than $\eta/3$. If $g_m \cap A = \varnothing$, set $\psi = \operatorname{Id}_M$. If not, locate a Euclidean patch W' in W and a standard n-cell B in W' containing g_m in its interior (Corollary 21.9A). Regard B as a cone from some point $c \in \operatorname{Int} B$ over ∂B. Apply Proposition 21.13 to obtain a minor modification θ of W' ($\theta = f_A^{-1}f_Y$) and starlike-equivalent subsets X and X_A of W' in B [here X corresponds to the set $f_A^{-1}(X)$ described in Proposition 21.13, while X_A corresponds to $f_A^{-1}(X_A)$] satisfying

(i) $\theta(g_m) \subset X$ and $\operatorname{dem} X = \operatorname{dem} g_m + 1 \leq n - 2$,
(ii) $X \cap A \subset X_A$ and $\operatorname{dem} X_A \leq n - 2 + \operatorname{dem} A + 1 - n < \operatorname{dem} A$;

in so doing it should be noted that the starlike structure of X_A is compatible with that of X, under some (suppressed) homeomorphism of W' to E^n carrying X and X_A to starlike sets. In addition, θ can be regulated to move points only a little, so we suppress θ as well. Just as recalled earlier in proving Proposition 1, based on the methods of Section 8 we know how to compress X_A near the cone point c under the composition of a finite sequence $f_1, f_2, \ldots, f_{k-1}$ of $(\eta/3)$-homeomorphisms. Aided by the inductive hypothesis, one can mimic the argument from the general case in Proposition 1, interspersing homeomorphisms $\psi_j\,(j = 1, \ldots, k - 2)$ supported in $W - f_j \cdots f_1(g_m)$, so that, for each $g \in G$ in $W\,(g \neq g_m)$,

$$\operatorname{diam} \psi_j f_j \cdots \psi_1 f_1(g) < \eta,$$

and, if $\psi_j f_j \cdots \psi_1 f_1(g) \cap f_j \cdots f_1(X_A) \neq \varnothing$,

$$\text{diam } \psi_j f_j \cdots \psi_1 f_1(g) < \eta/3.$$

Then the composition $\psi = f_{k-1} \psi_{k-2} f_{k-2} \cdots \psi_1 f_1$ satisfies $\psi(g_m) \cap A = \varnothing$, because $\psi(g_m) = f_{k-1} \cdots f_1(g_m) \subset X - X_A$, and diam $\psi(g) < \eta$ for the other $g \in G$ such that $\psi(g) \subset W$. These two features imply the stated conclusion of Lemma 3. ∎

Theorem 2 has several interesting consequences, among them Corollary 2B below, which represents a powerful improvement on Theorem 8.1.

Corollary 2A. *If G is a countable cell-like usc decomposition of S^n such that each $g \in H_G$ has embedding dimension $\leq n - 3$, then G is shrinkable.*

Corollary 2B. *If G is a countable usc decomposition of S^n such that each $g \in G$ is a flat cell of dimension $\leq n - 3$, then G is shrinkable.*

Corollary 2C. *If G is a countable usc decomposition of S^n such that each $g \in G$ is a contractible subpolyhedron of S^n having dimension $\leq n - 3$, then G is shrinkable.*

EXERCISES

1. Let G denote a cell-like decomposition of E^n such that H_G forms a null sequence and dem $N_G \leq k$, where $2k + 2 \leq n$. Show that for each $\varepsilon > 0$ there exists an ε-amalgamation $\tilde{G}$ of G such that $H_{\tilde{G}}$ forms a null sequence of starlike-equivalent sets.
2. Fill in details for the proof of Lemma 3.

23. SHRINKING SPECIAL $(n - 3)$-DIMENSIONAL DECOMPOSITIONS

Using a simple filtration device and the 0-dimensional shrinking theorem of the preceding section, Edwards [4] artfully synthesized a related $(n - 3)$-demensional shrinking theorem, which in turn he exploited to derive a 1-LCC shrinking theorem conjectured by Cannon [6].

The next lemma portrays that filtration, the instrument eventually put into operation to measure progress toward successful approximation by a homeomorphism.

Lemma 1. *Each n-dimensional σ-compact metric space Q^n can be filtered into σ-compact subsets $Q^n, Q^{n-1}, ..., Q^i, ..., Q^0$ such that dim $Q^i \leq i$, $Q^i \supset Q^{i-1}$, and dim$(Q^i - Q^{i-1}) \leq 0$.*

Proof. (By downward induction on i.) The σ-compact space Q^i has a countable basis $\{b_j\}$ of open sets whose frontiers (in Q^i) have dimension strictly less than that of Q^i. The union Q^{i-1} of the sets Fr b_j then is σ-compact, $\dim Q^{i-1} < \dim Q^i$, and $\dim(Q^i - Q^{i-1}) \le 0$. ∎

Theorem 2. *If G is a cell-like decomposition of the PL n-manifold M such that* dem $N_G \le n - 3$ *and M/G is finite-dimensional, then G is shrinkable.*

Remark. Due to the potential existence of cell-like dimension-raising decompositions, the hypothesis dem $N_G \le n - 3$ by itself is not known to imply $\dim(M/G) < \infty$.

Proof. By Theorem 17.7 $\dim(M/G) \le n$. Filter M/G into σ-compact subsets $Q^n = M/G, ..., Q^i, ..., Q^0$ satisfying the conclusions of Lemma 1. Name an open set W containing N_G, and consider the decomposition map $\pi: M \to M/G$. For simplicity, assume M is compact. Say that a map $f: X \to Y$ is 1-1 *over a subset A of Y if $f \,|\, f^{-1}(A)$ is 1-1.*

Claim. *For $i = 0, 1, ..., n$ and for each $\varepsilon > 0$ there exists a cell-like map $F_{i+1}: M \to M/G$ satisfying*

- (a) $\rho(F_{i+1}, \pi) < (i + 1)\varepsilon$,
- (b) F_{i+1} *is* 1-1 *over Q^i,*
- (c) dem $N_{F_{i+1}} \le n - 3$, *and*
- (d) $F_{i+1} \,|\, M - W = \pi \,|\, M - W$.

Once this claim is established, the proof will be complete, for certainly F_{n+1} will be a homeomorphism approximating π. The claim itself is proved by induction on i. The inductive step coincides essentially with that for $i = 0$, so we shall concentrate on the initial step.

To begin, we express Q^0 as the countable union of compact 0-dimensional sets Z_j $(j = 1, 2, ...)$. We shall find cell-like maps f_j $(j = 0, 1, ...)$, with $f_0 = \pi$, of M onto M/G satisfying the following conditions;

- (1) $\rho(f_{j+1}, f_j) < \varepsilon/2^{j+1}$;
- (2) f_j is 1-1 over $\bigcup_{k=1}^{j} Z_k$;
- (3) there exists a PL triangulation T_j of M with mesh less than $\frac{1}{2}^j$ $(j > 0)$ whose 2-skeleton $T_j^{(2)}$ misses $N_{f_j} \subset f_j^{-1}(\pi(N_G))$;
- (4) $f_{j+k} \,|\, (M - W) \cup T_j^{(2)} \cup f_j^{-1}(Z_j) = f_j \,|\, (M - W) \cup T_j^{(2)} \cup f_j^{-1}(Z_j)$ for $k \in \{1, 2, ...\}$;
- (5) for $x \in M - (T_j^{(2)} \cup f_j^{-1}(Z_j))$ and $k \in \{0, 1, ...\}$, $\rho(f_{j+k+1}(x),$ $f_{j+k}(x)) \le \dfrac{1}{4^k} \rho(f_{j+k}(x), f_j(T_j^{(2)}) \cup Z_j)$.

Before describing their construction, we describe their purpose. Condition (1), of course, guarantees that $\{f_j\}$ is a Cauchy sequence of maps converging to a map F_1 close to π, and F_1 necessarily will be cell-like by Theorem 17.4. Condition (2) requires f_j to be 1-1 over Z_j, condition (4) maintains the same action over Z_j by subsequent maps, and condition (5) provides the crucial controls preventing any other points of M, besides those of $f_j^{-1}(Z_j)$, from being sent to Z_j under F_1. Consequently, F_1 will be 1-1 over Q^0. Similarly, condition (3) identifies a finely meshed triangulation whose 2-skeleton $T_j^{(2)}$ misses the nondegeneracy set of f_j, condition (4) maintains the same action on $T_j^{(2)}$ by subsequent maps, and condition (5) prevents any other points of M from being sent to $F_1(T_j^{(2)}) = f_j(T_j^{(2)})$ under F_1. As a result, the nondegeneracy set of F_1 will be contained in $M - \bigcup T_j^{(2)}$, causing the nondegeneracy set of F_1 to have embedding dimension $\leq n - 3$, by Theorem 21.9. In summary, construction of such maps $\{f_j\}$ will establish the claim for the case $i = 0$.

To perform the first stage construction, observe that the decomposition G_1 induced over Z_1 by $f_0 = \pi$, where G_1 consists of the sets $f_0^{-1}(z)$, $z \in Z_1$, plus the remaining singletons from M, is a 0-dimensional, cell-like usc decomposition of M such that dem $N_{G_1} \leq n - 3$ (because $N_{G_1} \subset N_G$). According to Theorem 22.2, G_1 is strongly shrinkable. After choosing a triangulation T_1 of M with mesh less than $\frac{1}{2}$ and with 2-skeleton missing N_G (see Theorem 21.9), we obtain a map θ_1 of M onto itself, the end of some pseudo-isotopy, such that θ_1 realizes G_1, θ_1 moves no point outside $W - T_1^{(2)}$, and θ_1 is limited by the inverse image (under $f_0 = \pi$) of an open cover of M/G by sets having diameter less than $\varepsilon/2$.

We define f_1 as $f_0\theta_1^{-1}$, which obviously is a well-defined map close to f_0. Its inverse, $f_1^{-1} = \theta_1 f_0^{-1}$, discloses that f_1 is 1-1 over a subset of M/G when f_0 is, or, more formally, that

$$N_{f_1} \subset f_1^{-1} f_0(N_G) = f_1^{-1}(\pi(N_G)).$$

Furthermore, since each nondegenerate set $f_1^{-1}(q)$, $q \in M/G$, is homeomorphic via θ_1 to $f_0^{-1}(q)$, $f_1^{-1}(q)$ is cell-like. By construction of G_1 and θ_1, f_1 is 1-1 over Z_1. In particular, f_1 is a cell-like map satisfying conditions (1), (2), and (3).

The second stage construction is typical of the iteration. The decomposition G_2 induced over Z_2 by f_1 is an 0-dimensional cell-like decomposition of M such that

$$\text{dem } N_{G_2} \leq \text{dem } N_{f_1} \leq n - 3$$

(a fact verified later as Lemma 3). Again by Theorem 22.2, G_2 is strongly shrinkable. After choosing a triangulation T_2 having mesh less then $\frac{1}{4}$ and having 2-skeleton missing $N_{f_1} \supset N_{G_2}$, we obtain a map θ_2 of M onto itself, the end of another pseudo-isotopy, such that θ_2 realizes G_2, θ_2 fixes points outside $W - (T_1^{(2)} \cup T_2^{(2)} \cup f_1^{-1}(Z_1))$, and θ_2 is $\mathfrak{U}_2$-close to Id_M, where $\mathfrak{U}_2$ is the inverse image under f_1 of an open cover of M/G by sets having diameter less than $\varepsilon/4$. In addition, θ_2 is allowed to move no point outside $f_1^{-1}(Y_2)$, where Y_2 is an open set containing $f_1(N_{G_2})$ in $M/G - (Z_1 \cup f_1(T_1^{(2)}))$ whose components y_2 satisfy

$$\text{diam } y_2 \leq \tfrac{1}{4}\rho(\text{Cl } y_2, Z_1 \cup f_1(T_1^{(2)})).$$

Define f_2 as $f_1\theta_2^{-1}$. Conditions (1), (2), and (3) hold as before. Condition (4) follows quickly from the limitations on θ_2, and condition (5) follows from the additional limitation at the end.

By continuing in similar fashion, we build the maps $\{f_j\}$, establishing the claim for $i = 0$. It may be worth mentioning here that breaking Q^0 apart into the Z_j's is indispensable: although the decomposition induced by f_0 over Q^0, like that over Z_1, partitions M into singletons and a 0-dimensional collection of cell-like sets (whose union even has the appropriate embedding dimension!), this decomposition is not likely to be usc.

Now, on to the inductive step. Suppose that F_i is a cell-like map of M onto M/G satisfying statements (a) through (d) in the claim. The new goal is to produce another cell-like map $F_{i+1}: M \to M/G$ close to F_i and 1-1 over Q^i. Toward that end, express Q^i as a countable union of compact i-dimensional sets $Z_1, Z_2, \ldots$. Exactly like the case $i = 0$, we find cell-like maps f_j of M onto M/G ($j \in \{0, 1, \ldots\}$), with $f_0 = F_i$, satisfying conditions (1) through (5) listed there, except for condition (2), which is upgraded to:

(2′) f_j is 1-1 over $Q^{i-1} \cup (\bigcup_{k=1}^{j} Z_k)$.

The key for starting, and for iterating, is that the decomposition G_1 induced over Z_1 by $f_0 = F_i$ is 0-dimensional, because $f_0(N_{G_1}) \subset Z_1 - Q^{i-1} \subset Q^i - Q^{i-1}$. This exposes the principal benefit of the filtration. We shrink out G_1 via θ_1, define f_1 as $f_0\theta_1^{-1}$, and continue the iteration, proceeding precisely through the same drill as in the case $i = 0$, to complete the proof of the inductive step. This finishes the proof of the Claim and of Theorem 2, modulo the following omitted result about embedding dimension.

Lemma 3. *Suppose G is a cell-like decomposition of a (compact) PL n-manifold M such that* dem $N_G \leq n - 3$, *Z is a closed subset of M/G such that the decomposition $G(Z)$ induced over Z is strongly shrinkable, $\theta: M \to M$ is a map realizing $G(Z)$ which equals the end p_1 of some pseudo-isotopy $p_t: M \to M$, and f is the natural map $\pi\theta^{-1}$. Then* dem $N_f \leq n - 3$.

Proof.   Recall from the proof of Theorem 2, or recheck, that $\theta(N_G) \supset N_f$. To determine a typical 2-skeleton in $M - N_f$, start with a triangulation T having very small mesh and having 2-skeleton $T^{(2)} \subset M - N_G$. Then $\theta(T^{(2)}) \subset M - \theta(N_G) \subset M - N_f$. We shall prove that $\theta(T^{(2)})$ is the 2-skeleton of some triangulation of M having small mesh [since θ is just a map, $\theta(T)$ itself will not serve as a triangulation].

Fix $\varepsilon > 0$. Suppose T was chosen with mesh so fine that diam $p_t(\sigma) < \varepsilon/5$ for all $\sigma \in T$ and all $t \in I$. Choose a value $s \in (0, 1)$ close enough to 1 that

$$\text{diam } p_s(g^*) < \varepsilon/5 \qquad \text{for each} \quad g^* \in G(Z),$$

and

$$\rho(\theta p_s^{-1}, \text{Id}_M) < \varepsilon/5.$$

Because $p_s(G(Z))$ is strongly shrinkable, there exists an $(\varepsilon/5)$-map $\psi: M \to M$ fixed on $p_s(T^{(2)})$ and realizing $p_s(G(Z))$. Then $\theta p_s^{-1}\psi^{-1}$ is a $(2\varepsilon/5)$-homeomorphism of M carrying the tamely embedded polyhedron $\psi p_s(T^{(2)}) = p_s(T^{(2)})$ onto $\theta(T^{(2)})$, which shows $\theta(T^{(2)})$ to be the 2-skeleton of the triangulation $(\theta p_s^{-1}\psi^{-1})p_s(T)$, whose mesh is less than ε. ∎

Embedded in the proof of Theorem 2 are motion controls, pertaining to a special case, forcing maps f_j 1-1 over a closed subset $Z_1 \cup \cdots \cup Z_j$ to converge to a map 1-1 over the union of the Z_j's. Because such controls will be called for repeatedly, under more general circumstances, it is convenient now to dispose of the details, once and for all.

Proposition 4. *Let G denote a cell-like decomposition of an n-manifold M, and for $j = 1, 2, \ldots$, let Z_j denote a closed subset of M/G such that the decomposition $G(Z_j)$ induced over Z_j is shrinkable. Then $\pi: M \to M/G$ can be approximated, arbitrarily closely, by a cell-like map $F: M \to M/G$ that is 1-1 over $\bigcup_j Z_j$.*

Proof. For simplicity again focus on the case of a compact manifold M. Fix $\varepsilon > 0$. Following now-standard practice, we secure F as a limit of cell-like maps $\{f_i\}$, with $f_0 = \pi$. The relevant controls include some obvious ones:

(1) $\rho(f_j, f_{j+1}) < \varepsilon/2^{j+1}$,
(2) f_j is 1-1 over $\bigcup_{i=1}^{j} Z_i$; and
(3) $f_{j+k}|f_j^{-1}(Z_j) = f_j|f_j^{-1}(Z_j)$ for $k \in \{1, 2, \ldots\}$.

In addition, after f_j has been obtained, later maps must be regulated so no points of $M - f_j^{-1}(Z_j)$ are sent to Z_j by F. This is the slightly more troublesome control; it can be attained by requiring (for $k = 1, 2, \ldots$)

(4) for $p \in M - f_j^{-1}(Z_j)$, $\rho(f_{j+k}(p), f_j(p)) < \frac{1}{2}\rho(f_j(p), Z_j)$.

The first step is the easiest. Since $G(Z_1)$ is shrinkable, M admits a self-map θ_1 realizing $G(Z_1)$, limited by

$$\rho(f_0\theta_1^{-1}, f_0) = \rho(\pi\theta_1, \pi) < \varepsilon/2.$$

The map $f_1 = f_0 \theta_1^{-1}$ satisfies all four conditions above (the last two vacuously).

More complicated is the second step, aimed at shrinking out the decomposition induced over Z_2 by f_1. That decomposition coincides with $\theta_1 G(Z_2)$, by virtue of the identity

$$f_1^{-1}(z) = \theta_1 f_0^{-1}(z) = \theta_1 \pi^{-1}(z).$$

According to Corollary 13.2B, $\theta_1 G(Z_2)$ is shrinkable; furthermore, by Theorem 13.3 it is ideally shrinkable. As a result, M admits a self-map θ_2 realizing $\theta_1 G(Z_2)$ (the decomposition induced over Z_2 by f_1), reducing to the identity on $f_1^{-1}(Z_1)$, and limited by the regulations

$$\rho(f_1 \theta_2^{-1}, f_1) < \varepsilon/4 \quad \text{and} \quad \rho(f_1 \theta_2^{-1}(p), f_1(p)) \le \tfrac{1}{4}\rho(f_1(p), Z_1).$$

Then the map $f_2 = f_1 \theta_2^{-1} = f_0 \theta_1^{-1} \theta_2^{-1}$ satisfies conditions (1) through (4). The composition $\theta_2 \theta_1$, it should be noted, realizes the decomposition $G(Z_1 \cup Z_2)$ induced over $Z_1 \cup Z_2$.

Inductively, one should presume the existence of maps $f_1, \ldots, f_j$ satisfying conditions (1)–(4) so that, in addition, $f_i = f_0 \lambda_i^{-1}$, where $\lambda_i \colon M \to M$ is a map realizing the decomposition induced over $Z_1 \cup \cdots \cup Z_i$. With this extra hypothesis one can construct a map f_{j+1} satisfying the expanded induction hypothesis, by imitating what was done in the second step. ∎

Next, the dimension constraints found in Theorem 2 undergo a notable metamorphosis, shifting from a limitation on embedding dimension in the source to a more advantageous one on dimension in the quotient.

Theorem 5 (1-LCC shrinking). *If G is a closed-$(n - 3)$-dimensional cell-like decomposition of a (PL) n-manifold M, $n \ge 5$, for which $\pi(\bar{N}_G)$ is 1-LCC embedded in M/G, then G is shrinkable.*

Proof. The similarity between this theorem and Theorem 2 suggests a workable strategy: approximate $\pi \colon M \to M/G$ by a cell-like map $F \colon M \to M/G$ such that dem $N_F \le n - 3$, for then such an F (and, therefore, π) is a near-homeomorphism.

The desired map F materializes as the limit of cell-like maps whose nondegeneracy sets avoid successively finer skeleta. Specifically, for $\varepsilon > 0$ and $j \in \{1, 2, \ldots\}$ we produce a triangulation T_j of M with mesh less than $1/j$ and a cell-like map f_j of M onto M/G satisfying (where $f_0 = \pi$)

(1) $\rho(f_j, f_{j-1}) < \varepsilon/2^j$;
(2) $f_j(T_j^{(2)}) \cap \pi(\bar{N}_G) = \varnothing$;
(3) f_j is 1-1 over $\pi(M - \bar{N}_G)$; and
(4) each $T_j^{(2)}$ has a closed neighborhood W_j in $M - f_j^{-1}\pi(\bar{N}_G)$ such that $f_{j+k} \mid W_j = f_j \mid W_j$ for $k = 1, 2, \ldots$.

Consequently, the map $F = \lim f_j$ will be a cell-like map of M onto M/G and, because $F \mid W_j = f_j \mid W_j$, it will be 1-1 over each $f_j(T_j^{(2)})$, forcing dem $N_F \leq n - 3$.

Since the construction of f_1 from $f_0 = \pi$ is just a special case, we assume $T_1, \ldots, T_j$ and $f_1, \ldots, f_j$ have been obtained, subject to conditions (1) through (4). We select a triangulation T_{j+1} of M having mesh less than $1/(j + 1)$. The difficulty to be faced is the possibility that $T_{j+1}^{(2)} \cap f_j^{-1} \pi(\bar{N}_G) \neq \varnothing$, and the point is that such unpleasantness can be remedied by a homeomorphism ψ of M pushing $T_{j+1}^{(2)}$ off $f_j^{-1} \pi(\bar{N}_G)$, fixed on the previously defined sets W_i, and controlled to make $f_{j+1} = f_j \psi$ be close to f_j.

Readers well versed in radial engulfing techniques should readily perceive how to obtain ψ. As an alternative, we fill in a few details for $n \geq 6$ based not on engulfing but on the Bing–Kister theorem.

Subdivide $T_{j+1}^{(2)}$ to get a subcomplex whose underlying point set C lies in $T_{j+1}^{(2)} - f_j^{-1} \pi(\bar{N}_G)$ and contains each $W_i \cap T_{j+1}^{(2)}$ in its interior. Use the 1-LCC (and the implicit 0-LCC) condition to approximate $f_j \mid T_{j+1}^{(2)}$ by a map $\mu: T_{j+1}^{(2)} \to M/G$ such that $\mu \mid C = f_j \mid C$ and $\mu(T_{j+1}^{(2)}) \subset M/G - \pi(\bar{N}_G)$. Recall how the approximate lifting theorem (Theorem 16.7) allows μ to be regarded as $f_j \alpha$, where $\alpha: T_{j+1}^{(2)} \to M$ is a map for which $\alpha \mid C =$ inclusion. Invoke general position methods to adjust α slightly to a PL (with respect to T_{j+1}) embedding having the same properties. In addition, require $f_j \alpha$ to be so close to $f_j \mid T_{j+1}^{(2)}$ that there exists a short homotopy $h_t: T_{j+1}^{(2)} \to M/G$ between them, where $h_t \mid C = f_j \mid C$, $h_t(T_{j+1}^{(2)} - C) \cap W_i = \varnothing$, and each track $h(\{z\} \times I)$ has diameter less than $\varepsilon/2^{j+2}$ (Corollary 14.6E). Apply Theorem 16.8 to obtain a homotopy $H_t: T_{j+1}^{(2)} \to M$ between the inclusion and α such that $H_t \mid C =$ inclusion and $f_j H_t$ has the same properties as h_t, and then cover the nonstationary tracks $H(\{z\} \times I)$ by open subsets U_z of M for which diam $f_j(U_z) < \varepsilon/2^{j+2}$. Use Theorem 21.7 to secure a PL homeomorphism $\psi: M \to M$ such that $\psi \mid T_{j+1}^{(2)} = \alpha$, $\rho(f_j \psi, f_j) < \varepsilon/2^{j+1}$, and ψ equals the identity on each W_i. Finally, define f_{j+1} as $f_j \psi$ and name a closed neighborhood W_{j+1} of $T_{j+1}^{(2)}$ in $M - f_{j+1}^{-1} \pi(\bar{N}_G)$. ∎

Corollary 5A. *Suppose G is a cell-like decomposition of a (PL) n-manifold M, $n \geq 5$, and suppose Z is a closed $(n - 3)$-dimensional subset of M/G that is 1-LCC embedded there. Then the decomposition $G(Z)$ induced over Z is shrinkable.*

The proof is an exercise.

Corollary 5B. *Suppose G is a cell-like decomposition of a (PL) n-manifold M, $n \geq 5$, and suppose for $j \in \{1, 2, \ldots\}$ Z_j is a closed $(n - 3)$-dimensional subset of M/G that is 1-LCC embedded there. Then $\pi: M \to M/G$ can be approximated, arbitrarily closely, by a cell-like map $F: M \to M/G$ that is 1-1 over $\bigcup_j Z_j$.*

See also Proposition 4. It should be noted that in order to prove Corollary 5B one need never invoke (as was done in Proposition 4) the concept of ideally shrinkable decomposition. Instead, one can follow the structural outline provided by the proof of Proposition 4, at the second step using Corollary 5A, applied to the decomposition of the noncompact manifold $M - f_1^{-1}(Z_1)$ induced over $Z_2 - Z_1$ by f_1, to realize that decomposition by a map θ_2. It must be regulated so that, besides controls comparable to those given in Proposition 4, θ_2 extends via the identity on $f_1^{-1}(Z_1)$ to a map θ_2 of M onto itself. The extended map satisfies *all* the conditions required in the course of Proposition 4.

EXERCISES

1. Suppose G is a cellular usc decomposition of a (compact) n-manifold M. Show that $\pi: M \to M/G$ can be approximated by a cell-like map $F: M \to M/G$ such that $F(N_F)$ contains no open subset of M/G.
2. Prove Corollary 5A.

24. THE DISJOINT DISKS PROPERTY AND THE CELL-LIKE APPROXIMATION THEOREM

Like an elaborate puzzle, Edwards's cell-like approximation theorem involves many pieces, most of which have been laid out heretofore. The final one concerns a fundamental general position property. It has antecedents in the work of R. H. Bing [3] (seen here as the property used in Section 9 to diagnose nonshrinkable decompositions), in the property of Section 19 about inessential spanning, and in a mismatch condition first introduced by W. T. Eaton [1]. It surfaced explicitly in results of J. W. Cannon [6] preceding Edwards's work.

A metric space X has the *disjoint disks property*, abbreviated as DDP, if for any two maps $\mu_1, \mu_2: B^2 \to X$ and for each $\varepsilon > 0$, there exist approximating maps $\mu_1', \mu_2': B^2 \to X$ such that $\rho(\mu_i', \mu_i) < \varepsilon$ for $i \in \{1, 2\}$ and $\mu_1'(B^2) \cap \mu_2'(B^2) = \varnothing$.

The first result displays part of the strength of the DDP.

Proposition 1. *Let X denote a locally compact, separable ANR. The following statements are equivalent:*

(a) *X has the disjoint disks property.*

(b) *Each map $\mu: B^2 \to X$ can be approximated, arbitrarily closely, by an embedding $\lambda: B^2 \to X$.*

(c) *Each map μ of a finite 2-complex P into X can be approximated, arbitrarily closely, by an embedding $\lambda: P \to X$.*

(d) *Each map μ of a finite 2-complex P into X can be approximated by an embedding $\lambda: P \to X$ such that $\lambda(P)$ is 0-LCC and 1-LCC in X.*

Proof. Clearly (d) $\Rightarrow$ (c) $\Rightarrow$ (b). We show first that (b) $\Rightarrow$ (a).

Consider maps $\mu_1, \mu_2: B^2 \to X$ in the only case of interest, where $\mu_1(B^2) \cap \mu_2(B^2) \neq \varnothing$. Find disjoint 2-cells D_1 and D_2 in B^2 such that B^2 retracts to the union of $D_1 \cup D_2$ and an arc α running between them (see Fig. 24-1). Name homeomorphisms $\theta_i: D_i \to B^2$ and define $\mu: D_1 \cup D_2 \to X$ as $\mu \mid D_i = \mu_i \theta_i$ ($i = 1, 2$). Extend μ over α by having $\mu \mid \alpha$ trace out a path connecting $\mu(\partial\alpha)$ in the pathwise-connected set $\mu_1(B^2) \cup \mu_2(B^2)$, and then extend μ over the rest of B^2 by first retracting to $D_1 \cup \alpha \cup D_2$. Since μ can be approximated by embeddings $\lambda: B^2 \to X$, we can obtain disjoint approximations λ_1, λ_2 to μ_1, μ_2 by setting $\lambda_i = \lambda(\theta_i)^{-1}$.

(a) $\Rightarrow$ (b). The 2-cell B^2 contains a countable collection $\{(D_k, D_k')\}$ such that, for each k, D_k and D_k' are disjoint 2-cells in B^2 and the collection separates the points of B^2; that is, given any two distinct points $p, q \in B^2$, there exists a pair (D_k, D_k') for which $p \in D_k$ and $q \in D_k'$. Let $\mathcal{C}$ denote the space of all maps of B^2 into X, with the sup-norm metric, and let

$$\mathcal{O}_k = \{\mu \in \mathcal{C} \mid \mu(D_k) \cap \mu(D_k') = \varnothing\}.$$

Since the collection $\{(D_k, D_k')\}$ separates points, each $\lambda \in \bigcap_k \mathcal{O}_k$ is an embedding. Because X has a complete metric, $\mathcal{C}$ is a Baire space. Hence, in order to show why each $\mu \in \mathcal{C}$ can be approximated by an embedding $\lambda \in \bigcap \mathcal{O}_k$, it suffices to prove that $\mathcal{O}_k$ is open and dense in $\mathcal{C}$.

B^2

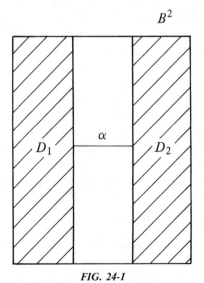

FIG. 24-1

That $\mathcal{O}_k$ is open in $\mathcal{C}$ should be transparent; that $\mathcal{O}_k$ is dense in $\mathcal{C}$ requires some explication. To do that, consider $\mu \in \mathcal{C}$. With the aid of homeomorphisms e, e' of B^2 onto D_k, D_k', respectively, one can produce a pair of maps, μe and $\mu e'$, from B^2 to X. Transferring back to the domain of μ, one finds, using the DDP, that the map $\mu \,|\, D_k \cup D_k'$ can be approximated by $\mu': D_k \cup D_k' \to X$ for which $\mu'(D_k) \cap \mu'(D_k') = \varnothing$. Moreover, since X is an ANR, Corollary 14.6E (actually, a variation of Corollary 14.6E to account for the noncompactness of X) implies that μ' can be obtained so close to $\mu \,|\, D_k \cup D_k'$ that μ' extends to a map $B^2 \to X$ approximating $\mu: B^2 \to X$. As a result, $\mathcal{O}_k$ is dense.

The implication (a) $\Rightarrow$ (c) is established by a similar argument.

For the remaining implication, we need a way to detect the 1-LCC property.

Lemma 2. *Suppose X is a locally compact ANR, $\mathcal{C}$ is the space of all maps $B^2 \to X$, $\mathfrak{D} = \{\alpha_i\}$ is a countable dense subset of $\mathcal{C}$, and A is a closed subset of X in $X - \bigcup_i \alpha_i(B^2)$. Then A is 1-LCC in X.*

Proof. Focus on a neighborhood U of an arbitrary point $a \in A$. By local contractibility, a has another neighborhood V that contracts in U. Then any map $f: \partial B^2 \to V - A$ extends to a map $F: B^2 \to U$, and the denseness of $\mathfrak{D}$ in $\mathcal{C}$ yields $\alpha_i \in \mathfrak{D}$ so close to F that $\alpha_i(B^2) \subset U$ and $f = F \,|\, \partial B^2$ is homotopic to $\alpha_i \,|\, \partial B^2$ in $U - A$. In other words, f is homotopic to $\alpha_i \,|\, \partial B^2$, which in turn is null homotopic in $\alpha_i(B^2)$, and the range of this homotopy lies in $U - A$. ∎

(a) $\Rightarrow$ (d). Let $\mathcal{C}$ denote the space of maps $B^2 \to X$. Since X has a countable basis, $\mathcal{C}$ has a countable dense subset $\{f_i\}$ (Dugundji [1, p. 265]).

Exactly as shown earlier, in the space $\mathfrak{M}(P, X)$ of maps $P \to X$, the subset $\mathcal{E}$ of embeddings is dense and is the intersection of open subsets $\mathcal{O}_k$ of $\mathfrak{M}(P, X)$. Thus, a given $\mu \in \mathfrak{M}(P, X)$ can be approximated by an embedding $\lambda_0 \in \mathcal{E}$. Applying the above to the 2-complex equal to the disjoint union of P and B^2, one sees that every pair of maps defined on P and B^2 can be approximated by disjoint embeddings of P and B^2. In particular, the maps λ_0 and f_1 can be approximated by disjoint embeddings $\lambda_1: P \to X$ and $f_1': B^2 \to X$. Similarly, λ_1 and f_2 can be approximated by disjoint embeddings $\lambda_2: P \to X$ and $f_2': B^2 \to X$, with λ_2 so close to λ_1 that $\lambda_2(P) \cap f_1'(B^2) = \varnothing$. Continuing in this fashion, one can produce embeddings $\lambda_i: P \to X$ and $f_i': B^2 \to X$ such that

$$\lambda_i(P) \cap f_j'(B^2) = \varnothing \qquad (j \in \{1, 2, ..., i\}).$$

Three aspects of this demand special care for the final outcome. First of all, the space $\mathcal{E}$, being a G_δ-subset of the complete metric space $\mathfrak{M}(P^2, X)$, has a complete metric; the successive embeddings λ_i must approximate their predecessors λ_{i-1} sufficiently well that $\{\lambda_i\}$ forms a Cauchy sequence in this metric so then $\{\lambda_i\}$ will converge to some $\lambda_\infty \in \mathcal{E}$ close to both $\lambda_0 \in \mathcal{E}$ and $\mu \in \mathfrak{M}(P, X)$. Second, after λ_i has been determined, the later elements in this sequence must be obtained so close to λ_i that $\lambda_\infty(P) \cap f_i'(B^2) = \emptyset$ for all i. Third, f_i' must be chosen close enough to f_i that $\{f_i'\}$ is another dense subset of $\mathcal{C}$.

Lemma 2 attests that $\lambda_\infty(P)$ is 1-LCC in X. Indeed, $\lambda_\infty(P)$ is also 0-LCC. (If this is not clear, it should be clear how to modify the construction to ensure its validity.) ∎

At last we are prepared to prove the fundamental result of decomposition theory, the culmination of Edwards's program.

Theorem 3 (cell-like approximation). *Suppose G is a cell-like decomposition of an n-manifold M, where $n \geq 5$. Then G is shrinkable if and only if M/G is finite-dimensional and has the disjoint disks property.*

Proof. One implication is trivial. For the other our concern, by long-established custom, centers on the case of compact M. To attain complete generality we should work in Euclidean patches containing parts of the nondegeneracy set, which exist by virtue of the cellularity of G (resulting from engulfing in topological manifolds), but to minimize technicalities we suppose M is a PL n-manifold. To minimize other technicalities, we also presume $n \geq 6$. We shall describe how to approximate the decomposition map $\pi: M \to M/G$ by a homeomorphism $F_3: M \to M/G$ within 3ε of π.

After listing the countably many finite 2-complexes P_i and recalling that the space of all maps $P_i \to M/G$ is separable, we apply Proposition 1 to determine a countable collection $A_1, A_2, \dots$ of compact sets in M/G, where each A_j is a 1-LCC subset of M/G homeomorphic to some P_i and, for any map μ of some P_i to M/G, there is a homeomorphism of P_i onto some A_k that approximates μ. By Corollary 23.5A the decomposition $G(A_j)$ induced over each A_j is shrinkable. According to Corollary 23.5B, there exists a cell-like map $F_1: M \to M/G$ such that $\rho(F_1, \pi) < \varepsilon$ and F_1 is 1-1 over $\bigcup A_j$.

In the next step we will approximate F_1 by another cell-like map $F_2: M \to M/G$, with $\rho(F_2, F_1) < \varepsilon$, so that the nondegeneracy set N_2 of F_2 has embedding dimension $\leq n - 3$. Then by Theorem 23.2 we will be able to approximate F_2 via a homeomorphism $F_3: M \to M/G$.

Consequently, all that remains is the "next step": approximating F_1 by a (cell-like) map F_2 for which dem $N_2 \leq n - 3$. (This is the only step where the restriction from $n \geq 5$ to $n \geq 6$ matters.) To get started, we determine a sequence of triangulations $T_1, T_2, \dots$ of M with meshes going to zero. Since $n \geq 6$, we can perform a general position adjustment so the various 2-skeleta R_k (where $R_k = T_k^{(2)}$) are pairwise disjoint. We shall obtain F_2 as a limit map resulting from a sequence of maps $f_k\colon M \to M/G$ to which the usual epsilonic restrictions pertain [that is, for $f_0 = F_1$, $\rho(f_{k+1}, f_k) < \varepsilon/2^{k+1}$], with the primary achievement of f_k ($k > 0$) being that f_k is 1-1 over $f_k(R_k)$, and with subsequent controls incorporated to guarantee that, in the limit,

$$F_2 \mid R_k = f_k \mid R_k \qquad \text{and} \qquad F_2^{-1}(f_k(R_k)) = R_k.$$

Then the nondegeneracy set N_2 of F_2 will lie in $M - \bigcup R_k$, ensuring dem $N_2 \leq n - 3$.

The basic maneuver is set forth in what follows.

Lemma 4. *Let R be a 2-complex in M, $f\colon M \to M/G$ a cell-like map that is 1-1 over $\bigcup A_j$, and $\delta > 0$. Then there exists a homeomorphism $h\colon M \to M$ such that $\rho(fh, f) < \delta$, h reduces to the identity outside $f^{-1}(N(f(R); \delta))$, and fh is 1-1 over $fh(R) \cup (\bigcup A_j)$.*

Proof. By the construction of the subsets A_j, there exists an embedding $\lambda\colon R \to \bigcup A_j$ close enough to $f \mid R$ to provide a $(\delta/4)$-homotopy $\psi_t\colon R \to M/G$ with $\psi_0 = f \mid R$ and $\psi_1 = \lambda$. This homotopy lifts (approximately) to a homotopy $\tilde{\psi}_t\colon R \to M$ between the inclusion and the embedding $f^{-1}\lambda\colon R \to M$. Since $\lambda(R)$ is 1-LCC in M/G and f is 1-1 over $\lambda(R)$, $f^{-1}\lambda(R)$ is 1-LCC in M (cf. the proof of Proposition 18.4). The fundamental taming theorem of J. L. Bryant and C. L. Seebeck III [1] then implies not only that $f^{-1}\lambda(R)$ is tame but, when their work is combined with a result of H. Gluck [1], that $f^{-1}\lambda(R)$ is ε-tame, meaning that there exists a small homeomorphism h^* on M supported near $f^{-1}\lambda(R)$ such that $h^*f^{-1}\lambda(R)$ is a subpolyhedron of M (Bryant–Seebeck [1]). In the situation at hand we allow h^* to move points only in $f^{-1}(N(f(R); \delta))$ and to move them so little that $\rho(fh^*, f) < \delta/2$ and $\tilde{\psi}_t\colon R \to M$ admits a modification to $\tilde{\psi}'_t\colon R \to M$ between the inclusion and $h^*f^{-1}\lambda$ for which $f\tilde{\psi}'_t\colon R \to M/G$ is a $(\delta/4)$-homotopy. The Bing–Kister result (Theorem 21.7) provides another homeomorphism $h'\colon M \to M$ supported in $f^{-1}(N(f(R); \delta))$ and satisfying $h' \mid R = h^*f^{-1}\lambda$ and $\rho(fh', f) < \delta/2$. Define h as $(h^*)^{-1}h'$. It follows that $\rho(fh, f) < \delta$ and that h is supported in $f^{-1}(N(f(R); \delta))$. Moreover, fh is clearly 1-1 over $\bigcup A_j$; in particular, fh is 1-1 over

$$fh(R) = f(h^*)^{-1}h'(R) = f(h^*)^{-1}h^*f^{-1}\lambda(R) \subset \bigcup A_j.$$

This establishes the lemma. ∎

To complete the proof of Theorem 3, we apply Lemma 4 for the map $f = F_1$, polyhedron R_1, and number $\delta = \varepsilon/2$ to obtain h_1. Because the map $f_1 = fh_1$ is 1-1 over $\bigcup A_j$, we can apply the lemma again. Generally, having found a map $f_k\colon M \to M/G$ (cell-like and 1-1 over $\bigcup A_j$), we apply Lemma 4 to f_k, R_k, and a small number $\delta_k < \varepsilon/2^{k+1}$ to obtain a homeomorphism

h_{k+1}, and then we set $f_{k+1} = f_k h_{k+1}$. The only technical concern in performing the successive iterations is the choice of δ_k's: it must be designed to guarantee that on R_k later maps agree with $f_k \mid R_k$ and, in addition, that no other points of M are sent to (any) $f_k(R_k)$ by the limiting cell-like map $F_2: M \to M/G$. Such technicalities were addressed in the proof of Proposition 23.4; the corresponding ones here are left to the reader. ■

Remarks. What makes the restriction $n \geq 6$ imperative in the preceding is the application of the Bing–Kister result (Theorem 21.7) for proving Lemma 4. It cannot be used for rearranging 2-complexes in 5-manifolds—the potential for linking of 2-spheres in E^5 makes this transparent. Unlike the proof of Theorem 23.5, the difficulty cannot be circumvented through engulfing methods. An argument for the 5-dimensional case developed by J. J. Walsh involves careful constructions with regular homotopies to approximate F_1 by a cell-like map F_2 with nondegeneracy set of embedding dimension 2, by virtue of avoiding a countable collection of singular 2-cells whose defining maps are dense in the space of all maps of B^2 into the given 5-manifold.

Knowing the equivalence between finite-dimensionality and the ANR property in this setting, we record formally an alternative statement of the cell-like approximation theorem.

Theorem 3′. *Suppose G is a cell-like decomposition of an n-manifold M, where $n \geq 5$. Then $\pi: M \to M/G$ can be approximated by homeomorphisms if and only if M/G is an ANR having the disjoint disks property.*

Corollary 3A. *If $f: M^m \to N^n$ is a proper cell-like surjection between manifolds and both $m, n \geq 5$, then f is a near-homeomorphism.*

Originally this corollary was derived by L. C. Siebenmann [1], who used strikingly different methods. The analog for $n = 3$ was proved by Armentrout [6] and for $n = 4$ by Quinn [2].

As another consequence, reasonable cell-like decompositions are stably shrinkable.

Corollary 3B. *If G is a cell-like decomposition of an n-manifold M such that M/G is finite-dimensional, then $G \times E^5$ is a shrinkable decomposition of $M \times E^5$. In particular, $(M/G) \times E^5$ is homeomorphic to $M \times E^5$.*

Proof. $(M/G) \times E^5$ satisfies the DDP because E^5 does. ■

Improvements to Corollary 3B are discussed in Section 26. Under certain restrictions on G, one can confirm a vital, best-possible stabilization result due to Cannon [6].

Corollary 3C. *If G is a closed-$(n - 2)$-dimensional cell-like decomposition of an n-manifold M, where $n \geq 4$, then $G \times E^1$ is a shrinkable decomposition of $M \times E^1$. In particular, $(M/G) \times E^1$ is homeomorphic to $M \times E^1$.*

Proof. Clearly $(M/G) \times E^1 = (M \times E^1)/(G \times E^1)$ is a finite-dimensional, cell-like image of the $(n + 1)$-manifold $M \times E^1$. Thus, it suffices to verify that $(M/G) \times E^1$ has the DDP.

Consider maps $\mu_1, \mu_2: B^2 \to (M/G) \times E^1$. Let $p_1: (M/G) \times E^1 \to M/G$ and $p_2: (M/G) \times E^1 \to E^1$ denote the projection maps, and let Z denote the closure in M/G of $\pi(N_G)$. By hypothesis, $\dim Z \leq n - 2$, so Z must be nowhere dense and 0-LCC in M/G. As a result, the maps $p_1\mu_1$, $p_1\mu_2: B^2 \to M/G$ can be modified to send increasingly dense 1-skeleta of B^2 into $(M/G) - Z$, yielding in the limit approximations $m_1, m_2: B^2 \to M/G$ such that $m_i^{-1}(Z)$ is a compact 0-dimensional set K_i $(i = 1, 2)$. Furthermore, for $i = 1, 2$, the maps $p_2\mu_i: K_i \to E^1$ can be approximated by disjoint embeddings $\beta_i: K_i \to E^1$, where β_i is so close to $p_2\mu_i \,|\, K_i$ that β_i extends to a map $\gamma_i: B^2 \to E^1$ close to $p_2\mu_i: B^2 \to E^1$ (see Corollary 14.6E).

Notice that the approximations $\mu_1', \mu_2': B^2 \to (M/G) \times E^1$ to μ_1, μ_2 given by $\mu_i'(b) = \langle m_i(b), \gamma_i(b) \rangle$ satisfy

$$\mu_1'(B^2) \cap \mu_2'(B^2) \subset [(M/G) - Z] \times E^1.$$

Since this subspace is a manifold of dimension at least 5, one can perform a final general position adjustment to produce disjoint approximations. ∎

Corollary 3D. (double suspension theorem). *If H^n is any homology n-sphere (meaning a compact n-manifold whose integral homology is isomorphic to that of S^n), then its double suspension, $\Sigma^2 H^n$, is homeomorphic to S^{n+2}.*

Proof. The crucial item concerns the manifold nature of $\Sigma^2 H^n$. The result about being the $(n + 2)$-sphere will follow from the consequence of the generalized Schönflies theorem that, whenever the suspension of a compactum yields a k-manifold, the manifold is actually the k-sphere.

When $n = 2$, $H^2 = S^2$, so we assume $n \geq 3$. The nicest situation arises when H^n is known to bound some (compact) contractible $(n + 1)$-manifold W. Let W^* be the manifold obtained by attaching a collar $H^n \times [0, 1)$ to W along $\partial W = H^n \leftrightarrow H^n \times \{0\}$. According to Corollary 3C, $(W^*/G_W) \times E^1$ is an $(n + 2)$-manifold. Since each point of the suspension circle, the set of potential nonmanifold points in $\Sigma^2 H^n$, has a neighborhood homeomorphic to $(W^*/G_W) \times E^1$, $\Sigma^2 H^n$ is an $(n + 2)$-manifold.

For $n \geq 4$ each PL homology n-sphere H^n bounds a contractible (PL) $(n + 1)$-manifold (Kervaire [1]), and the Kirby–Siebenmann results on PL structures [1] reveal that when $n \geq 5$ every H^n admits a PL triangulation.

Furthermore, M. H. Freedman [1] recently showed that every H^3 bounds a (not necessarily PL) contractible 4-manifold. Thus, the approach above works in most cases.

Generally, Cannon [5] and Edwards [3] independently have constructed cell-like decompositions G of S^{n+2} for which $S^{n+2}/G \approx \Sigma^2 H^n$. Such decompositions are shrinkable, because $\Sigma^2 H^n$ satisfies DDP, essentially by the argument given for Corollary 3C (see also Exercise 9). In Section 40 we will describe a related procedure revealing why $\Sigma^2 H^n$ is the cell-like image of an $(n + 2)$-manifold. ∎

Corollary 3E. *For $n \geq 4$ E^n contains a compact subset X that is non-cell-like (and noncellular) but $(E^n/G_X) \times E^1 \approx E^{n+1}$.*

Proof. Any compact polyhedral subset X of E^n which is not simply connected but which is homologically trivial will do. Then, by duality, the boundary of a regular (polyhedral) neighborhood V of X is necessarily a homology $(n - 1)$-sphere H^{n-1}, and the special point in E^n/G_X has a neighborhood homeomorphic to the open cone over H^{n-1}. It follows from what was established in Corollary 3D that $(E^n/G_X) \times E^1$ is an $(n + 1)$-manifold. Showing that this manifold actually represents E^{n+1} is left to the ingenious reader. ∎

The last two corollaries set forth conditions in a more classical mode, under which a given decomposition is shrinkable.

A closed-0-dimensional decomposition G of an n-manifold M is said to be *simple* provided it has a defining sequence $S = \{S_1, ..., S_i, ...\}$ [explicitly, S_i consists of a (locally) finite number of compact n-manifolds-with-boundary whose union $|S_i|$ covers N_G, $|S_{i+1}| \subset \text{Int}|S_i|$, and the components of $\bigcap |S_i|$ are elements of G] such that, for each index i and each $s \in S_i$, there exist n-cells B_1 and B_2 with $B_1 \cup B_2 \subset s$ and, for every $s' \in S_{i+1}$ with $s' \subset s$, either $s' \subset \text{Int } B_1$ or $s' \subset \text{Int } B_2$. This property of decompositions was introduced, in a more restrictive setting, by J. Neuzil [1].

Corollary 3F. *If G is a simple (necessarily cellular) usc decomposition of an n-manifold M, where $n \geq 5$, then G is shrinkable.*

The proof is an exercise. Standing in conspicuous contrast to Corollary 3F is the Bing minimal example, Example 9.6, a simple but nonshrinkable decomposition of E^3.

D. G. Wright [3] developed the next corollary, as well as the variation to it given as Exercise 12.

Corollary 3G. *If G is a countable, cellular decomposition of an n-manifold M, $n \geq 5$, where M contains an $(n - 3)$-dimensional closed subset X intersecting every $g \in H_G$, then G is shrinkable.*

Proof. Consider any two maps $\mu_1, \mu_2: B^2 \to M/G$, with lifts $q_i: B^2 \to M$ such that $\mu_i = \pi q_i$ ($i = 1, 2$). The first lift q_1 can be modified slightly to avoid the $(n - 3)$-demensional set X, causing those $g \in H_G$ touching $q_1(B^2)$ to have diameter at least $\varepsilon > 0$, where ε denotes the distance between $q_1(B^2)$ and X. By Proposition 23.4 μ_2 can be adjusted so as to miss $\pi(N_{G(\geq \varepsilon)})$. The images under the altered $\mu_1 = \pi q_1$ and μ_2 intersect only at points of the n-manifold $(M/G) - \pi(X)$, and a final general position adjustment there shows M/G satisfies the DDP.

EXERCISES

1. If G is a UV^1 decomposition of a compact metric space X such that X/G is an ANR with the DDP, prove that each $g \in G$ satisfies the cellularity criterion.
2. Suppose X is a CANR with the following disjoint k-cells property: any two maps of B^k into X can be approximated, arbitrarily closely, by maps $B^k \to X$ having disjoint images. Show that each map μ of a finite k-complex K into X can be approximated by an embedding $\lambda: K \to X$ for which $\lambda(K)$ is i-LCC, $i \in \{0, 1, ..., k\}$.
3. Suppose X is a CANR having the disjoint k-cells property. Show that each map of a compact k-dimensional metric space C into X can be approximated by an embedding.
4. Suppose G is a usc decomposition of a compact, connected n-manifold M, where M/G consists of at least two points, and suppose Z is a countable subset of M. Show that $\pi: M \to M/G$ can be approximated by a map $f: M \to M/G$ such that $f \mid Z$ is 1–1.
5. If G is a cellular usc decomposition of a (compact) n-manifold M, show that $\pi: M \to M/G$ can be approximated by a cell-like map $F: M \to M/G$ such that dem $N_F \leq n - 1$.
6. Suppose G is a cell-like decomposition of a compact PL n-manifold M, where $n \geq 4$, such that M/G (is an ANR and) satisfies the disjoint 1-cells property.
   (a) For any finite 1-dimensional subpolyhedron P of M and $\varepsilon > 0$, find a cell-like map $f: M \to M/G$ such that $\rho(f, \pi) < \varepsilon$, $f \mid M - \pi^{-1}(N(\pi(P); \varepsilon)) = \pi \mid M - \pi^{-1}(N(\pi(P); \varepsilon))$, and $f \mid P$ is 1–1. (*Hint:* In the space of all maps $M \to M/G$, let $\mathcal{C}$ denote the closure of $\{\pi h \mid h: M \to M$ is a homeomorphism supported in $\pi^{-1}(N(\pi(P); \varepsilon))\}$. Use the Bing–Kister result to prove that the desired maps form a dense G_δ-subset of $\mathcal{C}$.]
   (b) For any collection of pairwise disjoint, finite 1-dimensional subpolyhedra $P_1, P_2, ...$ of M and any $\varepsilon > 0$, find a cell-like map $F: M \to M/G$ such that $F \mid \bigcup P_i$ is 1–1.
   (c) Choose $P_1, P_2, ...$ as above so that the map $F: M \to M/K$ promised in (b) yields dem $F^{-1}(q) \leq n - 2$ for each $q \in M/G$.
7. Find an alternate result established by the proof of Corollary 3C, strong enough to imply the same conclusion for 0-dimensional cell-like decompositions G of M.
8. Prove Corollary 3F.

9. Suppose $f: M \to X \times E^1$ is a proper, cell-like mapping of an n-manifold M, $n \geq 5$, onto $X \times E^1$, where X contains a closed subset Y such that $\dim Y \leq n - 3$ and $(X - Y) \times E^1$ is an n-manifold. Show that $X \times E^1$ is an n-manifold.

10. Use the fact that each $(n - 1)$-sphere in E^n bounds a compact contractible subset of E^n to show that, if $f: E^n \to N^n$ is a proper map onto an n-manifold N^n such that at most a countable number of the sets $f^{-1}(x)$ are nondegenerate, then each $f^{-1}(x)$ is cell-like. Thus, for $n \geq 5$, f is a near-homeomorphism.

11. Suppose G is a cell-like decomposition of an ANR X such that X/G has the DDP, and suppose C is a closed subset of X/G. Then $X/G(C)$ has the DDP.

12. If G is a countable, cellular decomposition of a PL n-manifold M, $n \geq 5$, where M contains an $(n - 2)$-dimensional subpolyhedron P intersecting every $g \in H_G$, then G is shrinkable.

25. CELL-LIKE DECOMPOSITIONS OF 2-MANIFOLDS— THE MOORE THEOREM

In 1925 R. L. Moore [1] showed every usc decomposition of S^2 into nonseparating continua yields S^2 as the decomposition space. Translated into modern terminology, this amounted to showing that all cell-like decompositions of S^2 yield S^2. Although Moore never concerned himself with the matter, it has long been perceived that, indeed, such decompositions are shrinkable. That stronger result is the current subject.

Theorem 1. *Every cell-like usc decomposition G of a 2-manifold M is strongly shrinkable.*

There are several sound reasons for examining this low-dimensional situation. First, the theorem is intrinsically beautiful. Second, it holds immense historical significance; throughout most of the twentieth century, especially since the midpoint, it has served decomposition theory as a motivating force, as the ideal to be matched, and as a bench mark indicating the relative position of new results. Third, such an investigation offers a quick treatment of Theorem 1 unifying the 2-dimensional case and the higher-dimensional ones. In much the same style, but somewhat pared down, as the cell-like approximation theorem, this treatment provides, perhaps as an additional point in its favor, a legitimate forum in which to review the lengthy argument given for the latter.

A substantial advance toward Theorem 1 is backed by an analog of Theorem 22.2.

Lemma 2. *Every cell-like 0-dimensional decomposition G of a 2-manifold M is strongly shrinkable.*

Recapitulating some of Edwards's program, we first show how Lemma 2 implies Theorem 1. The two major parts in that program not recast this time are the disjoint disks property, which is disregarded for obvious reasons, and the amalgamation trick, which is ignored because in the present low-dimensional setting the failure of general position to instantaneously transform mapped arcs into embedded arcs makes it more difficult to establish amalgamation results than to prove the decomposition results they are designed to support. Reappearing are the filtration device and the method for improving maps $M \rightarrow M/G$, based on Lemma 2, over the successive sets comprising the filtration.

Proof that Lemma 2 implies Theorem 1. Theorem 17.9 attests that $\dim M/G \le 2$. Consequently, by Lemma 23.1 M/G can be filtered into σ-compact subsets Q^2, Q^1, Q^0 satisfying

$$M/G = Q^2 \supset Q^1 \supset Q^0 \supset Q^{-1} = \varnothing$$

where $\dim Q^i \le i$ and $\dim(Q^i - Q^{i-1}) \le 0$.

Write Q^0 as the countable union of compacta X_j. Apply Lemma 2 and Proposition 23.4 to approximate $\pi: M \rightarrow Q^2$ by a cell-like map f_0 that is 1-1 over Q^0.

Write Q^1 as the countable union of compacta Z_j. Since $Z_j - Q^0$ is 0-dimensional (for all j), the decomposition $G_f(Z_j)$ induced by f_0 over Z_j is cell-like and 0-dimensional. Again each $G_f(Z_j)$ is shrinkable by Lemma 2, so f_0 can be approximated by a cell-like map $f_1: M \rightarrow Q^2$ that is 1-1 over Q^1.

In exactly the same fashion f_1 can be approximated by a map $f_2: M \rightarrow Q^2$ that is 1-1 over Q^2 and hence is a homeomorphism, implying G is shrinkable. ∎

When tackling Lemma 2 a person can select from a variety of approaches. One method is to recognize M/G as a 2-manifold by invoking known characterizations. A more primitive method is to employ results from continua theory for extracting well-placed simple closed curves in M. Either approach gives rise to well-placed disks in M that serve as both generators and regulators for the required shrinking homeomorphisms. The latter method will be used here.

The heart of Lemma 2 is isolated in the following Claim. Verification that this Claim implies Lemma 2, which entails reconsideration of Theorem 8.9, is left as an exercise.

Claim. *For each $g \in H_G$ and open subset U of M containing g, there exists a 2-cell D such that $g \subset D \subset U$ and $\partial D \cap N_G = \varnothing$.*

Proof. Restrict U, if necessary, to make $M - U$ be connected (and nonvoid). Use cell-likeness to obtain a smaller neighborhood V of g such that every simple closed curve in Cl V is null homotopic in U, and apply Corollary 15.4C to find a 2-cell B with $g \subset$ Int $B \subset B \subset V$.

Let $J = \partial B$. Based on the fact that the complement in M/G of any 0-dimensional closed subset is 0-LC, determine a sequence of maps $f_i: J \to V - g$ starting with the inclusion f_0 and satisfying:

(1) f_{i+1} is homotopic to f_i,

(2) $f_i(J)$ misses all those nondegenerate elements of G having diameter at least 2^{-i}, and

(3) $\{f_i\}$ is a Cauchy sequence converging to a map $F: J \to$ Cl V for which $F(J) \cap N_G = \varnothing$.

Then $F(J)$ separates g from $M - U$ because each $f_i(J)$ does. It follows from Wilder [1, Theorem IV.6.7] that $F(J)$ contains a simple closed curve S separating g from $M - U$. By the planar Schönflies theorem, S bounds a 2-cell D in U with the desired properties. ∎

In the usual way Theorem 1 leads to the 2-dimensional version of the Armentrout–Quinn–Siebenmann approximation theorem (Corollary 24.3A):

Corollary 1A. *Proper cell-like maps between 2-manifolds can be approximated by homeomorphisms.*

EXERCISES

1. Show that Lemma 2 follows from the Claim made after it.

2. Let G denote a cell-like 0-dimensional decomposition of a compact 2-manifold M. Use the preceding results to show that for every $\varepsilon > 0$ there is an ε-amalgamation K of G into a null sequence cell-like decomposition.

V ──────────────────────────

SHRINKABLE
DECOMPOSITIONS

Applications of the cell-like approximation theorem are given in Chapter V, most frequently to the product of two cell-like decompositions. For products with Euclidean spaces, it is established that all finite-dimensional cell-like decompositions of manifolds are stably shrinkable, upon crossing with E^2; conditions under which the product just with E^1 is shrinkable are also studied, even in a low-dimensional case. About products with other spaces, work of Bass is reproduced showing the product of two finite-dimensional cell-like decompositions of manifolds having dimension greater than 1 to be invariably shrinkable.

Also described is a fruitful method, called spinning, for employing certain decompositions of one manifold to form related decompositions of higher-dimensional ones. Then the extent to which shrinkability of the one determines shrinkability of the other is analyzed, in virtually all cases.

26. PRODUCTS OF E^2 AND E^1 WITH DECOMPOSITIONS

Whenever G is a finite-dimensional, cell-like decomposition of E^n, then $G \times E^k$ is known to be a shrinkable decomposition of E^{n+k}, provided $k \geq 5$ (Corollary 24.3B). This is not the best possible result, as we are about to learn. Here we open an extensive investigation concerning the stable shrinkability of G. First, the coarse estimate $k \geq 5$ on the unrestricted shrinkability of $G \times E^k$ is improved to $k \geq 2$. The optimal estimate may well be $k = 1$; whether this is the case stands as a major unsettled issue. Subsequently, conditions on G implying the shrinkability of $G \times E^1$ are examined. Corollary 24.3C has already provided one—that G be closed-$(n - 2)$-dimensional. A variation to be established erases the word "closed" at the cost of one dimension.

The discussion starts with an analysis of the local separation (more precisely: nonseparation) properties of certain compacta in E^n/G, geared toward verifying the disjoint 1-cells property introduced in the Exercises following Section 24. Evoked by this is a class of spaces intimately related to cell-like decomposition spaces: the generalized manifolds. A locally compact (separable) metric space X is called a *generalized n-manifold* if X is a finite-dimensional ANR and if, for each $x \in X$, $H_k(X, X - \{x\})$ is isomorphic to $H_k(E^n, E^n - \{origin\})$ for all k (singular homology with integer coefficients). It is a consequence of the classical Vietoris–Begle mapping theorem (Begle [1]) that every finite-dimensional cell-like image of an n-manifold is a generalized n-manifold; another argument is forthcoming. Conversely, for $n > 4$ F. Quinn [3,4] has shown that nearly every generalized n-manifold is the cell-like image of some n-manifold. As a result, for $n \geq 5$, it seems likely that n-manifolds are characterized topologically as the generalized n-manifolds having the DDP.

Proposition 1. *Let G denote a cell-like decomposition of an n-manifold M and $x \in X = M/G$. Then $H_n(X, X - \{x\}) = \mathbb{Z}$ and $H_i(X, X - \{x\}) = 0$ $(i \neq n)$.*

Proof. First assume X is an ANR. Theorem 17.1 certifies that the natural maps $\pi: M \to X$ and $M - \pi^{-1}(x) \to X - \{x\}$ are homotopy equivalences, which therefore induce isomorphisms at the homology level. The algebraic five lemma (Spanier [1, p. 185]) discloses that $H_*(X, X - \{x\})$ is isomorphic to $H_*(M, M - \pi^{-1}(x))$. Moreover, $H_*(M, M - \pi^{-1}(x)) \cong H_*(E^n, E^n - \{0\})$, essentially because homology cannot distinguish between the cell-like $\pi^{-1}(x)$ and a point; more formally, $H_j(M, M - \pi^{-1}(x)) \cong \check{H}^{n-j}(\pi^{-1}(x))$, which is $\mathbb{Z}$ when $j = n$ and trivial otherwise, because $\pi^{-1}(x)$ is homotopically like a point (Corollary 15.4D).

The same conclusion holds even when X is not known to be an ANR. By Theorem 16.10 the natural maps $M \to X$ and $M - \pi^{-1}(x) \to X - \{x\}$ induce isomorphisms at the homotopy level, which, in spirit, seems stronger than what was needed above. The key for understanding how to obtain homology isomorphisms is to view singular chains as mapped complexes, with the boundary of a chain regarded as an identified subcomplex. Just as with the proof of Theorem 16.10, the strong approximate lifting theorem (16.18) is perfectly tailored for determining that these natural maps induce homology isomorphisms. ■

Corollary 1A. *If G is a cell-like decomposition of an n-manifold M and $\dim M/G < \infty$, then M/G is a generalized n-manifold.*

Generalized manifolds satisfy an Alexander–Lefschetz duality theorem, so they share the homological features of manifolds. Rather than engaging all the massive machinery of duality theory, we give an elementary derivation of some homological data.

Lemma 2. *Suppose X is a locally compact ANR such that, for some integer $r > 0$ and every $x \in X$, $H_i(X, X - \{x\}) \cong 0$ whenever $i \in \{0, 1, ..., r\}$. Then for each k-dimensional closed subset A of X, where $k \leq r$, $H_j(X, X - A) \cong 0$ whenever $j \in \{0, ..., r - k\}$.*

Proof. The argument proceeds by induction on dim A. Clearly, the result is valid when dim $A = -1$. Assuming it to be true for all closed subsets of dimension $< k$, we consider a k-dimensional closed subset A and some $z \in H_j(X, X - A)$, where $0 \leq j \leq r - k$. We shall show that $z = 0$.

At the core of the proof is the observation that, when A' and A'' are closed subsets of X for which $\dim(A' \cap A'') < k$, the Mayer–Vietoris sequence for the excisive couple of pairs $\{(X, X - A'), (X, X - A'')\}$ yields an inclusion-induced isomorphism α

$$0 \cong H_{j+1}(X, X - (A' \cap A'')) \longrightarrow H_j(X, X - (A' \cup A'')) \overset{\alpha}{\longrightarrow}$$

$$H_j(X, X - A') \oplus H_j(X, X - A'') \longrightarrow H_j(X, X - (A' \cap A'')) \cong 0$$

because of the inductive assumption, applied to $A' \cap A''$.

Identify a compact pair $(C, C') \subset (X, X - A)$ carrying a representative of z. Since $H_j(X, X - \{x\}) \cong 0$ for all $x \in X$, each $a \in A$ has a neighborhood U_a for which the image of z in $H_j(X, X - U_a)$ is trivial. Elementary dimension-theoretic properties lead to a cover $\{C_i \mid i = 1, 2, ..., m\}$ of $C \cap A$ by closed sets such that $\{C_i\}$ refines the cover $\{U_a \mid a \in C \cap A\}$, the interior relative to A of $\bigcup C_i$ contains $C \cap A$, and the frontier in A of each C_i has dimension $< k$. Define A_i as $\mathrm{Cl}(A - \bigcup_{j=i+1}^{m} C_j)$, where $i \in \{0, 1, ..., m\}$. Since A_0 does not intersect C, the image of z in $H_j(X, X - A_0)$ is trivial. Inductively, we presume the image of z in $H_j(X, X - A_{i-1})$ to be trivial. For applying the Mayer–Vietoris argument above, think of A' as A_{i-1} and of A'' as $\mathrm{Cl}(C_i - \bigcup_{s > i} C_s)$; note that $A' \cup A'' = A_i$ and

$$\dim(A' \cap A'') \leq \dim \bigcup_s \mathrm{Fr}\, C_s < k.$$

As a result, the homomorphism α is an isomorphism. By construction the image of z in $H_j(X, X - A'')$ is trivial, because its image is trivial in some $H_j(X, X - U_a)$ where $A'' \subset U_a$. Hence, α displays that the image of z in $H_j(X, X - (A' \cup A'')) = H_j(X, X - A_i)$ is trivial as well. Ultimately this shows that $z = 0$, since $A_m = A$. ∎

Corollary 2A. *Each k-dimensional closed subset A of a generalized n-manifold X, where $k \leq n - 2$, is 0-LCC.*

Proof. Consider a small pathwise-connected open neighborhood U of some point $a \in A$. From the exact homology sequence of the pair $(U, U - A)$, one finds

$$H_1(U, U - A) \to \tilde{H}_0(U - A) \to \tilde{H}_0(U) \cong 0,$$

and the left-hand term is trivial by Lemma 2. Thus, $\tilde{H}_0(U - A) \cong 0$ or, equivalently, $U - A$ is pathwise-connected. ∎

Proposition 3. *If G is a cell-like decomposition of an n-manifold M, $n \geq 3$, then $X = M/G$ has the disjoint 1-cells property.*

Proof. Consider a map $f_1: B^1 = [-1, 1] \to X$. Partition B^1 as

$$-1 = t_0 < t_1 < \cdots < t_{i-1} < t_i < \cdots < t_k = 1$$

so that each interval $[t_{i-1}, t_i]$ has small size (depending on f_1). Since Peano continua are arcwise-connected, each $f_1([t_{i-1}, t_i])$ contains an arc from $f_1(t_{i-1})$ to $f_1(t_i)$, whenever $f_1(t_{i-1}) \neq f_1(t_i)$. Now it is a straightforward matter to produce a map $f_1': B^1 \to X$ near f_1 such that $f_1' \mid [t_{i-1}, t_i]$ is either a constant mapping or an embedding defining an arc in $f_1([t_{i-1}, t_i])$. Consequently, the image $A = f_1'(B^1)$ is 1-dimensional.

Given another map $f_2: B^1 \to X$, we invoke Corollary 2A to determine a map f_2' approximating f_2 but with $f_2'(B^2) \subset X - A = X - f_1'(B^1)$. ∎

Corollary 3A. *Suppose G is a cell-like decomposition of an n-manifold M, $n \geq 3$. Then each map $f: P^1 \to M/G$ of a finite 1-complex P^1 can be approximated, arbitrarily closely, by an embedding $F: P^1 \to M/G$ for which $F(P^1)$ is 0-LCC.*

Proof. See also Exercise 24.2 or the proof of Proposition 24.1. ∎

Up to this juncture we have used both the disjoint disks property and a disjoint arcs property. The central results of this section involve an intermediate concept. We say that a metric space S has the *disjoint arc–disk property* (DADP) if for all maps $f: B^1 \to S$ and $m: B^2 \to S$ and for all $\varepsilon > 0$, there exist maps $f': B^1 \to S$ and $m': B^2 \to S$ such that $\rho(m', m) < \varepsilon$, $\rho(f', f) < \varepsilon$, and $f'(B^1) \cap m'(B^2) = \varnothing$.

Proposition 4. *For a locally compact metric space S, the following statements are equivalent:*

(a) *S has the DADP;*

(b) *Each map $f: B^1 \to S$ can be approximated, arbitrarily closely, by a map $f': B^1 \to S$ such that $f'(B^1)$ is nowhere dense, 0-LCC, and 1-LCC;*

(c) *Each map $m: B^2 \to S$ can be approximated, arbitrarily closely, by a map $m': B^2 \to S$ such that $m'(B^2)$ is nowhere dense and 0-LCC.*

This follows from Baire category arguments similar to those employed in proving Proposition 24.1.

Lemma 5. *Suppose S is an LC^1 metric space having the DADP, μ_1 and μ_2 are maps of B^2 to S, P is a finite 1-complex in B^2, and $\varepsilon > 0$. Then there exist maps $\mu_1', \mu_2': B^2 \to S$ such that $(\mu_i', \mu_i) < \varepsilon$ (i = 1, 2) and $\mu_1'(P) \cap \mu_2'(B^2) = \varnothing$.*

Proof. As an elementary consequence of the DADP, there exist maps $f: P \to S$ and $\mu_2': B^2 \to S$ approximating $\mu_1 \mid P$ and μ_2, respectively, where $f(P) \cap \mu_2'(B^2) = \varnothing$. When S is an ANR, f can be obtained so close to $\mu_1 \mid P$ that it extends to $\mu_1': B^2 \to S$ within ε of μ_1 (Corollary 14.6E); under the hypothesis that S is LC^1, one can extend f over the skeleta of some fine triangulation of B^2 for which P is a subcomplex to achieve the same conclusion. ∎

Upon stabilization of a given decomposition, each extra E^1-factor enhances the disjoint cells property held by the decomposition space.

Proposition 6. *If S is an LC^1 metric space having the DADP, then $S \times E^1$ has the DDP.*

Proof.   Consider maps $\mu_1, \mu_2: B^2 \to S \times E^1$ and $\varepsilon > 0$. Name a triangulation T of B^2 having mesh so small that diam $\mu_i(\sigma) < \varepsilon' = \varepsilon/17$ for each $\sigma \in T$ and $i \in \{1, 2\}$, and let P denote its 1-skeleton. Let $p: S \times E^1 \to S$ denote the projection map.

Since S has the DADP, Lemma 5 establishes the existence of maps $\mu_1', \mu_2': B^2 \to S \times E^1$ within ε' of μ_1, μ_2, whose projections to E^1 agree with those of μ_1, μ_2, and which satisfy

$$(*) \qquad\qquad p\mu_1'(P) \cap p\mu_2'(B^2) = \varnothing.$$

Repeating, while limiting the motion so as to maintain (*), one procures maps $\mu_1'', \mu_2'': B^2 \to S \times E^1$ satisfying $\rho(\mu_i'', \mu_i') < \varepsilon'$ and

$$(**) \qquad p\mu_1''(B^2) \cap p\mu_2''(P) = \varnothing = p\mu_1''(P) \cap p\mu_2''(B^2).$$

Note that diam $\mu_i''(\sigma) < 5\varepsilon'$ for each $\sigma \in T$.

Enumerate the 2-simplexes of T as $\sigma_1, \sigma_2, ..., \sigma_m$. For each σ_j choose a point $s_j \in E^1$ so that $\mu_1''(\sigma_j)$ is within $5\varepsilon'$ of $S \times \{s_j\}$, and then choose another point $t_j \in E^1$ so that $\mu_2''(\sigma_j)$ is within $5\varepsilon'$ of $S \times \{t_j\}$ but no t_j belongs to $\{s_j \mid j = 1, ..., m\}$. Now one can easily define μ_1''' and μ_2''' such that $\mu_i''' \mid P = \mu_i'' \mid P$ $(i \in \{1, 2\})$ and, for $j = 1, 2, ..., m$, $\mu_1'''(\sigma_j)$ is contained in

$$[p\mu_1''(\partial\sigma_j) \times (s_j - 5\varepsilon', s_j + 5\varepsilon')] \cup [p\mu_1''(\sigma_j) \times \{s_j\}],$$

a set of diameter $< 15\varepsilon'$, while $\mu_2'''(\sigma_j)$ is contained in

$$[p\mu_2''(\partial\sigma_j) \times (t_j - 5\varepsilon', t_j + 5\varepsilon')] \cup [p\mu_2''(\sigma_j) \times \{t_j\}].$$

The effect is suggested in Fig. 26-1.

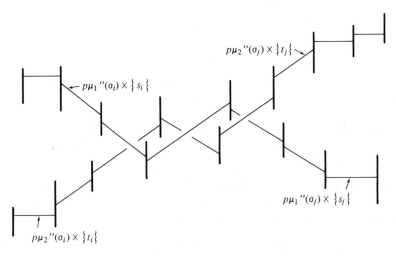

FIG. 26-1

It follows from (**) and the choice of points s_j and t_j that $\mu_1'''(B^2) \cap \mu_2'''(B^2) = \varnothing$. Moreover, $\rho(\mu_i''', \mu_i) < 17\varepsilon' = \varepsilon$. ∎

Proposition 7. *If G is a cell-like decomposition of an n-manifold M, $n \geq 3$, then $(M/G) \times E^1$ has the DADP.*

Proof. Let $X = M/G$ and let $p_X: X \times E^1 \to X$ and $p_E: X \times E^1 \to E^1$ denote the projection mappings. Given a map $f: B^1 \to X \times E^1$, we can apply Corollary 3A to approximate $p_X f$ by a 0-LCC embedding $e_X: B^1 \to X$. We shall prove that the product embedding $e: B^1 \to X \times E^1$, defined by $e(b) = \langle e_X(b), p_E f(b) \rangle$, gives a 1-LCC embedded arc.

First, however, we shall explain why the arc $A = e_X(B^1) \times \{0\}$ is 1-LCC embedded in $X \times E^1$. This is an application of the Seifert–van Kampen theorem, similar to the proof of Corollary 18.5A. Let $W \times (-\delta, \delta)$ be a neighborhood of $a \in A$. There exists a connected open subset U of X such that $U \times (-\delta, \delta)$ is another neighborhood of a and that every loop in U is null homotopic in W. By Corollary 2A, $U - e_X(B^1)$ is pathwise-connected. Every loop in $(U \times (-\delta, \delta)) - A$ can be expressed as a finite composition of loops from $(U \times (-\delta, \delta)) - (e_X(B^2) \times [0, \delta))$ and from $(U \times (-\delta, \delta)) - (e_X(B^1) \times (-\delta, 0])$. In either case, such loops are contractible in $W \times (-\delta, \delta)$. As a result, the image of $\pi_1((U \times (-\delta, \delta)) - A)$ in $\pi_1((W \times (-\delta, \delta)) - A)$ is trivial, and A is 1-LCC embedded at a.

Additionally, one should note the relatively obvious facts that A is nowhere dense and 0-LCC in $X \times E^1$.

The Klee trick (see the proof of Proposition 10.4) indicates that $e(B^1)$ and A are equivalently embedded in $X \times E^1$, via a homeomorphism θ of $X \times E^1$

to itself such that $p_X \theta = p_X$. Certainly then $e(B^1)$ is nowhere dense, 0-LCC, and 1-LCC in $X \times E^1$. According to Proposition 4, $X \times E^1$ has the DADP. ∎

Theorem 8. *If G is a cell-like decomposition of an n-manifold M, where $n \geq 3$, then $(M/G) \times E^2$ has the DDP.*

Propositions 6 and 7 combine to give Theorem 8. The cell-like approximation theorem of Section 24 then yields a fairly strong result concerning the stable shrinkability of G.

Corollary 8A. *If G is a cell-like decomposition of an n-manifold M, where $n \geq 3$, such that $\dim(M/G) < \infty$, then $G \times E^2$ is shrinkable and $(M/G) \times E^2$ is homeomorphic to $M \times E^2$.*

Whether $G \times E^1$ itself is always shrinkable is not known. With this in mind, we turn to an examination of some conditions on G implying $G \times E^1$ is shrinkable.

Theorem 9. *If G is an $(n - 3)$-dimensional cell-like decomposition of an n-manifold M, where $n \geq 4$, then $G \times E^1$ is shrinkable and $(M/G) \times E^1$ is homeomorphic to $M \times E^1$.*

Proof. In order to show that $(M/G) \times E^1$ has the DDP, we will prove that M/G has the DADP and will apply Proposition 7. In order to prove that M/G does have the DADP, we shall show how to approximate a given map $m: B^2 \to M/G$ by a map $m': B^2 \to M/G$ such that $\dim m'(B^2) \leq n - 2$. Then Corollary 2A will corroborate that $m'(B^2)$ is 0-LCC, and Proposition 4 will certify that M/G has the DADP.

For support in producing the desired approximation m', consider the complete metric space $\mathcal{C}$ of maps $B^2 \to M/G$. Determine open subsets $W_1, W_2, \ldots$ of M/G, with $\pi(N_G) \subset W_{i+1} \subset W_i$ for all i, such that $\dim(\bigcap W_i) \leq n - 3$, and construct straight line segments $L_1, L_2, \ldots$ in B^2 such that $\dim(B^2 - \bigcup L_i) = 0$. Define

$$\mathcal{O}_k = \{h \in \mathcal{C} : h \mid h^{-1}(M/G - W_k) \text{ is a } (1/k)\text{-map}\}$$

$$\mathcal{P}_k = \left\{h \in \mathcal{C} : h \mid \bigcup_{i=1}^{k} L_i \text{ is a } (1/k)\text{-map}\right\}.$$

Routine arguments show both $\mathcal{O}_k$ and $\mathcal{P}_k$ to be open subsets of $\mathcal{C}$. Moreover, each is dense: that $\mathcal{O}_k$ is dense follows when $n \geq 5$ because each h can be approximated by the projection under π of an embedding into M and when $n = 4$, at least for the apparently worst possible situation where N_G is dense in M, because each map $h \in \mathcal{C}$ can be approximated by the projection of an immersion into M whose points of self-intersection all lie in N_G; that $\mathcal{P}_k$ is dense follows from Corollary 3A and the proof of Lemma 5. By the Baire category theorem, each $m \in \mathcal{C}$ can be approximated by a map $m' \in (\bigcap \mathcal{O}_k) \cap (\bigcap \mathcal{P}_k)$. Each point inverse under m' is 0-dimensional (because it is the union of at most one point from $\bigcup L_i$ together with a subset of $B^2 - \bigcup L_i$), and the image $S_2(m')$ of the nondegeneracy set (i.e., $S_2(m') = \{x \in m'(B^2) \mid (m')^{-1}(x) \neq \text{point}\}$) is contained in $\bigcap W_i$ and, therefore, has dimension at most $n - 3$. According to a result of H. Freudenthal [1]

$$\dim m'(B^2) \leq \max\{\dim B^2, 1 + \dim S_2(m')\} \leq n - 2.$$

As explained previously, this implies that M/G has the DADP. ∎

Whenever G is a closed $(n - 2)$-dimensional cell-like decomposition, it is fairly easy to obtain a map $m': B^2 \to M/G$, close to a given map $m: B^2 \to M/G$, such that $\dim m'(B^2) \leq n - 2$. Thus, the first part of the argument just presented provides an alternative proof of Corollary 24.3D, which attests that $G \times E^1$ is shrinkable.

Even when G is a closed-$(n - 1)$-dimensional cell-like decomposition of E^n, whether $G \times E^1$ must be shrinkable remains unresolved. Nonetheless, it is shrinkable provided $\pi(N_G)$ lies in a subset with its own inherently well-behaved structure—namely in a generalized $(n - 1)$-manifold or in an $(n - 1)$-complex (Theorem 12). The crux of this argument involves a verification of the DDP outlined in a separate lemma, Lemma 11, which has other applications later on; before embarking on that argument, we treat an important special case. The proof is brief, its techniques minimal, and its structure similar to that of Lemma 11.

Theorem 10. *If G is a cell-like decomposition of E^{n-1}, regarded as $E^{n-1} \times \{0\} \subset E^n$, $n \geq 5$, such that E^{n-1}/G is finite-dimensional, then the trivial extension G^T of G over E^n is shrinkable.*

Proof. It suffices to show that E^n/G^T has the DDP, so consider maps $\mu_1, \mu_2: B^2 \to E^n/G^T$. For notational simplicity, let $\pi: E^n \to E^n/G^T$ denote the obvious map, $X = \pi(E^{n-1} \times \{0\})$, $Y_+ = \pi(E^{n-1} \times [0, \infty))$ and $Y_- = \pi(E^{n-1} \times (-\infty, 0])$.

First, improve μ_1, μ_2 to maps μ'_1, μ'_2 such that each $(\mu'_i)^{-1}(X)$ is a 1-dimensional compactum A_i (or, if preferred, a 1-manifold) by taking μ'_i to be the image under π of a nice map $B^2 \to E^n$. Set $T_i = (\mu'_i)^{-1}(Y_+)$ and $L_i = (\mu'_i)^{-1}(Y_-)$ for $i = 1, 2$.

Next, approximate $\mu'_i \mid A_i$ by $\mu''_i: A_i \to X$ with $\mu''_1(A_1) \cap \mu''_2(A_2) = \varnothing$. The capability of performing this step is substantiated by Corollary 3A or Exercise 24.2. Moreover, this can be done with μ''_i so close to $\mu'_i \mid A_i$ that

μ_i'' extends to maps $\mu_i''\colon T_i \to Y_+$ and $\mu_i''\colon L_i \to Y_-$ for which the combined μ_i'', defined on B^2, is close to μ_i'.

The final step depends on the simple verification that X is 0-LCC and 1-LCC in both Y_+ and Y_-. Consequently, μ_i'' can be approximated by μ_i''' such that $\mu_i''' \mid A_i = \mu_i'' \mid A_i$, $\mu_i'''(B^2 - A_i) \cap X = \varnothing$, and $\mu_1'''(B^2 - A_1) \cap \mu_2'''(B^2 - A_2) = \varnothing$, by first exploiting these LCC properties to obtain $\mu_i'''(T_i - A_i) \subset Y_+ - X$ and $\mu_i'''(L_i - A_i) \subset Y_- - X$, and by trading then on the manifold nature of $Y_+ - X$ and $Y_- - X$ to adjust further so that the two images of $B^2 - A_i$ are disjoint. Because $\mu_i'''(A_i) \subset X$ and $\mu_1'''(A_1) \cap \mu_2'''(A_2) = \varnothing$, it follows that $\mu_1'''(B^2) \cap \mu_2'''(B^2) = \varnothing$. ∎

In [6] Cannon states a result much like the following promised improvement of Theorem 10.

Lemma 11. *Suppose Y is a generalized n-manifold, $n \geq 5$; Y_1 and Y_2 are closed subsets of Y, each satisfying the DDP, whose intersection X is a generalized $(n - 1)$-manifold, with $Y - X$ being a manifold; and each pair of maps $f_e\colon B^2 \to Y_e$ $(e = 1, 2)$ can be approximated arbitrarily closely by maps $f_e'\colon B^2 \to Y_e$ having disjoint images. Then Y satisfies the DDP.*

Proof. Consider $\mu_1, \mu_2\colon B^2 \to Y$. Since X is nowhere dense in the generalized n-manifold Y, these maps can be adjusted to μ_1', μ_2' where each $(\mu_i')^{-1}(X)$ is a 1-dimensional compactum A_i.

As before, it is possible to approximate the two maps $\mu_i' \mid A_i$ by disjoint embeddings μ_i'' of A_i in X. The crucial operation in the proof involves the advance identification of a dense collection of nice 2-cell images in Y_1 and Y_2, permitting a subsequent extension of μ_i'' over B^2 in which the image of $B^2 - A_i$ is diverted around $\mu_1''(A_1) \cup \mu_2''(A_2)$. In Theorem 10, where X was 1-LCC embedded, this was no problem; in the current situation, this is a real issue.

Claim 1. *For $e = 1, 2$, the space $\mathcal{C}_e$ of all maps $B^2 \to Y_e$ has two (disjoint) countable dense subsets $\{\lambda_{ej}\}$, $\{\lambda_{ej}'\}$ such that each λ_{ej} and each λ_{ej}' is an embedding and no two of $\lambda_{1j}(B^2)$, $\lambda_{1k}'(B^2)$, $\lambda_{2r}(B^2)$, $\lambda_{2s}'(B^2)$ intersect.*

The proof of the Claim, which is an exercise, depends on the hypothesized disjoint disks properties and an argument like that of Proposition 24.1.

Claim 2. *If U is an open subset of B^2 and $\mu\colon \mathrm{Cl}\, U \to Y_e$ is a map, then μ can be approximated by a map $\mu'\colon \mathrm{Cl}\, U \to Y_e$ such that $\mu' \mid \mathrm{Fr}\, U = \mu \mid \mathrm{Fr}\, U$ and*

$$\mu'(U) \cap X \subset \bigcup [\lambda_{ej}(B^2) \cap X].$$

Proof. Since Y is a generalized manifold and, therefore, satisfies duality (Wilder [1]), it follows that $Y_e - X$ is 0-LC at each point of X. Choose a triangulation T of U with mesh tending toward zero near Fr U, carefully approximate μ by μ' with $\mu'(T^{(1)}) \subset Y_e - X$, and then adjust the map on the various 2-simplexes to achieve the desired conclusion. This disposes of Claim 2. ∎

Here is the primary difference between this argument and the one given for Theorem 10. Each $\lambda_{ej}(B^2) \cap X$, $\lambda'_{ej}(B^2) \cap X$ is a 2-dimensional closed subset of X; hence, each is 0-LCC embedded there (by Corollary 2A). As a result, the maps $\mu'_i \mid A_i$ can be approximated by disjoint embeddings μ''_i of A_i in X, missing every $\lambda_{ej}(B^2)$ and every $\lambda'_{ej}(B^2)$. Of course, μ''_i extends to a map $\mu''_i : B^2 \to Y$ with $\mu''_i((\mu'_i)^{-1}(Y_e)) \subset Y_e$ for $e = 1, 2$ and with μ''_i close to μ'_i.

Apply Claim 2 to approximate μ''_i by μ'''_i such that $\mu'''_i \mid A_i = \mu''_i \mid A_i$ and

$$\mu'''_1(B^2 - A_1) \cap X \subset \bigcup_{e=1}^{2} \bigcup_j [\lambda_{ej}(B^2) \cap X],$$

$$\mu'''_2(B^2 - A_2) \cap X \subset \bigcup_{e=1}^{2} \bigcup_j [\lambda'_{ej}(B^2) \cap X].$$

This yields that $\mu'''_1(B^2) \cap \mu'''_2(B^2) \cap X = \varnothing$. As usual, a final general position adjustment in the manifold $Y - X$ completes the proof. ∎

Theorem 12. *If G is a cell-like decomposition of an n-manifold M, $n \geq 4$, for which $\pi(N_G)$ is contained in a generalized $(n-1)$-manifold X embedded in M/G as a closed subset, then $G \times E^1$ is a shrinkable decomposition of $M \times E^1$.*

Proof. Clearly $(M \times E^1)/(G \times E^1)$ is finite-dimensional; the more substantial matter is to detect that it satisfies the DDP.

The problem being essentially local, we can assume X to be connected, and because of the duality properties arising from a local orientation in M/G, X to separate M/G into two components, with closures Y_1 and Y_2.

The argument rests on the elementary observation that, since each Y_i is an ANR (Theorem 14.7) and $Y_i - X$ is 0-LCC, given any dense subset D of E^1, one can approximate each map of B^2 in $Y_i \times E^1$ by another map of B^2 into

$$((Y_i - X) \times E^1) \bigcup (X \times D).$$

The presence of disjoint dense subsets of E^1 makes it easy to check that the cover of $(M/G) \times E^1$ by $Y_1 \times E^1$ and $Y_2 \times E^1$ has the piecewise disjoint disks properties required for applying Lemma 11. ∎

Corollary 12A. *Suppose G is a cell-like decomposition of an n-manifold M, $n \geq 4$; X is a generalized $(n - 1)$-manifold embedded in M/G as a closed subset; and G(X) is the decomposition of M induced over X. Then $G(X) \times E^1$ is a shrinkable decomposition of $M \times E^1$.*

Corollary 12B. *If G is a cell-like decomposition of an n-manifold M, $n \geq 4$, for which $\pi(N_G)$ is contained in an $(n - 1)$-complex P topologically embedded in M/G as a closed subset, then $G \times E^1$ is shrinkable.*

Proof. Underlying the $(n - 2)$-skeleton of P is a (closed) subset A, and $P - A$ is an $(n - 1)$-manifold. The decomposition $G(A)$ of M induced over A by π is shrinkable, by Corollary 24.3C. Consequently, the natural map of $M \times E^1$ onto $(M/G) \times E^1$ is approximable by a cell-like map ψ 1–1 over both $A \times E^1$ and $((M/G) - P) \times E^1$. Deleting $\psi^{-1}(A \times E^1)$ from the domain and $A \times E^1$ from the range, one finds from Theorem 12 that the restricted ψ can be approximated by a homeomorphism ψ' of $(M \times E^1) - \psi^{-1}(A \times E^1)$ onto $((M/G) - A) \times E^1$, subject to controls forcing its extension over $\psi^{-1}(A \times E^1)$ via ψ to be a homeomorphism close to the original map. ∎

Corollary 12C. *Suppose G is a cell-like decomposition of an n-manifold M, $n \geq 4$; P is an $(n - 1)$-complex embedded in M/G as a closed subset; and G(P) is the decomposition of M over P. Then $G(P) \times E^1$ is a shrinkable decomposition of $M \times E^1$.*

A generalized n-manifold X is said to be *locally encompassed by manifolds* if each point $x \in X$ has arbitrarily small neighborhoods U_x whose frontiers in X are $(n - 1)$-manifolds. If such a space X is the cell-like image of an n-manifold $(n \geq 4)$, the next result attests that $X \times E^1$ is a manifold. Unfortunately, not all generalized $(n - 1)$-manifolds have this property; there are examples of cell-like decompositions G of S^n, due to Bing and Borsuk [1] and also to S. Singh [1] for the case $n = 3$, generalized to $n > 3$ by D. G. Wright [1], Singh [2, 3], and later by J. J. Walsh and the author [1], such that S^n/G contains no 2-cell and certainly, therefore, no $(n - 1)$-manifold.

The proof of Theorem 13 is lengthy and laborious. Those willing to skip ahead may prefer to look first at Section 30, particularly the proof of Lemma 30.2, as an easier warm-up.

Theorem 13. *If G is a cell-like decomposition of an n-manifold M, $n \geq 4$, such that M/G is locally encompassed by manifolds, then $G \times E^1$ is a shrinkable decomposition of $M \times E^1$.*

Proof. That M/G is locally encompassed by manifolds certainly implies $\dim(M/G) \leq n$. Thus, to establish this theorem, the essential part is to show that $(M/G) \times E^1$ satisfies the DDP. The argument is delicate and drawn out, so we break it down into seven fairly short steps.

Fix maps μ_1, $\mu_2 \colon B^2 \to (M/G) \times E^1$ and a rational positive number ε.

Step 1. *An alteration of the decomposition map.* Because M/G is locally encompassed by manifolds, it contains a countable collection $\{N_i\}$ of compact $(n-1)$-manifolds such that each point of M/G has arbitrarily small neighborhoods with frontiers equal to some N_i. By Corollary 12A the decompositions $G(N_i) \times E^1$ are shrinkable, and by Theorem 10 the trivial extensions to $M \times E^1$ of the decompositions $G \times \{t\}$ defined on $M \times \{t\}$ are shrinkable for all $t \in E^1$. According to Proposition 23.4, the decomposition map $M \times E^1 \to (M/G) \times E^1$ can be approximated by a cell-like map $p \colon M \times E^1 \to (M/G) \times E^1$ that is 1-1 over all the sets $N_i \times E^1$ and over $(M/G) \times \{q\}$, for every rational number q. This puts $p(N_p)$ in the product of $((M/G) - \bigcup N_i)$ with the irrationals, so there exist 0-dimensional F_σ sets C and C' in M/G and E^1, respectively, such that $p(N_p) \subset C \times C'$.

Step 2. *Modifications of the map μ_e.* Determine a triangulation T of B^2 such that diam $\mu_e(\sigma) < \varepsilon$ for all $\sigma \in T$ and $e = 1, 2$. Since $(M/G) \times E^1$ satisfies the DADP, just as in the proof of Proposition 6 we can adjust μ_1 and μ_2 so that

$$\mu_1(B^2) \cap \mu_2(T^{(1)}) = \varnothing = \mu_1(T^{(1)}) \cap \mu_2(B^2);$$

taking each of these adjusted maps to be the image under p of disjoint embeddings in $M \times E^1$, we may assume that $\mu_1(B^2) \cap \mu_2(B^2) \subset C \times C'$.

Noting that $C \times E^1$ is 1-dimensional and σ-compact, which implies it is 0-LCC in $(M/G) \times E^1$, we can modify the μ_e's further so that $\mu_e(T^{(1)}) \cap (C \times E^1) = \varnothing$, without affecting the set $S = \mu_1(B^2) \cap \mu_2(B^2)$. Finally, since $C \times \{t\}$ is 1-LCC in $(M/G) \times E^1$ for each $t \in E^1$, we may assume, in addition, that for $e = 1, 2$ and k an integer

$$\mu_e(B^2) \cap (C \times \{k\varepsilon/2\}) = \varnothing.$$

Step 3. *Finiteness considerations.* Choose a neighborhood W^* of C in M/G whose components each have diameter less than $\varepsilon/2$, and define W as $W^* \times (E^1 - \bigcup_k \{k\varepsilon/2\})$. Every point $s \in S = \mu_1(B^2) \cap \mu_2(B^2)$ then has a neighborhood V_s in W of the form $V_s = U_s \times J_s$, where $J_s = (k_s \cdot \varepsilon/2, (k_s + 1) \cdot \varepsilon/2)$ for some integer k_s and where U_s is a connected open subset of M/G with closure in W^*, having manifold frontier, and satisfying

(A₀) $(U_s \times E^1) \cap (\mu_1(T^{(1)}) \cup \mu_2(T^{(1)})) = \varnothing$,

(B₀) $(\text{Fr } U_s \times E^1) \cap \mu_1(B^2) \cap \mu_2(B^2) = \varnothing$,

(C₀) $(U_s \times \text{Fr } J_s) \cap (\mu_1(B^2) \cup \mu_2(B^2)) = \varnothing$.

Cover S by a finite collection $\{V_1, \ldots, V_r\}$ of such sets V_s.

Step 4. *A reduction.* For $j = 1, \ldots, r$ we intend to describe maps $\mu_{1,j}$ and $\mu_{2,j}$ of B^2 to $(M/G) \times E^1$ agreeing with μ_1 and μ_2 over $((M/G) \times E^1) - W$ and satisfying the analogs of (A_0), (B_0), and (C_0) above, as well as the two conditions below:

$$\mu_{e,j}(\mu_e^{-1}(W)) \subset W, \quad \text{and} \quad \mu_{1,j}(B^2) \cap \mu_{2,j}(B^2) \subset S - \left(\bigcup_{i=1}^{j} V_i \right).$$

Of course, in the terminal case $j = r$ we will have the required disjoint approximations.

This can be accomplished by verifying the following implicit inductive step: if f_1 and f_2 are maps of B^2 into $(M/G) \times E^1$ such that

$$(B_{i-1}) \qquad f_1(B^2) \cap f_2(B^2) \cap \left[\bigcup_{m=1}^{r} (\mathrm{Fr}\ U_m \times E^1) \right] = \varnothing$$

$$(C_{i-1}) \qquad [f_1(B^2) \cup f_2(B^2)] \cap \left[\bigcup_{m=1}^{r} (U_m \times \mathrm{Fr}\ J_m) \right] = \varnothing,$$

where $V_m = U_m \times J_m$, and that

$$f_1(B^2) \cap f_2(B^2) \subset S - \bigcup_{j=1}^{i-1} V_j,$$

then there exist maps $F_1, F_2 : B^2 \to (M/G) \times E^1$ such that for $e = 1, 2$

$$F_e^{-1}(W) = f_e^{-1}(W), \qquad F_e \,|\, B^2 - F_e^{-1}(W) = f_e \,|\, B^2 - f_e^{-1}(W),$$

$$[F_1(B^2) \cup F_2(B^2)] \cap \left[\bigcup_{m=1}^{r} (U_m \times \mathrm{Fr}\ J_m) \right] = \varnothing$$

$$F_1(B^2) \cap F_2(B^2) \subset (f_1(B^2) \cap f_2(B^2)) - V_i \subset S - \left(\bigcup_{j=1}^{i} V_j \right).$$

Step 5. *Elimination of intersections from* V_i. Choose a point $t_i \in J_i$ such that

$$[f_1(B^2) \cup f_2(B^2)] \cap (U_i \times \{t_i\}) = \varnothing$$

[see (C_{i-1})]. For $e = 1, 2$ define $Z_e = f_e^{-1}(\mathrm{Cl}\ V_i)$ and

$$Y_e = f_e^{-1}(\mathrm{Fr}\ V_i) = f_e^{-1}((\mathrm{Fr}\ U_i) \times J_i).$$

By Theorem 14.7, $((\mathrm{Fr}\ U_i) \times J_i) \cup (U_i \times \{t_i\})$ is an ANR, so by Borsuk's homotopy extension theorem (14.6), the map $f_1 \,|\, Y_1$ extends to

$$m_1 : Z_1 \to ((\mathrm{Fr}\ U_i) \times J_i) \cup (U_i \times \{t_i\}),$$

since $f_1 \mid Y_1$ is homotopic to one extendable over Z_1 (namely f_1 followed by the "projection" of Cl $U_i \times J_i$ to Cl $U_i \times \{t_i\}$). Now choose $s_i \in J_i$ such that

$$[m_1(Z_1) \cup f_2(B^2)] \cap (U_i \times \{s_i\}) = \varnothing.$$

As above, $f_2 \mid Y_2$ extends to (see Fig. 26-2)

$$m_2: Z_2 \to ((\text{Fr } U_i) \times J_i) \cup (U_i \times \{s_i\}).$$

Summarizing, we have produced maps $m_e: Z_e \to (\text{Cl } U_i) \times J_i$ for $e = 1, 2$ that extend over the rest of B^2 via f_e, and the images of which intersect nowhere in V_i, although they may intersect at points of $(\text{Fr } U_i) \times J_i$.

Step 6. *General position improvements.* To circumvent difficulties peculiar to the case $n = 4$, we specialize to the case $n \geq 5$. For $e = 1, 2$ let $R_e = m_e^{-1}((\text{Fr } U_i) \times J_i)$. Because Fr $U_i \times E^1$ is an n-manifold of dim ≥ 5 and because the sets $U_i \times \{s_i\}$ and $U_i \times \{t_i\}$ are disjoint, the maps m_e of R_e in $((\text{Fr } U_i) \times J_i) \cup (U_i \times \{s_i, t_i\})$ admit general position modifications, affecting no points of Fr R_e, hence extending over $Z_e - R_e$ via m_e, so that $m_1(Z_1 - Y_1) \cap m_2(Z_2 - Y_2) = \varnothing$. Define maps $f_e': B^2 \to (M/G) \times E^1$ as m_e on Z_e and as f_e elsewhere. Then any $s \in f_1'(B^2) \cap f_2'(B^2)$ belongs either to

$$[S \cap f_1(B^2) \cap f_2(B^2)] - V_i \subset S - \bigcup_{j=1}^{i} V_j$$

or to

$$[f_1(Y_1) \cap m_2(Z_2 - Y_2)] \cup [m_1(Z_1 - Y_1) \cap f_2(Y_2)]$$

[see (B_{i-1})]. By expelling points of the second kind, without adding any others, we will achieve our goal.

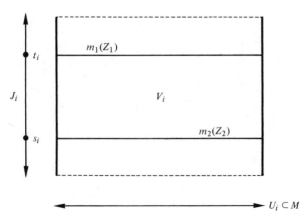

FIG. 26-2

Step 7. *Final modifications.* It follows from condition (B_{i-1}) that $m_1(Y_1) \cap m_2(Y_2) = \varnothing$. Thus, for $e = 1, 2$ B^2 contains disjoint open subsets P_e and Q_e, as depicted in Fig. 26-3, satisfying

$$Y_e \subset P_e \subset (f_e')^{-1}(W \cap [((M/G) - U_i) \times J_i]),$$

$$(m_e \mid Z_e - Y_e)^{-1}(m_1(Z_1) \cap m_2(Z_2)) \subset Q_e \subset Z_e - Y_e,$$

$$f_1'(P_1) \cap f_2'(P_2) = \varnothing = f_1'(Q_1) \cap f_2'(Q_2).$$

Note that any point $b \in B^2$ for which

$$f_e'(b) \in [f_1(Y_1) \cap m_2(Z_2 - Y_2)] \cup [m_1(Z_1 - Y_1) \cap f_2(Y_2)]$$

belongs to $P_e \cup Q_e$.

Since $\operatorname{Fr} U_i$ is an $(n-1)$-manifold separating M/G, properties of generalized manifolds guarantee that U_i and $(M/G) - \operatorname{Cl} U_i$ are 0-LC at $\operatorname{Fr} U_i$. After choosing disjoint dense subsets D, D' of $E^1 - \{s_i, t_i\}$, we exploit once again the observation used to prove Theorem 12 to obtain maps $F_e: B^2 \to (M/G) \times E_1$ such that

$$F_e \mid B^2 - (P_e \cup Q_e) = f_e' \mid B^2 - (P_e \cup Q_e),$$

$$F_e(P_e) \subset (W \cap [((M/G) - \operatorname{Cl} U_i) \times J_i)] \cup [\operatorname{Fr} U_i) \times (D \cap J_i)],$$

$$F_e(Q_e) \subset (U_i \times J_i) \cup [(\operatorname{Fr} U_i) \times (D' \cap J_i)],$$

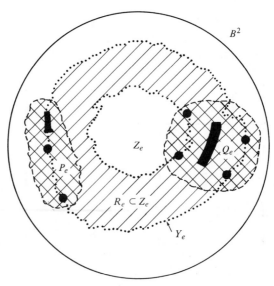

FIG. 26-3

all of which can be arranged with minor care in the approximation so that, in addition,

$$F_1(P_1) \cap F_2(P_2) = \varnothing = F_1(Q_1) \cap F_2(Q_2),$$

$$F_1(P_1 \cup Q_1) \cap m_2(B^2 - (P_2 \cup Q_2)) = \varnothing,$$

$$m_1(B^2 - (P_1 \cup Q_1)) \cap F_2(P_2 \cup Q_2) = \varnothing.$$

The disjointness of D and D' then implies that

$$F_1(P_1 \cup Z_1) \cap F_2(P_2 \cup Z_2) = \varnothing.$$

Consequently, the maps F_1 and F_2 satisfy

$$F_1(B^2) \cap F_2(B^2) \subset f_1'(B^2) \cap f_2'(B^2),$$

and they eradicate intersections of the second kind identified in Step 6. Routine auditing of the entries will confirm that these maps fulfill the requirements of the reduction designated in Step 4. ∎

Corollary 13A. *Suppose G is a cell-like decomposition of an n-manifold M, $n \geq 4$, and $\{P_i\}$ is a countable collection of closed subsets of M/G such that each P_i is either an $(n - 1)$-complex or a generalized $(n - 1)$-manifold and that each point of $(M/G) - \bigcup P_i$ is locally encompassed by manifolds. Then $G \times E^1$ is a shrinkable decomposition of $M \times E^1$.*

EXERCISES

1. Suppose G is a cell-like decomposition of an n-manifold M, $n \geq 5$, such that M/G is an ANR and suppose k is an integer such that every map $\mu: B^2 \to M/G$ can be approximated by $\mu': B^2 \to M/G$ for which $\dim \mu'(B^2) \leq k$. Show that $\pi: M \to M/G$ can be approximated by a cell-like map $F: M \to M/G$ such that $\dim F(N_F) \leq k$.

2. Suppose G is a cell-like decomposition of E^n such that E^n/G has the following disjoint point–disk property: any two maps $B^0 \to E^n/G$, $B^2 \to E^n/G$ can be approximated by maps with disjoint images. Show that the set of points in E^n/G having cellular preimages is dense in E^n/G.

3. Suppose G is a cell-like decomposition of an n-manifold M, $n \geq 3$. Show that $(M/G) \times E^1$ has the following disjoint disk triples property: any three maps $\mu_i: B^2 \to M/G$ $(i = 1, 2, 3)$ can be approximated by maps $\mu_i': B^2 \to M/G$, where $\rho(\mu_i', \mu_i) < \varepsilon$ for $i \in \{1, 2, 3\}$, such that $\mu_1'(B^2) \cap \mu_2'(B^2) \cap \mu_3'(B^2) = \varnothing$.

4. If G is a cell-like decomposition of an n-manifold M, $n \geq 4$, such that $\text{dem } N_G \leq n - 2$, prove that $(M/G) \times E^1$ has the DDP.

5. For any cell-like decomposition G of an n-manifold M, $n \geq 3$, every map $f: B^2 \to M/G$ can be approximated by a map $F: B^2 \to M/G$ such that $\dim F^{-1}(x) \leq 0$ for all $x \in M/G$.

6. Prove Claim 1 made in the proof of Lemma 11.

7. Establish Lemma 11 without the hypothesis that $Y - X$ is a manifold.

8. Suppose S is a locally compact ANR and U is an open subset of S whose frontier F is an ANR such that U is 0-LC at each point of F. Let (Z, Z') be a pair of closed subsets of B^2 ($Z' \subset Z$), f a map of (Z, Z') into $(Y \times E^1, F \times E^1)$, where $Y = U \cup F$, and D any dense subset of E^1. Show that f can be approximated by a map f' of (Z, Z') into $(Y \times E^1, F \times E^1)$ such that $f' \mid Z' = f \mid Z'$ and $f'(Z - Z') \subset [(U \times E^1) \cup (F \times D)]$.

9. Prove Corollary 13A.

10. If G is a finite-dimensional cell-like decomposition of an n-manifold M, $n \geq 4$, show that the natural map $M \times E^1 \to (M/G) \times E^1$ can be approximated by a cell-like map p for which $\dim p(N_p) \leq 1$.

27. PRODUCTS OF E^1 WITH DECOMPOSITIONS OF E^3

According to Theorem 17.12, each cell-like decomposition G of E^3 is finite-dimensional. Therefore, Corollary 26.8A reveals $G \times E^2$ to be a shrinkable decomposition of E^5. Nothing in Section 26 sheds much light on the shrinkability of the product $G \times E^1$, however, because the cell-like approximation theorem does not apply. It is noteworthy that Theorem 23.4, with no lower bound on the dimensions in which it applies, can be used in E^4 for studying $G \times E^1$. As an initiation into systematic investigation of such product decompositions, we analyze the first nontrivial case, where G is a closed-0-dimensional.

Theorem 1. *If G is a closed-0-dimensional cell-like decomposition of S^3, then $(S^3/G) \times E^1$ is homeomorphic to $S^3 \times E^1$. In particular, $G \times E^1$ is a shrinkable decomposition of $S^3 \times E^1$.*

Theorem 1 was derived independently in 1974 by R. D. Edwards and R. T. Miller [1] and by C. P. Pixley and W. T. Eaton [1]. Later an alternative proof was given by J. W. Cannon [4]. Their clever, diverse *ad hoc* techniques, predating Edwards's work on the cell-like approximation theorem, presently seem less mainstream than those to be laid out here, which exploit several philosophical and substantive facets of what has been developed thus far.

By Corollary 20.4A it suffices to prove the ensuing reduction of Theorem 1.

Proposition 2. *If X is a cell-like subset of E^3, then $G_X \times E^1$ is a shrinkable decomposition of $E^3 \times E^1 = E^4$.*

At the outset one can simplify the problem presented in Proposition 2 by selecting, in whatever way seems beneficial, a preferred cell-like set $X^* \subset E^3$ for which $E^3 - X = E^3 - X^*$. A convenient simplification involves the notion of split handle pair, introduced in Cannon [4]. A pair (H_1, H_2) of (3-dimensional) cubes with handles is called a *split handle pair* provided H_2 has twice as many handles as H_1, which run through the handles of H_1 as shown in Fig. 27-1. We will make use of the following (Cannon [4, Lemma]):

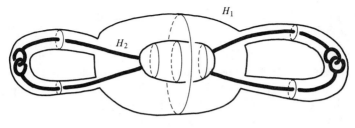

FIG. 27-1

Lemma 3. *Let X be a cell-like set in E^3 and U an open set containing X. Then there exists a split handle pair (H_1, H_2) and a PL embedding $F: H_1 \to E^3$ such that $U \supset F(H_1) \supset F(\text{Int } H_2) \supset X$.*

We say that a cell-like subset X^* of E^3 is in *standard position* if there exists a sequence of PL cubes with handles $H_1, H_2, H_3, \ldots$ such that for $j \in \{1, 2, \ldots\}$:

   (1) $H_{j+1} \subset \text{Int } H_j$;
   (2) $X^* = \bigcap_j H_j$;
   (3) (H_{2j-1}, H_{2j}) is a split handle pair;
   (4) H_{2j-1} is the union of a 3-cell having diameter $< 1/j$ and of 1-handles having cross-sectional diameter $< 1/j$; and
   (5)   the linking pairs of handles in H_{2j} each lie in a 3-cell in H_{2j-1} having diameter less than $1/j$.

This rigid linking pattern will expedite the shrinking argument eventually sought here. For now, observe that Property (4) prompts the familiar dem $X \le 1$.

It is permissible to treat only cell-like sets X^* in standard position, for by Lemma 3 every cell-like set X in E^3 is equivalent to such an X^*, the weak interpretation of equivalence being that $E^3 - X^*$ is homeomorphic to $E^3 - X$. Better yet, first given a measure of closeness, one can obtain X^* and then can approximate the original decomposition map $E^3 \to E^3/G_X$ by a map $f: E^3 \to E^3/G_X$ whose only nondegenerate preimage is X^*.

Throughout the rest of this section we shall employ the following notation: X, for a cell-like subset of E^3 in standard position; Q, for $(E^3/G_X) \times E^1$; and f_0, for the natural map $E^4 = E^3 \times E^1 \to Q$.

The result stated below is the main lemma. It has applications to other, more general 4-space decomposition problems.

Lemma 4. *Let X be a cell-like subset of E^3 in standard position. Then there exists a sequence of triangulations $\{T_j\}$ of E^4, with mesh $T_j \to 0$, and there exists a sequence of cell-like maps $\{f_j: E^4 \to Q\}$ satisfying:*

(a) $\lim f_j = f$ is a proper cell-like map (close to f_0);
(b) f is 1-1 over $Q - f_0(X \times E^1)$; and
(c) f is 1-1 over $\bigcup_j f(T_j^{(2)})$.

Proof that Lemma 4 implies Proposition 2.   Consider the decomposition $G_f = \{f^{-1}(q) \mid q \in Q\}$ induced by f. Conclusion (a) of Lemma 4 implies that G_f is cell-like and conclusion (c) implies that its nondegeneracy set has embedding dimension at most 1. Consequently, Theorem 23.2 attests that G_f is shrinkable, so there exists a homeomorphism $h \colon E^4 \to E^4/G_f$ approximating the decomposition map p. As customary, set $F = fp^{-1}h$, which is a homeomorphism $E^4 \to Q$. With controls on the choice of f ensured by conclusion (a), one can obtain F close to f_0, showing that $G_X \times E^1$ is shrinkable. ∎

A *prismatic triangulation* of $E^4 = E^3 \times E^1$ is a cell-complex subdivision of E^4 expressible as $T = T^* \times T^{**}$, where T^* and T^{**} denote triangulations of E^3 and E^1, respectively. For Lemma 2, working with prismatic triangulations is equally acceptable as with the usual simplicial ones. The advantage of the former stems from the 1-demensionality of X, indicating that the 1-skeleton of T^* can be taken to miss X; this puts the only portion of the entire 2-skeleton of T meeting $X \times E^1$ in those levels determined by the 0-skeleton of T^{**}.

The argument for the technical lemma stated next embodies the crucial geometric features needed to prove Lemma 4. The rest of the proof of Lemma 4 is dirty detail.

Lemma 5.   *Let $\varepsilon > 0$ and let $T = T^* \times T^{**}$ be a prismatic triangulation of E^4 with $T^{*(1)} \cap X = \varnothing$. Then there exist an ε-homeomorphism $g \colon E^4 \to E^4$, an isotopy $\theta_t \colon E^4 \to E^4$, and an index $k \in \mathbb{Z}_+$ satisfying:*

(a) $\theta_t \mid [E^3 - N(X; \varepsilon)] \times E^1 = \text{Id}$,
(b) $\theta_t \mid E^3 \times [E^1 - N(T^{**(0)}; \varepsilon)] = \text{Id}$,
(c) $\theta_t \mid g(T^{(2)}) = \text{Id}$,
(d) $\{t \in E^1 \mid g(T^{(2)}) \cap \theta_1(H_k \times \{t\}) \neq \varnothing\}$ is nowhere dense;
(e) *for all* $t \in E^1$, $g(T^{(2)}) \cap \theta_1(H_k \times \{t\}) = g(T^{(2)}) \cap (H_k \times \{t\})$;
(f) $g(T^{(2)}) \cap \theta_1(H_k \times \{t\}) \neq \varnothing$ *implies* diam $\theta_1(H_{k+1} \times \{t\}) < \varepsilon$; and
(g) *for each* $t \in E^1$ *with* $g(T^{(2)}) \cap (H_k \times \{t\}) \neq \varnothing$, *there exists a neighborhood* U_t *of t in E^1 for which $\theta_1 \mid H_{k+1} \times U_t$ is a product embedding.*

Proof.   Choose an odd integer $k \in \mathbb{Z}_+$ so that H_k consists of a small (diam $< \varepsilon/3$) 3-cell and thin (cross-sectional diam $< \varepsilon/3$) handles.

Approximate the inclusion $T^{(2)} \to E^4$ by a level-preserving homeomorphism g' of $E^4 = E^3 \times E^1$ to itself such that $g'(T^{(2)}) \cap (H_k \times \{t\})$, where nonvoid, consists of mutually exclusive disks in the handles and away

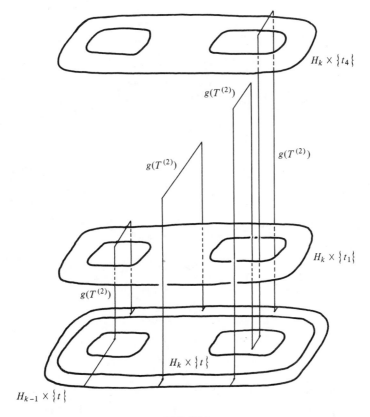

FIG. 27-2

from the linked overlaps present in $H_{k+1} \times \{t\}$. Keeping E^3-coordinates fixed, approximate g' by $g: E^4 \to E^4$ such that $g(T^{(2)}) \cap (H_k \times \{t'\})$, when nonvoid, is a single disk. See Fig. 27-2. Perform these adjustments in a uniform way to cause

$$\{t \in E^1 \mid g(T^{(2)}) \cap (H_k \times \{t\}) \neq \varnothing\}$$

to be a countable discrete set.

The prearranged linking pattern now allows quick description of the isotopy θ_t that shrinks the aforementioned countable collection of level sets $H_{k+1} \times \{t\}$ to small size, fixing $g(T^{(2)})$. Unlink the handles of such an $H_{k+1} \times \{t\}$ and replace them, in unlinked fashion, back in $H_k \times \{t\}$; then shrink the image in $H_k \times \{t\}$ near the single disk $g(T^{(2)}) \cap (H_k \times \{t\})$. This can easily be accomplished, subject to (a), (b), and (c), with θ_1 behaving like a product map near the level sets $H_{k+2} \times \{t\}$ where $g(T^{(2)}) \cap (H_k \times \{t\}) \neq \varnothing$. See Fig. 27-3. ∎

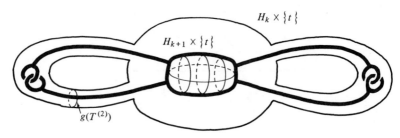

Before applying θ_t

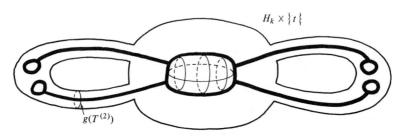

Handles of $H_{k+1} \times \{t\}$ unlinked and replaced

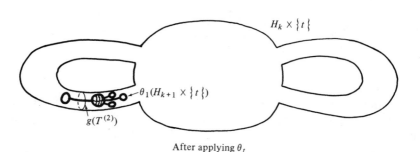

After applying θ_t

FIG. 27-3

With standard epsilonic and convergence controls superimposed, repeated applications of Lemma 5 give rise to the next result indicating that f_0 can be approximated by a map $f_1 : E^4 \to Q$ which is 1–1 over the image of a given 2-skeleton.

Lemma 6. *Let* $T = T^* \times T^{**}$ *be a prismatic triangulation of* E^4 *with* $T^{*(1)} \cap X = \varnothing$ *and let* $\delta > 0$. *Then there exist a* δ-*homeomorphism* $g : E^4 \to E^4$ *and a pseudo-isotopy* ψ_t *of* E^4 *(with* $\psi_0 = \mathrm{Id}$*) satisfying:*

(a) $\psi_t | [E^3 - N(X; \delta)] \times E^1 = \mathrm{Id}$;

(b) $\psi_t | E^3 \times [E^1 - N(T^{**(0)}; \delta)] = \mathrm{Id}$;

(c) *each nondegenerate inverse set under ψ_1 is also an inverse set of f_0 (giving it the form $X \times \{t\}$)*;

(d) *for some closed 0-dimensional subset C of E^1, $X \times C$ contains the nondegenerate inverses under ψ_1*;

(e) $\psi_1(X \times \{t\}) \cap g(T^{(2)}) \neq \varnothing$ *implies* $\psi_1(X \times \{t\}) = $ point; *and*

(f) *ψ_t is ultimately stationary away from the nondegeneracy set (that is, for each compact set K in E^4 missing the nondegeneracy set of ψ_1, there exists $t(0) \in [0, 1)$ such that $\psi_t | K = \psi_{t(0)} | K$ for all $t \in [t(0), 1]$).*

The following straightforward application of Lemma 6 arranges a more pronounced coherence with the main lemma.

Lemma 7. *For each $\varepsilon > 0$ there exists a triangulation T_1 of E^4 having mesh less than ε and there exists a cell-like map $f_1: E^4 \to Q$ such that*

(a) $\rho(f_0, f_1) < \varepsilon$;

(b) $f_1 | [E^3 - N(X; \varepsilon)] \times E^1 = f_0 | [E^3 - N(X; \varepsilon)] \times E^1$;

(c) *for some closed 0-dimensional subset C_1 of E^1, $X \times C_1$ contains the nondegenerate inverses under f_1; and*

(d) f_1 *is 1-1 over* $f_1(T^{(2)}) \cup f_0((E^3 - X) \times E^1)$.

Proof. Name a prismatic triangulation T of E^4 with $T = T^* \times T^{**}$ such that mesh $T < \varepsilon/3$ and $T^{*(1)} \cap X = \varnothing$. Apply Lemma 6 to obtain g and ψ_t, using $\delta \in (0, \varepsilon/3)$ so small that the δ-neighborhood of any inverse set of f_0 is mapped to an ε-set in Q. Let $\psi = \psi_1$.

Set $T_1 = g(T)$ and $f_1 = f_0(\psi)^{-1}$. Conclusion (c) of Lemma 6 implies f_1 is well defined and continuous. Conclusions (a) and (b) and the choice of δ imply that conclusion (a) here holds; conclusion (a), that (b) here holds; and conclusion (d), that (c) here holds.

Since $f_1^{-1} = \psi f_0^{-1}$, f_1 is 1-1 over each point $q \in Q$ over which f_1 is 1-1, substantiating part of (d). For the other part, consider $z \in T_1^{(2)} = g(T^{(2)})$. If $f_1^{-1}f_1(z)$ is not a point, $f_0^{-1}f_1(z)$ is not a point either, which implies it is a set $X \times \{t\}$. Then

$$z \in (X \times \{t\}) \cap g(T^{(2)})$$

and $\psi(X \times \{t\}) = \psi f_0^{-1} f_1(z)$ is a point by conclusion (e) of Lemma 6. As a result, f_1 is 1-1 over $f_1(T_1^{(2)})$, as required. ∎

Remarks about the proof of Lemma 4. Lemma 7 inaugurates the construction of the sequence of triangulations and cell-like maps called for in Lemma 4. Continuation of the process is not quite standard nor quite automatic, for the map f_1 destroys parts of the product geometry prevalent in f_0. Before dropping the subject, we indicate some critical aspects of the repetitions.

To be able to exploit uniform continuity, we expand the Euclidean factors into their natural compactifications so as to have $X \subset E^3 \subset S^3$ and $E^1 \subset S^1$, and we work in $S^3 \times S^1$ instead of $E^3 \times E^1$.

Name a small number $\varepsilon_2 \in (0, \varepsilon/2)$ and focus on the pseudo-isotopy ψ_t used to define $f_1 = f_0 \psi^{-1}$. There exists $\delta > 0$ such that for $A \subset S^3 \times S^1$ with diam $A < \delta$, diam $\psi_t(A) < \varepsilon_2$. Choose a prismatic triangulation $\hat{T} = \hat{T}^* \times \hat{T}^{**}$ of $S^3 \times S^1$ such that mesh $\hat{T} < \delta/3$, $\hat{T}^{*(1)} \cap X = \varnothing$, and $\hat{T}^{**(0)} \cap C_1 = \varnothing$. (This displays the primary purpose of C_1—it helps contrive a prismatic triangulation whose 2-skeleton avoids the sets shrunk out by ψ_1).

Find $\eta > 0$ for which $N(\hat{T}^{(2)}; \eta) \cap (X \times C_1) = \varnothing$ and $(N(X; \eta) \times N(\hat{T}^{**(0)}; \eta)) \cap \hat{T}^{(2)} = \varnothing$. Apply Lemma 6 again with triangulation $\hat{T}$ and positive number $\min\{\delta/3, \eta\}$ to obtain g_2 and $\hat{\psi}_t$. Then mesh $g_2(\hat{T}) < \delta$. Also, $g_2(\hat{T}^{(2)}) \cap (X \times C_1) = \varnothing$, so $f_0^{-1} f_0 g_2(\hat{T}^{(2)})$ is a compact set K missing $X \times C_1$, which contains the nondegeneracy set of ψ_t. Due to that constraint, conclusion (f) of Lemma 6 gives $t(0) \in (0, 1)$ such that $\psi_{t(0)} | K = \psi | K$.

Set $T_2 = \psi_{t(0)} g_2(\hat{T})$. Since mesh $g_2(\hat{T}) < \delta$, the earlier choice of δ yields mesh $T_2 < \varepsilon_2$. Further, let $\hat{\psi} = \hat{\psi}_1$ and $f_2 = f_0 \hat{\psi}^{-1} \psi^{-1}$.

It should be clear that epsilonic restrictions can be installed to get at (a) and (b) of Lemma 4. The most absorbing conclusion there is (c). To see why f_2 is 1-1 over $f_2(T_2^{(2)})$, note that $f_2 | T_2^{(2)} = f_2 | \psi_{t(0)} g_2(\hat{T}^{(2)})$ and $f_2 \psi_{t(0)} | g_2(\hat{T}^{(2)}) = f_0 \hat{\psi}^{-1} \psi^{-1} \psi_{t(0)} | g_2(\hat{T}^{(2)}) = f_0 \hat{\psi}^{-1} | g_2(\hat{T}^{(2)})$ by the choice of $t(0)$ above. Hence f_2 is 1-1 over $f_2(T_2^{(2)})$ for precisely the same reasons that f_1 is 1-1 over $f_1(T_1^{(2)})$. To see why f_2 is also 1-1 over $f_2(T_1^{(2)})$, check that the choice of η causes $\hat{\psi}_t$ to be stationary on $T_1^{(2)}$, implying f_2 and f_1 coincide there. This suggests that f will agree with f_1 on $T_1^{(2)}$, but for the ultimate conclusion one must begin adding other controls, as in the proof of Proposition 23.4, to guarantee that the limit map f will agree with f_1 over $f_1(T_1^{(2)})$.

The next step in the process is a repeat of the one just described, with the composition $\psi_t \hat{\psi}_t$ playing the role of ψ_t. Successive steps proceed similarly. This completes our outline toward a proof of Lemma 4. ■

Notes. Modifications of the techniques just described are made in Daverman–Row [1] to show that $G \times E^1$ is a shrinkable decomposition of E^4 provided G is a 0-dimensional cell-like decomposition of E^3 and in Daverman–Preston [2] to do the same provided G is a cell-like decomposition of E^3 for which N_G has embedding dimension 1.

28. SPUN DECOMPOSITIONS

Among the various nonshrinkable cellular decompositions of E^3, the ones described in Section 9 hold extraordinary appeal because of their specific, uncomplicated, pictorial description. Notably, in light of the proliferating collection of positive results concerning shrinkability of cell-like decompositions in higher dimensions, we have not yet encountered a single nonshrinkable cellular decomposition there, let alone one having an equally specific and uncomplicated description. As we saw in the preceding sections, the most obvious technique, taking the product of E^1 with a cell-like decomposition of E^{n-1}, has never produced an interesting example. A more fruitful technique is presented in this section; it provides fairly specific examples which are cellular but nonshrinkable.

The method, that of spinning decompositions, was first detailed in a general form by J. W. Cannon [1]. He passes credit for the idea to E. Cheeseman, who described to him a wild Cantor set in E^4 obtained by spinning a 3-dimensional solid horned sphere (see Corollary 9B), but others such as L. L. Lininger [1] had exploited spinning earlier for similar purposes.

At the onset the point of view is based on generalized polar coordinates, where points of E^{k+1} are represented in terms of direction and distance from the origin. Explicitly, one can regard E^{k+1} as

$$([0, \infty) \times S^k)/\mathcal{G}_k = (E^1_+ \times S^k)/\mathcal{G}_k,$$

where $\mathcal{G}_k$ denotes the decomposition (not cell-like) whose only nondegenerate element is $\{0\} \times S^k$. Taking products with E^{n-k-1}, one can regard E^n as

$$E^n = (E^{n-k-1} \times E^1_+ \times S^k)/\mathcal{R}_k = (E^{n-k}_+ \times S^k)/\mathcal{R}_k,$$

·where $\mathcal{R}_k$ denotes the trivial extension over $E^{n-k}_+ \times S^k$ of the relation

$$\{\{x\} \times S^k \mid x \in E^{n-k-1} \times \{0\}\}.$$

This is easiest to visualize when $k = 1$, for then one can imagine rotating E^{n-1}_+ about its "edge" $E^{n-2} \times \{0\}$; when $n = 3$, this coincides with rotating a half-plane about a linear axis.

In similar fashion, passing to one-point compactifications, one can view S^n as $(B^{n-k} \times S^k)/\mathfrak{I}_k$, where $\mathfrak{I}_k$ denotes the trivial extension over B^{n-k} of $\{\{b\} \times S^k \mid b \in \partial B^{n-k}\}$. Due to compactness features, we prefer working in S^n; nevertheless, comparable properties usually hold in the related representation of E^n.

A useful notational item is the map $\psi_k : S^n \to B^{n-k}$ induced from the diagram

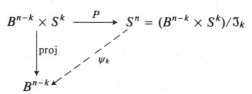

It is 1-1 over ∂B^{n-k}, and its other point-inverses are k-spheres. The set $\psi_k^{-1}(\partial B^{n-k})$ serves as the *distinguished $(n - k - 1)$-sphere* about which an $(n - k)$-cell is spun to sweep out S^n.

If X is a subset of B^{n-k}, then by the *k-spin of X* we mean the subset $\text{Sp}^k(X)$ of S^n defined as $P(X \times S^k) = \psi_k^{-1}(X)$. For instance, in case X is an arc in B^{n-k} such that $X \cap \partial B^{n-k}$ is an endpoint e of X, then $\text{Sp}^k(X)$ is topologically

$$(X \times S^k)/G_{\{e\} \times S^k} \approx ([0, 1] \times S^k)/G_{\{0\} \times S^k},$$

which is a $(k + 1)$-cell.

Lemma 1   *Let X denote a compact subset of B^{n-k} intersecting ∂B^{n-k}. If both X and $X \cap \partial B^{n-k}$ are cell-like, then $\text{Sp}^k(X)$ is cell-like; conversely, if $\text{Sp}^k(X)$ is cell-like, then X is cell-like.*

Proof.   Suppose first that X and $X \cap \partial B^{n-k}$ are cell-like. As a special case, assume that $X \cap \partial B^{n-k}$ is a single point z. Let U be an open subset of S^n containing X. Choose an open subset V of B^{n-k} containing X such that $P(V \times S^k) \subset U$. Since X is cell-like and V is an ANR, there exists a map $\theta_t: X \to V$ starting at the inclusion, fixing z throughout, and ending at the constant $\theta_1: X \to \{z\}$. The map $\tilde{\theta}_t$ induced by $\theta_t \times \text{Id}$ functions as the required contraction of $P(X \times S^k)$ in $P(V \times S^k) \subset U$.

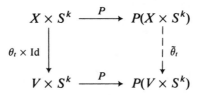

In the general case, either find a contraction of X in V that keeps $X \cap \partial B^{n-k}$ in $V \cap \partial B^{n-k}$ throughout, or use the argument above to prove that $P(X \times S^k)/G_{P((X \cap \partial B^{n-k}) \times S^k)}$ is cell-like and infer that $P(X \times S^k)$ itself must be.

Next suppose that $\text{Sp}^k(X)$ is cell-like. Fix a point $*$ of S^k and let $j: B^{n-k} \to S^n$ be the natural map

$$B^{n-k} \to B^{n-k} \times * \hookrightarrow B^{n-k} \times S^k \xrightarrow{\ P\ } S^n.$$

Then $\psi_k j = \text{Id}_{B^{n-k}}$. Let U be a neighborhood in B^{n-k} of X. The set $\text{Sp}^k(U) = P(U \times S^k)$ is a neighborhood in S^n of $\text{Sp}^k(X)$. Since $\text{Sp}^k(X)$ is cell-like, there exists a contraction h_t of $\text{Sp}^k(X)$ in $\text{Sp}^k(U)$, and $\psi_k h_t j$ provides a contraction of X in U.   ∎

Proposition 2.   *Suppose X is a cell-like proper subset of B^{n-k} intersecting ∂B^{n-k} nontrivially. Then X satisfies the cellularity criterion in B^{n-k} if and only if $\text{Sp}^k(X) = P(X \times S^k)$ satisfies the cellularity criterion in S^n.*

Proof. Suppose first that $Sp^k(X)$ satisfies the cellularity criterion in S^n. Consider a neighborhood U of X in B^{n-k}. Find another neighborhood V such that each loop in $Sp^k(V) - Sp^k(X)$ is contractible in $Sp^k(U) - Sp^k(X)$. Let $j: B^{n-k} \to S^n$ denote an embedding, as before, such that $\psi_k j = \text{Id}_{B^{n-k}}$. Given a loop $f: \partial B^2 \to V - X$ one can produce an extension $F: B^2 \to U - X$ by extending $jf: \partial B^2 \to Sp^k(V) - Sp^k(X)$ to $\tilde{F}: B^2 \to Sp^k(U) - Sp^k(X)$ and defining F as $\psi_k \tilde{F}$.

Next suppose that X satisfies the cellularity criterion, and consider a neighborhood W of $Sp^k(X)$. Determine a neighborhood U of X in B^{n-k} for which $Sp^k(U) \subset W$. By hypothesis U contains another neighborhood V of X such that loops in $V - X$ are null homotopic in $U - X$. We shall show that loops in $Sp^k(V) - Sp^k(X)$ are null homotopic in $Sp^k(U) - Sp^k(X)$. Our argument applies only when $k \geq 1$; the case $k = 0$ is left to the reader.

Consider $f: \partial B^2 \to Sp^k(V) - Sp^k(X)$. Since the distinguished sphere $Sp^k(\partial B^{n-k})$ has dimension $\leq n - 2$, we can perform a short preliminary homotopy [in $Sp^k(V) - Sp^k(X)$] moving $f(\partial B^2)$ off $Sp^k(\partial B^{n-k})$. Because $f(\partial B^2)$ then misses the image of the nondegeneracy set of P, P induces a map

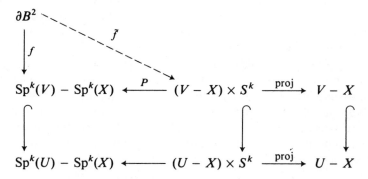

$\tilde{f}: \partial B^2 \to (V - X) \times S^k$ such that $P\tilde{f} = f$. The prearranged neighborhoods give rise to a map $F: B^2 \to U - X$ extending $\text{proj} \circ \tilde{f}$. It is convenient to regard F as a homotopy $F: S^1 \times [0, 1] \to U - X$ between $\text{proj} \circ \tilde{f}$ and a constant map. Since X is cell-like, it does not separate any open subset of S^{n-k}, making possible a modification of the constant mapping so its image rests in $U \cap \partial B^{n-k}$. Homotopy lifting properties then yield another map $\tilde{F}: S^1 \times [0, 1] \to (U - X) \times S^k$ preserving commutativity in the following diagram. Although $\tilde{F}(S^1 \times \{1\})$ is not likely to be a point (when $k = 1$; when $k > 1$ such difficulties dissolve), $P\tilde{F}(S^1 \times \{1\})$ must be a point, and $P\tilde{F}$ provides a homotopy between $f = P\tilde{f}$ and a constant map, with range in $P((U - X) \times S^k) \subset W - Sp^k(X)$.

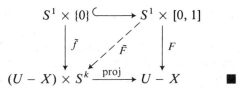

Corollary 2A. *Suppose X is a cell-like subset of B^{n-k} that satisfies the cellularity criterion in B^{n-k} and intersects ∂B^{n-k} in a cell-like set. Then $\mathrm{Sp}^k(X)$ is cellular in S^n.*

See also Lemma 1 and Theorem 18.5.

Corollary 2B. *If X is a compact subset of B^{n-k} such that $X \cap \partial B^{n-k}$ is cell-like and $\mathrm{Sp}^0(X)$ is cellular in S^{n-k}, then $\mathrm{Sp}^k(X)$ is cellular in S^n.*

Corollary 2C. *Suppose X is a compact subset of B^m such that $X \cap \partial B^m$ is cell-like. Then $\mathrm{Sp}^k(X)$ is cellular in S^{m+k} if and only if $\mathrm{Sp}^n(X)$ is cellular in S^{m+n}.*

The method initially uncovered for concocting noncellular cell-like sets in S^n ($n > 3$) involves suspensions: the iterated suspension ΣX of a given cell-like set X in S^k ($k < n$) having nonsimply-connected complement resides naturally as a noncellular subset of $\Sigma S^k \approx S^n$ (Exercise 18.6). A new method involves spins: if an arc A directly connects a given noncellular, cell-like subset X' of Int B^k to ∂B^k, then the k-spin of $A \cup X'$ is a noncellular subset of S^n.

Equipped now with precepts for recognizing spins that are cellular, we revert to studying decompositions. A usc decomposition G of B^{n-k} is *admissible* if each $g \in H_G$ intersects ∂B^{n-k} in a cell-like (nonempty) set. Given an admissible decomposition G of B^{n-k}, we define the associated *k-fold spun decomposition* (of S^n), often referred to simply as the *k-spin* of G, denoted by $\mathrm{Sp}^k(G)$, as the trivial extension of $\{P(g \times S^k) \mid g \in H_G\}$. Admissible cell-like decompositions of B^{n-k} are natural forerunners, though not the most general ones, of cell-like spun decompositions of S^n.

Proposition 3. *Let G denote an admissible usc decomposition of B^{n-k}. Then $\mathrm{Sp}^k(G)$ is a cell-like decomposition if and only if G is cell-like. Furthermore, in case G is cell-like, $\mathrm{Sp}^k(G)$ is a cellular decomposition if and only if each $g \in H_G$ satisfies the cellularity criterion in B^{n-k}.*

This is an immediate consequence of Lemma 1 and Proposition 2, plus, of course, Theorem 18.5.

The next lemma introduces more notation. Its proof is obvious.

Lemma 4. *If G is an admissible decomposition of B^{n-k}, then ψ_k induces a map $\tilde{\psi}_k: S^n/\mathrm{Sp}^k(G) \to B^{n-k}/G$ that is 1-1 over $\pi(\partial B^{n-k})$ and that is conjugate to the projection $Y \times S^k \to Y$ over $Y = B^{n-k}/G - \pi(\partial B^{n-k})$. In particular, $\tilde{\psi}_k \tilde{\pi} = \pi \psi_k$.*

Proposition 5. *Suppose G is an admissible decomposition of B^{n-k}, $n - k \geq 5$, such that $S^n/\mathrm{Sp}^k(G)$ is homeomorphic to S^n. Then B^{n-k}/G has the disjoint disks property.*

Proof. Consider two maps $f_1, f_2 \colon B^2 \to B^{n-k}/G$ and $\varepsilon > 0$. Define an embedding $\tilde{j} \colon B^{n-k}/G \to S^n/\mathrm{Sp}^k(G)$ as $\tilde{j} = \tilde{\pi} j \pi^{-1}$, so that the diagram below is commutative,

$$
\begin{array}{ccc}
B^{n-k} & \xrightarrow{\ \pi\ } & B^{n-k}/G \\
\Big\downarrow{\scriptstyle j} & & \Big\downarrow{\scriptstyle \tilde{j}} \\
S^n & \xrightarrow{\ \tilde{\pi}\ } & S^{n-k}/\mathrm{Sp}^k(G)
\end{array}
$$

Then $\tilde{\psi}_k \tilde{j} = \tilde{\psi}_k \tilde{\pi} j \pi^{-1} = \pi \psi_k j \pi^{-1} = \mathrm{Id}_{B^{n-k}/G}$. Since $S^n/\mathrm{Sp}^k(G) \approx S^n$ by hypothesis, the maps $\tilde{j} f_1$ and $\tilde{j} f_2$ can be approximated by disjoint embeddings $F_1, F_2 \colon B^2 \to S^n$, subject to the requirement that $\rho(\tilde{\psi}_k F_i, \tilde{\psi}_k \tilde{j} f_i) < \varepsilon$ ($i = 1, 2$). Although $\tilde{\psi}_k F_1(B^2)$ and $\tilde{\psi}_k F_2(B^2)$ could intersect, they cannot intersect at any point of $\pi(\partial B^{n-k})$, over which $\tilde{\psi}_k$ is 1-1. Hence, a final general position adjustment in the $(n - k)$-manifold $(B^{n-k}/G) - \pi(\partial B^{n-k})$ will remove all intersections. ∎

The dimension restriction $n - k \geq 5$ is significant only for the final general position modification. In case $n \geq 5$ but $n - k$ is unrestricted, the initial part of the argument establishes the following.

Proposition 6. *Suppose G is an admissible decomposition of B^{n-k} such that $S^n/\mathrm{Sp}^k(G)$ is homeomorphic to S^n, $n \geq 5$. Then for any two maps f_1, f_2 of B^2 to B^{n-k}/G and any $\varepsilon > 0$ there exist maps $\tilde{f}_1, \tilde{f}_2 \colon B^2 \to B^{n-k}/G$ such that $\rho(\tilde{f}_i, f_i) < \varepsilon$ ($i = 1, 2$) and*

$$
\tilde{f}_1(B^2) \cap \tilde{f}_2(B^2) \cap \pi(\partial B^{n-k}) = \varnothing.
$$

When n is low-dimensional, it is less easy to find nice maps of B^2 into S^n, but with apparently stronger hypotheses on the decomposition of B^{n-k}, it is still possible to obtain the same conclusion as in Proposition 6.

Proposition 7. *Suppose G is an admissible cell-like decomposition of B^m such that $\mathrm{Sp}^0(G)$ is shrinkable. Then any two maps f_1, f_2 of B^2 into B^m/G can be approximated by maps $\tilde{f}_1, \tilde{f}_2 \colon B^2 \to B^m/G$ such that*

$$
\tilde{f}_1(B^2) \cap \tilde{f}_2(B^2) \cap \pi(\partial B^m) = \varnothing.
$$

Proof. One can naturally express S^m as the union of two embedded m-cells, $j_1(B^m)$ and $j_2(B^m)$, where $j_i(\partial B^m) = P(\partial B^m \times S^0)$ and $\tilde{\psi}_0 \tilde{\pi} j_i = \pi \colon B^m \to B^m/G$ ($i = 1, 2$). Given maps $f_1, f_2 \colon B^2 \to B^m/G$, one can lift them to maps $F_1, F_2 \colon B^2 \to B^m$ that satisfy $\rho(f_i, \pi F_i) < \varepsilon/2$ and then modify the lifts slightly so that $F_i(B^2) \subset \text{Int } B^m$. The two copies $j_1 F_1(B^2)$ and $j_2 F_2(B^2)$ are disjoint subsets of S^m, having some distance δ between them.

Determine $\gamma > 0$ so that $\tilde{\psi}_0$ sends γ-subsets of $S^m/\text{Sp}^0(G)$ to $\varepsilon/2$-subsets of B^m/G. Since $\text{Sp}^0(G)$ is shrinkable, there exists a homeomorphism h of S^m to itself such that

$$\text{diam } h(\text{Sp}^0(g)) < \delta \qquad \text{for each} \quad g \in H_G$$

and

$$\rho(\tilde{\pi}h, \tilde{\pi}) < \gamma$$

It follows that $h^{-1}j_1 F_1(B^2)$ and $h^{-1}j_2 F_2(B^2)$ meet no common element of $\text{Sp}^0(G)$; in other words, $\tilde{\pi}h^{-1}j_1 F_1(B^2) \cap \tilde{\pi}h^{-1}j_2 F_2(B^2) = \varnothing$. Define $\tilde{f}_i$ as $\tilde{\psi}_0\tilde{\pi}h^{-1}j_i F_i$. Since $\tilde{\psi}_0$ is 1–1 over $\pi(\partial B^m)$, $\tilde{f}_1(B^2) \cap \tilde{f}_2(B^2) \cap \pi(\partial B^m) = \varnothing$.

$$
\begin{array}{ccc}
S^m & \xrightarrow{\ \tilde{\pi}\ } & S^m/\text{Sp}^0(G) \\
\ \downarrow{\scriptstyle \psi_0} & & \ \downarrow{\scriptstyle \tilde{\psi}_0} \\
B^m & \xrightarrow{\ \pi\ } & B^m/G
\end{array}
$$

Furthermore,

$$\rho(\tilde{f}_i, f_i) \le \rho(\tilde{f}_i, \pi F_i) + \rho(\pi F_i, f_i) \le \rho(\tilde{\psi}_0\tilde{\pi}h^{-1}j_i F_i, \tilde{\psi}_0\tilde{\pi}j_i F_i) + \varepsilon/2$$

$$\le \rho(\tilde{\psi}_0\tilde{\pi}h^{-1}, \tilde{\psi}_0\tilde{\pi}) + \varepsilon/2 < \varepsilon/2 + \varepsilon/2 = \varepsilon. \qquad \blacksquare$$

Example 1. *A nonshrinkable cellular decomposition of S^n $(n > 3)$.* It is the $(n - 3)$-spin of a decomposition G of B^3 suggested by Fig. 28-1, and it represents a variation of Example 9.2 with minor modifications causing it to be admissible. The nonshrinkability of $\text{Sp}^{n-3}(G)$ follows from properties of Example 9.2 that conflict with the necessary condition of Propositions 6 and 7.

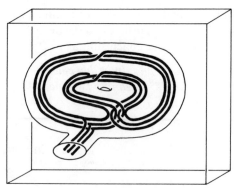

FIG. 28-1

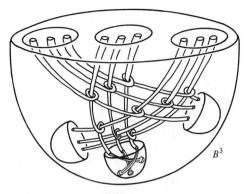

FIG. 28-2

Example 2.   Another nonshrinkable cellular decomposition of S^n ($n > 3$).
This is the $(n - 3)$-spin of an example sometimes called the three-toed sloth,
constructed by L. O. Cannon [1] and suggested by Fig. 28-2. Of historical
interest, it was the first such example obtained by spinning. Cannon's
argument that Example 2 does not satisfy the conclusions of Propositions 6
and 7 is much subtler than that for Example 1.

A partial converse to Proposition 6, the main positive result of this section
provides a sufficient condition for a spun (admissible) decomposition to be
shrinkable. Stated below is a technical lemma used in the proof.

Lemma 8.   *Suppose G is a cell-like decomposition of B^m such that any two
maps $f_1, f_2: B^2 \to B^m/G$ can be approximated by maps $\tilde{f}_1, \tilde{f}_2$ such that*

$$\tilde{f}_1(B^2) \cap \tilde{f}_2(B^2) \cap \pi(\partial B^m) = \varnothing.$$

*Then for any map $f: B^2 \to B^m/G$ and any finite open cover $\mathcal{W}$ of $\pi(\partial B^m)$
there exist approximations $\tilde{f}: B^2 \to B^m/G$ to f such that, for each $W \in \mathcal{W}$,*

$$(W \cap \pi(\partial B^m)) - \tilde{f}(B^2) \neq \varnothing.$$

Proof.   Exercise.

Theorem 9.   *Suppose G is an admissible cell-like decomposition of B^{n-k},
$n \geq 5$, such that $\dim(B^{n-k}/G) < \infty$. Then $\mathrm{Sp}^k(G)$ is shrinkable if and only
if every pair of maps f_1, f_2 from B^2 to B^{n-k}/G can be approximated by maps
$\tilde{f}_1, \tilde{f}_2: B^2 \to B^{n-k}/G$ such that*

$$\tilde{f}_1(B^2) \cap \tilde{f}_2(B^2) \cap \pi(\partial B^{n-k}) = \varnothing.$$

Proof.   The forward implication is Proposition 6.

To establish the other implication, one can simply show that $S^n/\mathrm{Sp}^k(G)$
is a finite-dimensional space satisfying the DDP. The finite-dimensionality

is relatively automatic—$S^n/\mathrm{Sp}^k(G)$ splits into a closed piece homeomorphic to $\pi(\partial B^{n-k})$ and another open piece homeomorphic to $[\pi(B^{n-k}) - \pi(\partial B^{n-k})] \times S^k$, each of which is finite-dimensional, implying that the union is.

For notational convenience, set $Z = \tilde{\psi}_k^{-1}\pi(\partial B^{n-k}) = \tilde{\pi}P(\partial B^{n-k} \times S^k)$. Since $Z \supset \tilde{\pi}(N_{\mathrm{Sp}^k(G)})$ it will suffice to show that any two maps $\mu_1, \mu_2 \colon B^2 \to S^n/\mathrm{Sp}^k(G)$ can be approximated by maps $\tilde{\mu}_1, \tilde{\mu}_2$ such that $\tilde{\mu}_1(B^2) \cap \tilde{\mu}_2(B^2) \cap Z = \varnothing$. Details are given only for the case $k \geq 1$; the remaining case is left to the reader.

Name an open cover $\mathcal{W}$ of $\pi(\partial B^{n-k}) = \tilde{\psi}_k(Z)$ in B^{n-k}/G by connected open subsets so small that diam $\tilde{\psi}_k^{-1}(W) < \varepsilon/2$ for each $W \in \mathcal{W}$. Determine a closed neighborhood Q of $\pi(\partial B^{n-k})$ small enough that $\mathcal{W}$ covers Q. Find $\eta > 0$ so that any η-subset of Q is contained in some $W \in \mathcal{W}$. Finally, determine $\delta > 0$ so that the image, under $\tilde{\psi}_k$, of any δ-subset of $S^n/\mathrm{Sp}^k(G)$ has diameter smaller than $\eta/3$.

Given maps $\mu_1, \mu_2 \colon B^2 \to S^n/\mathrm{Sp}^k(G)$, subdivide B^2 with a small mesh triangulation T and approximate the maps by μ_1', μ_2' so that $\mu_i'(T^{(1)}) \cap Z = \varnothing$, diam $\mu_i'(\sigma) < \delta$ for each $\sigma \in T$, and $\mu_i'(\sigma) \cap Z \neq \varnothing$ implies $\mu_i'(\sigma) \subset \tilde{\psi}_k^{-1}(Q)$ ($i = 1, 2$). Then set $C_i = \bigcup\{\sigma \in T \mid \mu_i'(\sigma) \cap Z = \varnothing\}$.

By hypothesis, the two maps $\tilde{\psi}_k\mu_1', \tilde{\psi}_k\mu_2' \colon B^2 \to B^{n-k}/G$ can be approximated by maps whose images have no intersection in common with $\pi(\partial B^{n-k}) = \tilde{\psi}_k(Z)$. Since B^{n-k}/G is an ANR, there exists approximations f_1, f_2 to $\tilde{\psi}_k\mu_1', \tilde{\psi}_k\mu_2'$, respectively, such that for $i = 1, 2$,

$$f_i \mid C_i = \tilde{\psi}_k\mu_i' \mid C_i, \qquad \rho(f_i, \tilde{\psi}_k\mu_i') < \eta/3,$$

$$f_i(\sigma) \subset Q \text{ for every } \sigma \in T \text{ such that } \mu_i'(\sigma) \cap Z \neq \varnothing,$$

$$f_1(B^2) \cap f_2(B^2) \cap \tilde{\psi}_k(Z) = \varnothing.$$

Furthermore, it is an elementary consequence of Lemma 8 that such maps f_i can be obtained so that $f_i(B^2) \cap \tilde{\psi}_k(Z)$ contains no $W \cap \tilde{\psi}_k(Z)$, $W \in \mathcal{W}$. [Here $\mathcal{W}$ should be redefined as a finite cover, every element of which meets $\tilde{\psi}_k(Z)$.]

All that remains to be done is a controlled "lifting" of the maps $f_i \mid \sigma$, similar to that performed in proving Proposition 3. For $i = 1, 2$ set $\tilde{\mu}_i = \mu_i'$ on those simplexes σ of T for which $\sigma \subset C_i$. With the remaining 2-simplexes σ of T, regard $f_i \mid \sigma$ as a homotopy between $f_i \mid \partial\sigma$ and a constant map. The crucial step is to determine this constant map as a mapping into $\pi(\partial B^{n-k}) = \tilde{\psi}_k(Z)$. This can be done because $B^{n-k}/G - \pi(\partial B^{n-k})$ is 0-LC at each point of $\pi(\partial B^{n-k})$; thus, a path in any $W \in \mathcal{W}$ from an arbitrary point to $z \in W \cap \tilde{\psi}_k(Z)$ can be modified to a new path meeting $\tilde{\psi}_k(Z)$ in (at most) its endpoints. It is also significant that, for the simplexes σ in question, diam $f_i(\sigma) < \eta$, so that by the initial choice of numbers some $W_\sigma \in \mathcal{W}$

contains $f_i(\sigma)$. The homotopy of $F_i \mid \partial\sigma$ we choose will operate in W_σ and will end, not at an arbitrary constant map, but in one that sends $\partial\sigma$ to a point of $\tilde{\psi}_k(Z)$ not in the other image of B^2, and the adjustments from the earlier homotopy to this one produce no additional points of intersection with $\tilde{\psi}_k(Z)$ besides this point of constancy.

Now we claim that there exists a map $\tilde{\mu}_i: \sigma \to S^n/\mathrm{Sp}^k(G)$ such that $\tilde{\mu}_i \mid \partial\sigma$ $= \mu_i' \mid \partial\sigma$ and $\tilde{\psi}_k\tilde{\mu}_i(\sigma)$ is so close to $f_i(\sigma)$ that the new image also misses the other image of B^2. Clearly there exists a map $F_i: \partial\sigma \times I \to \pi^{-1}(W)$ so that $\pi F_i(\partial\sigma \times I)$ is close enough to $f_i(\partial\sigma \times I)$ to miss the other image and that $F_i(\partial\sigma \times \{1\}) \subset \partial B^{n-k}$. Exactly as in Proposition 2, there exists a lift $\tilde{F}_i: \partial\sigma \times I \to \psi_k^{-1}\pi^{-1}(W)$ agreeing with $\tilde{\mu}_i'$ on $\partial\sigma \times \{0\}$, now thought of as $\partial\sigma$. Then $\tilde{\mu}_i = \tilde{\pi}\tilde{F}_i$ acts as the desired map on σ.

$$
\begin{array}{ccccc}
\psi_k^{-1}\pi^{-1}(W) & \xrightarrow{\;\tilde{\pi}\;} & \tilde{\psi}_k^{-1}(W) & \xleftarrow{\;\mu_i'\;} & \partial\sigma \\
{\scriptstyle \psi_k}\downarrow & & {\scriptstyle \tilde{\psi}_k}\downarrow & & \downarrow \\
B^{n-k} \supset \pi^{-1}(W) & \xrightarrow{\;\pi\;} & W & \xleftarrow{\;f_i\;} & \partial\sigma \times I
\end{array}
$$

To finish, maps $\tilde{\mu}_i$ must be defined on the appropriate simplexes $\sigma \in T$ successively, constantly measuring disjointness properties at Z, in like fashion. Ultimately, this provides the required maps $\tilde{\mu}_1, \tilde{\mu}_2: B^2 \to S^n/\mathrm{Sp}^k(G)$ such that $\tilde{\mu}_1(B^2) \cap \tilde{\mu}_2(B^2) \cap Z = \varnothing$. ∎

Corollary 9A. *Suppose G is an admissible cell-like decomposition of B^m such that $\mathrm{Sp}^0(G)$ is a shrinkable decomposition of S^m, and suppose $m + k \geq 5$. Then $\mathrm{Sp}^k(G)$ is shrinkable.*

Proof. See Proposition 7 for the requisite disjoint disk property. ∎

Corollary 9B. *If G is the decomposition of B^3 into points and arcs described by the defining sequence suggested in Fig. 28-3 (whose 0-spin is the Bing decomposition given as Example 9.1), then for all $k \geq 2$ $\mathrm{Sp}^k(G)$ is a shrinkable decomposition of S^{3+k}. Consequently, $\tilde{\pi}(\mathrm{Cl}\, N_{\mathrm{Sp}^k(G)})$ is a Cantor set wildly embedded in S^{3+k}.*

Proof. In Section 9 it was shown that $\mathrm{Sp}^0(G)$ is shrinkable and that $\mathrm{Cl}\, N_G$ has nonsimply connected complement in B^3. A global variation to the argument set forth in the first half of Proposition 2 (see Exercise 4) demonstrates that $\mathrm{Cl}\, N_{\mathrm{Sp}^k(G)}$ [as well as $\tilde{\pi}(\mathrm{Cl}\, N_{\mathrm{Sp}^k(G)})$] also has nonsimply connected complement, which implies the wildness of the Cantor set $\tilde{\pi}(\mathrm{Cl}\, N_{\mathrm{Sp}^k(G)})$ in $S^{3+k}/\mathrm{Sp}^k(G) \approx S^{3+k}$. ∎

Remarks. A shrinking of the spun Bing decomposition $\mathrm{Sp}^k(G)$ can be carried out explicitly, even when $k = 1$, by manual techniques like those of Section 9. Lininger [1] first observed this; Edwards [3] depicted with rich

B^3

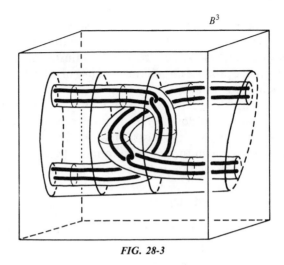

FIG. 28-3

detail how to do it when $k = 1$ in his early work on the double suspension problem. Accordingly, the spun Bing decomposition $\mathrm{Sp}^k(G)$ is shrinkable for all $k \geq 0$, and wild Cantor sets can be found in S^n whenever $n \geq 3$ (earlier results give wild cells of dim $k \geq 1$ in S^n whenever $n \geq 3$).

Theorem 9 is noteworthy for what it does not do. It does not attest that $\mathrm{Sp}^k(G)$ is shrinkable if and only if $\mathrm{Sp}^0(G)$ is shrinkable. By Corollary 9A the latter is sufficient to imply the former (at least for $n \geq 5$); however, it is not necessary, and that fact accounts for some of the technical variations demanded in this section. If one modifies Bing's minimal example (Example 9.6) slightly, as in Example 1 of this section, to obtain an admissible decomposition G_M of B^3, one finds from Theorem 9 or from Corollary 24.3F that $\mathrm{Sp}^k(G_M)$ is shrinkable for all $k \geq 2$. Similarly, if Bing's dogbone decomposition G_D is appropriately described in $B^3 \subset S^3$, then one finds from Theorem 9 that $\mathrm{Sp}^k(G_D)$ is shrinkable ($k \geq 2$); in fact, $\mathrm{Sp}^1(G_D)$ is also known to be shrinkable (Cannon–Daverman [2, Theorem 8.3]). On the other hand, neither $\mathrm{Sp}^0(G_M)$ nor $\mathrm{Sp}^0(G_D)$ is shrinkable.

EXERCISES

1. A compact subset C of an m-manifold-with-boundary M is said to be *cellular-at-the-boundary* if, for each open subset U of M containing C, there exists an embedding $e: B^{m-1} \times [0, 1] \to U$ such that

$$C \cap \partial M \subset e(\mathrm{Int}\, B^{n-1} \times \{0\}) \subset \partial M \qquad \text{and} \qquad C \subset e(\mathrm{Int}\, B^{m-1} \times [0, 1)).$$

Show geometrically, with reference to the cellularity criterion, that if the compact subset C of B^m is cellular-at-the-boundary in B^m, then $\mathrm{Sp}^k(C)$ is cellular in S^{m+k}, for each $k \geq 0$.

2. Prove the analogue of Proposition 6 in case $n = 4$, under the assumption that dim $\pi(\partial B^{n-k}) < 4$.
3. Prove Lemma 8.
4. If G is an admissible usc decomposition of B^{n-k} such that $S^n - N_{\mathrm{Sp}^k(G)}$ is simply connected, prove that $B^{n-k} - N_G$ is simply connected. Do the same when these nondegeneracy sets are replaced by their closures.
5. Let e be an embedding of $X \times I$ in B^m (X a Cantor set) such that $e(X \times \{0\}) = e(X \times I) \cap \partial B^m$ and that $B^m - e(X \times I)$ is 1-LC at each point of $e(X \times (0, 1))$, and let G denote the decomposition of B^m in which $H_G = \{e(\{x\} \times I) \,|\, x \in X\}$. Show that $\mathrm{Sp}^k(G)$ is a shrinkable decomposition of S^{m+k} provided $m \geq 4$ and $m + k \geq 5$.
6. Prove Theorem 9 in case $k = 0$.

29. PRODUCTS OF GENERALIZED MANIFOLDS

For $n \geq 3$ Euclidean $(n + 1)$-space admits exotic factorizations—E^4 is the product of E^1 with Bing's dogbone space; E^{n+1} is the product of E^1 with E^n modulo any arc; indeed, for essentially every known cell-like decomposition G of E^n, E^{n+1} is the product of E^1 with E^n/G.

Examples like these do not exhaust the list, however, because it is possible to express E^n ($n \geq 6$) as the product of two nonmanifolds. Generalizing the Andrews–Curtis result about the product of E^1 with E^n modulo an arc (Theorem 10.7), K. W. Kwun [1] showed that the product of E^m modulo an arc with E^n modulo an arc is always E^{m+n}. Since then, others have found additional nonmanifold decomposition spaces whose product is E^{m+n}. The definitive result was derived by C. D. Bass [2], based, as might be expected, on the cell-like approximation theorem. Bass proved that if G_m is a cell-like decomposition of an m-manifold M^m and G_n is a cell-like decomposition of an n-manifold N^n ($m \geq 3, n \geq 3$) such that both M^m/G_m and N^n/G_n are ANRs, then $G_m \times G_n$ is shrinkable, so $(M^m/G_m) \times (N^n/G_n)$ is homeomorphic to $M^m \times N^n$. The argument, of course, reduces to showing the product has the DDP, and that is the subject of this section.

As notational shortcuts throughout, we write M^m/G_m and N^n/G_n as X and Y, respectively, and we use $P_X: X \times Y \to X$ and $P_Y: X \times Y \to Y$ to denote the projection maps.

Lemma 1. *Suppose A is a nowhere dense, 0-LCC subset of X, $f: B^2 \to X \times Y$, C is a closed subset of B^2 such that $f(C) \cap (A \times Y) = \varnothing$, and $\varepsilon > 0$. Then there exists $F: B^2 \to X \times Y$ such that $\rho(F, f) < \varepsilon$, $F|C = f|C$, $F(B^2) \cap (A \times Y)$ is a 0-dimensional set K, and $P_Y|K$ is 1-1. Furthermore, if Z is a closed subset of $A \times Y$ such that, for every $a \in A$, $P_Y(Z \cap (\{a\} \times Y))$ is nowhere dense in Y, then Z is 1-LCC in $X \times Y$.*

Proof. Since $A \times Y$ is nowhere dense and 0-LCC in $X \times Y$, standard approximation methods yield a minor adjustment to f so that $f^{-1}(A \times Y)$ is a 0-dimensional set Q. The required map F, which, among other features, must send a set like Q onto something 0-dimensional, will be obtained as a limit of a convergent sequence $\{F_i\}$. We shall describe how to obtain the first approximation F_1 to f; the usual iterative details will be left to the reader.

Specifically, the goal is to find a triangulation T of B^2 having small mesh and a map $F_1 \colon B^2 \to X \times Y$ satisfying

(i) $\rho(F_1, f) < \varepsilon$;

(ii) $F_1 | C \cup T^{(1)} = f | C \cup T^{(1)}$;

(iii) for each 2-simplex $\sigma \in T$, $F_1(\sigma) \cap (A \times Y) \subset S_\sigma \times \{y_\sigma\}$, where S_σ is a small subset of X and where the points y_σ corresponding to distinct 2-simplexes are themselves distinct; and

(iv) $F_1(B^2) \cap Z = \varnothing$ (this automatically yields that Z is 1-LCC in $X \times Y$).

Each $q \in Q$ has a neighborhood U_q in $B^2 - C$ such that $f(U_q) \subset V_q \times W_q$, where V_q and W_q are open subsets of X and Y, respectively, W_q is contractible in some connected open subset W_q' of Y, $\operatorname{diam}(V_q \times W_q') < \varepsilon$, and $Z \cap (V_q \times O_q) = \varnothing$, where O_q is another open subset of W_q. [Starting with a neighborhood W_q' of $P_Y f(q)$, choose W_q, V_q, and O_q, and then U_q, in that order.]

Determine a triangulation T of B^2 whose 1-skeleton $T^{(1)}$ misses Q and whose mesh is so small that any $\sigma \in T$ touching Q is associated with some $q \in Q$ for which $U_q \supset \sigma$.

For $\sigma \in T$ such that $\sigma \cap Q = \varnothing$, define $F_1 | \sigma$ as $f | \sigma$. For the other 2-simplexes $\sigma \in T$, we intend to modify $f | \sigma$ slightly, dissecting it into a part with horizontal (X-direction) image and another part with vertical (Y-direction) image. In the situation at hand, where there exists $q \in Q$ such that $U_q \supset \sigma$, we choose a point $w_\sigma \in O_q \subset W_q$, doing this so that the w_σ's associated with different 2-simplexes are distinct, and we name a contraction ψ_t of W_q to w_σ in W_q'. For notational convenience, we also name a collar $c_\sigma = \partial\sigma \times I$ on $\partial\sigma$ in σ, with each $s \in \partial\sigma$ corresponding to $\langle s, 0 \rangle \in \partial\sigma \times I$, and a homeomorphism $h_\sigma \colon \sigma \to \operatorname{Cl}(\sigma - c_\sigma)$ respecting the collar structure, meaning that $h_\sigma(s) = \langle s, 1 \rangle \in c_\sigma$ for each $s \in \partial\sigma$. Then for $\tilde{s} \in \tilde{\sigma} = \operatorname{Cl}(\sigma - c_\sigma)$ we define $F_1(\tilde{s}) = P_X f(h^{-1}(\tilde{s})) \times \{w_q\}$ and for $\langle s, t \rangle \in c_\sigma = \partial\sigma \times I$ we define $F_1(\langle s, t \rangle) = P_X f(\langle s, 0 \rangle) \times \psi_t P_Y f(\langle s, 0 \rangle)$. It should be clear, because $f(\partial\sigma) \cap (A \times Y) = \varnothing$, that $F_1(c_\sigma) \cap (A \times Y) = \varnothing$ as well and that $F_1(\sigma) \cap (A \times Y) = F_1(\tilde{\sigma}) \cap (A \times Y)$ is contained in the product of $V_q \cap A$ with $\{w_\sigma\}$, which gives $F_1(B^2) \cap Z = \varnothing$, as required.

Lemma 2. *The space $X \times Y$ has the disjoint arc–disk property.*

Proof. Since each of X and Y has the disjoint 1-cells property, any map $\mu: B^1 \to X \times Y$ leads to $P_X\mu: B^1 \to X$ and $P_Y\mu: B^1 \to Y$ that can be approximated by embeddings $e_X: B^1 \to X$ and $e_Y: B^1 \to Y$. Set $A = e_X(B^1)$ and $Z = e(B^1)$, where $e(b) = \langle e_X(b), e_Y(b) \rangle$. For any $a \in A$, $P_Y(Z \cap (\{a\} \times Y))$ is a singleton, which certainly is nowhere dense in Y. By Lemma 1, $e(B^1)$ is 1-LCC. According to Proposition 26.4, $X \times Y$ has the DADP.

Theorem 3. *Suppose G_m is a cell-like decompositon of an m-manifold M^m ($m \geq 3$) and G_n is a cell-like decomposition of an n-manifold N^n ($n \geq 3$) such that the spaces $X = M^m/G_m$ and $Y = N^n/G_n$ are ANRs. Then $G_m \times G_n$ is a shrinkable decomposition of $M^m \times N^n$.*

Proof. It suffices to show that $X \times Y \approx (M^m \times N^n)/(G_m \times G_n)$ has the DDP. Toward that end, let $\alpha_0, \beta_0: B^2 \to X \times Y$. Choose a triangulation T of B^2 so that the diameters of both $\alpha_0(\sigma)$ and $\beta_0(\sigma)$ are small for every $\sigma \in T$.

Step 1. By Lemma 2 the maps α_0, β_0 can be approximated by maps α_1, β_1, respectively, such that $\alpha_1(T^{(1)}) \cap \beta_1(B^2) = \varnothing$. Symmetrically, the latter can be approximated by maps α_2, β_2 for which $\alpha_2(B^2) \cap \beta_2(T^{(1)}) = \varnothing$, but with controls limiting the motion to maintain the original disjointness property; that is,

$$\alpha_2(T^{(1)}) \cap \beta_2(B^2) = \varnothing = \alpha_2(B^2) \cap \beta_2(T^{(1)}).$$

Step 2. Since X has the disjoint 1-cells property, the maps α_2, β_2 have approximations α_3, β_3 such that $P_X\alpha_3 \,|\, T^{(1)}$ and $P_X\beta_3 \,|\, T^{(1)}$ are embeddings with disjoint images. These can be obtained subject to limitations ensuring again that

$$\alpha_3(T^{(1)}) \cap \beta_3(B^2) = \varnothing = \alpha_3(B^2) \cap \beta_3(T^{(1)}).$$

Step 3. Exactly as in the proof of Lemma 1, for each 2-simplex $\sigma \in T$ we can produce a collar $c_\sigma = \partial\sigma \times I$ on $\partial\sigma$ in σ (with each $s \in \partial\sigma$ corresponding to $\langle s, 0 \rangle \in c_\sigma = \partial\sigma \times I$) and also a 2-cell $\tilde{\sigma} = \mathrm{Cl}(\sigma - c_\sigma)$. Just as was done in that argument, we find maps α_4, β_4 close to α_3, β_3 (the degree of closeness dependent on the mesh of T) satisfying

(i) $\alpha_4 \,|\, T^{(1)} = \alpha_3 \,|\, T^{(1)}$ and $\beta_4 \,|\, T^{(1)} = \beta_3 \,|\, T^{(1)}$;
(ii) $\alpha_4(c_\sigma)$ is contained in a small subset of $P_X\alpha_3(\partial\sigma) \times Y$, while $\beta_4(c_\sigma)$ is contained in a small subset of $P_X\beta_3(\partial\sigma) \times Y$;
(iii) $\alpha_4(\tilde{\sigma}) = P_X\alpha_3(\sigma) \times \{w_{\sigma,\alpha}\}$ and $\beta_4(\tilde{\sigma}) = P_X\beta_3(\sigma) \times \{w_{\sigma,\beta}\}$; and
(iv) the sets $\{w_{\sigma,\alpha} \,|\, \text{2-simplexes } \sigma \in T\}$ and $\{w_{\sigma,\beta} \,|\, \text{2-simplexes } \sigma \in T\}$ are disjoint.

We shall refer to the sets $\alpha_4(c_\sigma)$, $\beta_4(c_\sigma)$ as walls and to the sets $\alpha_4(\tilde{\sigma})$, $\beta_4(\tilde{\sigma})$ as floors. It follows from step 2 and condition (ii) that no wall of $\alpha_4(B^2)$ meets a wall of $\beta_4(B^2)$, and it follows from conditions (iii) and (iv) above that no floor of $\alpha_4(B^2)$ meets a floor of $\beta_4(B^2)$. Thus, the only possible intersections occur where a wall of one meets a floor of the other.

Step 4. To improve the floors of $\alpha_4(B^2)$, we apply Lemma 1 and obtain an approximation α_5 to α_4 such that $\alpha_5|\bigcup c_\sigma = \alpha_4|\bigcup c_\sigma$ and $\alpha_5(B^2) \cap (P_X\beta_4(T^{(1)}) \times Y)$ is a compact 0-dimensional set K for which $P_Y|K$ is 1-1. The map α_5 can be obtained so close to α_4 that $\alpha_5(\bigcup \tilde\sigma)$ still misses the floors of $\beta_4(B^2)$. Since $P_X|\alpha_3(T^{(1)})$ and $P_X|\beta_3(T^{(1)})$ were made disjoint in Step 2, it should be clear that $K \cap (P_X\alpha_5(T^{(1)}) \times Y) = \varnothing$.

Step 5. Here we renovate the walls of $\beta_4(B^2)$. Let K_σ denote $K \cap (P_X\beta_4(\partial\sigma) \times Y)$, for each 2-simplex $\sigma \in T$. Then $\partial\sigma \times Y \approx S^1 \times Y$ is very nearly an $(n + 1)$-manifold. What sets $X \times Y$ apart from $X \times E^1$ (when $n = \dim Y > 1$) is that K_σ is 1-LCC in $P_X\beta_4(\partial\sigma) \times Y$. This fact, which does not hold in case $Y = E^1$, follows directly from the Seifert–van Kampen argument given in Proposition 26.7, based on the property that $P_Y|K_\sigma$ is 1-1. Previous modifications have determined β_4 so that $\beta_4(\partial c_\sigma) \cap \alpha_5(B^2) = \beta_4(\partial c_\sigma) \cap K_\sigma = \phi$. Consequently, because of the 1-LCC condition on K_σ, the various $\beta_4|c_\sigma$ can be approximated by maps $\beta_5|c_\sigma$ satisfying

$$\beta_5|\partial c_\sigma = \beta_4|\partial c_\sigma \quad \text{and} \quad \beta_5(c_\sigma) \subset (P_X\beta_4(\partial\sigma) \times Y) - K_\sigma.$$

Such maps combine to produce $\beta_5\colon B^2 \to X \times Y$ agreeing with β_4 on $\bigcup \tilde\sigma$. The walls of $\beta_5(B^2)$ now avoid $\alpha_5(B^2)$ completely.

Steps 6 and 7. Repeat steps 4 and 5 for the symmetric situation to improve, first, the floors of $\beta_5(B^2)$ and, then, the walls of $\alpha_5(B^2)$. Although α_5 no longer has floors in the same horizontal sense as α_4, its walls agree with those of α_4, and the modification of step 7 pertains to those walls. On the other hand, $\beta_5(B^2)$ has the same floors as $\beta_4(B^2)$ and it has walls quite like those of $\beta_4(B^2)$; nevertheless, the final modification (step 6) of $\beta_5(B^2)$ involves the common floors. After completing all seven steps, one will have eliminated all intersections between $\alpha_6(B^2)$ and $\beta_6(B^2)$.

This establishes that $X \times Y$ has the DDP. ∎

Corollary 3A. *If G_1, G_2, and G_3 are cell-like decompositions of positive-dimensional manifolds such that the three decomposition spaces are ANRs, then $G_1 \times G_2 \times G_3$ is shrinkable.*

EXERCISES

1. Let G_1 and G_2 denote admissible cell-like decompositions of B^m ($m \geq 1$) and B^n ($n \geq 1$), respectively, such that B^m/G_1 and B^n/G_2 are ANRs. Prove that $G_1 \times G_2$ is an admissible decomposition of $B^m \times B^n$. Then prove that $\mathrm{Sp}^k(G_1 \times G_2)$ is shrinkable whenever $m + n \geq 4$ and $k \geq 1$.
2. For $n \geq 3$ show that there exists an uncountable collection $\{X_\alpha\}$ of topologically distinct spaces such that each product $X_\alpha \times X_\alpha$ is homeomorphic to E^{2n}.

30. A MISMATCH PROPERTY IN LOCALLY SPHERICAL DECOMPOSITION SPACES

In order to apply the cell-like approximation theorem to a particular cell-like decomposition G of an n-manifold M, two problems must be addressed: the first and often harder is to detect the finite-dimensionality of M/G, and the second is to verify that M/G satisfies the DDP. In this section we investigate a property that immediately dispels the first problem. When G is a (cell-like) usc decomposition of an n-manifold M, we say that M/G is *locally spherical* if each $x \in M/G$ has arbitrarily small neighborhoods U_x whose frontiers Fr U_x are $(n - 1)$-spheres. Under such circumstances, when M/G is spherical, it clearly is a finite-dimensional space; whether it satisfies the DDP when $n \geq 5$ and G is cell-like remains unsolved. However, evidence from lower dimensions suggests that it might not, for S. Armentrout [5] has shown the decomposition space associated with his decomposition [7] of E^3 into points and straight line segments to be locally spherical despite being a nonmanifold.

The initial result pertaining to locally spherical, cell-like decompositions G of n-manifolds M $(n \geq 5)$ preceded the establishment of the cell-like approximation theorem. In 1973 J. W. Cannon [1, Theorem 62] showed that G is shrinkable if the $(n - 1)$-spheres Fr U_x promised by the sphericality hypothesis are 1-LCC embedded in M/G. Later, with the aid of Edwards's theorem, Daverman [7] obtained the same conclusion in case a one-sided 1-LCC property holds—if the neighborhoods U_x are 1-LC at points of Fr U_x.

Here such strict 1-LCC conditions about sides of the $(n - 1)$-spheres Fr U_x are exchanged for a more relaxed property. An $(n - 1)$-sphere S in the decomposition space M/G (associated, as usual, with a cell-like decomposition of a connected n-manifold M) is said to satisfy the *homotopy mismatch property* (HMP) if S contains disjoint subsets Q_1 and Q_2 such that each map $\mu_i: B^2 \to \mathrm{Cl}\, U_i$ can be approximated, arbitrarily closely, by a map $\mu_i': B^2 \to U_i \cup Q_i$ $(i = 1, 2)$, where U_1 and U_2 denote the components of $M/G - S$. The principal result to be proved is the following:

Theorem 1. *Suppose G is a cell-like decomposition of an n-manifold M, $n \geq 5$, such that each $x \in M/G$ has arbitrarily small neighborhoods U_x whose frontiers* Fr U_x *are $(n - 1)$-spheres satisfying the homotopy mismatch property. Then G is shrinkable.*

A corollary subsumes both of the earlier theorems by Cannon and Daverman about locally spherical decompositions. It serves as a significant extension of Theorem 8.9 and the result of Exercise 8.6 as well.

Corollary 1A. *Suppose G is a cell-like decomposition of an n-manifold M, $n \geq 5$, such that each $x \in M/G$ has arbitrarily small neighborhoods U_x whose*

*frontiers are $(n - 1)$-spheres and for which either U_x or $M/G - \text{Cl }U_x$ is
1-LC at every point of* Fr U_x. *Then G is shrinkable.*

Work of E. P. Woodruff [1,2] concerning decompositions of E^3 lends
motivation for Theorem 1. Bred in the same spirit as Theorem 8.9, her
emphasis was placed on the geometry surrounding the decomposition itself
rather than on the decomposition space. The operant hypothesis in
(Woodruff [1]) required each nondegenerate element to possess arbitrarily
small neighborhoods in E^3 whose frontiers were 2-spheres, possibly wildly
embedded in E^3, but missing all the nondegenerate elements. The proof of
Theorem 1 establishes a high-dimensional analog of her result.

Corollary 1B. *Suppose G is a cell-like decomposition of an n-manifold,
$n \geq 5$, such that each $g \in H_G$ has arbitrarily small neighborhoods V_g in M
whose frontiers are $(n - 1)$-spheres that miss N_G and satisfy the homotopy
mismatch property in M. Then G is shrinkable.*

In the 3-dimensional situation studied by Woodruff, Eaton's mismatch
theorem [1] reveals that each 2-sphere in E^3 satisfies the HMP, but in higher
dimensions $(n - 1)$-spheres in E^n do not necessarily satisfy it. That accounts,
in a way, for the appearance of HMP in Corollary 1B, even though it does
not appear in the statement of Woodruff's result. Whether Corollary 1B is
valid without the hypothesis about frontiers satisfying HMP has endured as
an open question.

The central features about the proof of Theorem 1 emerge in the following
result.

Lemma 2. *Suppose G is a cell-like decomposition of an n-manifold M,
$n \geq 5$, such that M/G is locally spherical and the $(n - 1)$-spheres* Fr U *satisfy
the HMP; f_1 and f_2, maps of B^2 to M/G; C, a closed subset of M/G such
that $C_0 = C \cap f_1(B^2) \cap f_2(B^2)$ is 0-dimensional; and W, an open subset of
M/G containing C_0. Then there exist maps $F_1, F_2: B^2 \to M/G$ satisfying*

(a) $F_1(B^2) \cap F_2(B^2) \cap C = \varnothing$,
(b) $F_i f_i^{-1}(W) \subset W$, *and*
(c) $F_i | B^2 - f_i^{-1}(W) = f_i | B^2 - f_i^{-1}(W)$ $(i = 1, 2)$.

Proof. We break the proof into several steps.

Step 1. Finiteness considerations. By hypothesis each point $c \in C_0$ has
an open neighborhood U_c whose closure is in W and whose frontier Fr U_c
is an $(n - 1)$-sphere satisfying the HMP. Since C_0 is compact we can extract
from the open cover $\{U_c \mid c \in C_0\}$ a finite subcover $\{U_j \mid 1 \leq j \leq r\}$ and trim
the latter to another cover $\{V_j \mid 1 \leq j \leq r\}$ of C_0 by open sets in M/G such
that $V_j \subset \text{Cl }V_j \subset U_j$.

Step 2.   A reduction.   Given maps $f_1, f_2 \colon B^2 \to M/G$ for which

$$f_1(B^2) \cap f_2(B^2) \cap C \subset \bigcup_{j=k}^{r} V_j,$$

we shall describe maps $F_1, F_2 \colon B^2 \to M/G$ such that

(a') $F_1(B^2) \cap F_2(B^2) \cap C \subset \displaystyle\bigcup_{j=k+1}^{r} V_j,$

(b') $F_i f_i^{-1}(W) \subset W$; and

(c') $F_i \mid B^2 - f_i^{-1}(W) = f_i \mid B^2 - f_i^{-1}(W)$ $(i = 1, 2).$

Repeated application (r times) of this reduced version will establish Lemma 2.

Step 3.   Eradication of $f_i(B^2)$ from U_k.   Define $Z_i = f_i^{-1}(\mathrm{Cl}\ U_k)$ and $Y_i = f_i^{-1}(\mathrm{Fr}\ U_k)$ $(i = 1, 2)$. The map $f_i \mid Y_i \colon Y_i \to \mathrm{Fr}\ U_k$ extends to a map $m_i \colon Z_i \to \mathrm{Fr}\ U_k$ $(i = 1, 2)$, since $\mathrm{Fr}\ U_k$ is a simply connected ANR. See Fig. 30-1.

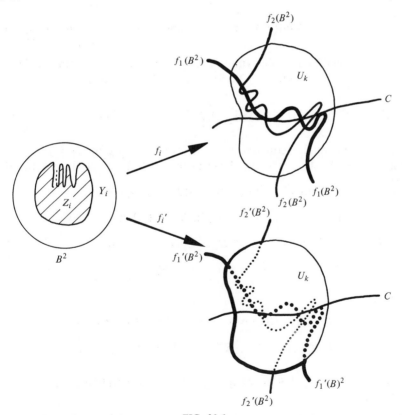

FIG. 30-1

Step 4. General position improvements. To circumvent special difficulties when $n = 5$, we consider only the case $n \geq 6$, leaving the extra case
to the reader. Because $\dim \mathrm{Fr}\, U_k \geq 5$, m_1 and m_2 admit general position
modifications, affecting no points of Y_1 or Y_2, so that $m_1(Z_1 - Y_1) \cap
m_2(Z_2 - Y_2) = \varnothing$. Define maps $f_i\colon B^2 \to M/G$ as m_i on Z_i and as f_i
elsewhere $(i = 1, 2)$. Then any $c \in f_1'(B^2) \cap f_2'(B^2) \cap C$ belongs either to

$$f_1(B^2) \cap f_2(B^2) - V_k \subset f_1(B^2) \cap f_2(B^2) \cap \left(\bigcup_{j=k+1}^{r} V_j \right)$$

or

$$C \cap ([f_1(Y_1) \cap m_2(Z_2 - Y_2)] \cup [m_1(Z_1 - Y_1) \cap f_2(Y_2)]).$$

Only the points of the second kind cause any further concern.

Step 5. Final improvements to m_i. Define sets

$$X_1 = Z_1 \cap m_1^{-1}(C \cap m_1(Z_1) \cap f_2(Y_2)) \quad \text{and} \quad X_1^* = X_1 - m_1^{-1}\left(\bigcup_{j=k+1}^{r} V_j \right)$$

and define sets X_2 and X_2^* symmetrically. Then X_i^* is a compact subset of
X_i and $X_i^* \subset Z_i - Y_i$ $(i = 1, 2)$, for if $x \in X_1^* \cap Y_1$, then

$$m_1(x) \in [C \cap m_1(Y_1) \cap f_2(Y_2)] - \bigcup_{j=k}^{r} V_j \subset [C \cap f_1(Y_1) \cap f_2(Y_2)] - \bigcup_{j=k}^{r} V_j,$$

which is empty by the reductive hypothesis of step 2. Define $T_1 = f_1^{-1} m_2(X_2^*)$
and $T_2 = f_2^{-1} m_1(X_1^*)$. It should be clear that $T_i \subset Y_i$ $(i = 1, 2)$ and thus that

$$m_1(X_1^*) \cap m_2(X_2^*) = f_1(T_1) \cap f_2(T_2) = \varnothing.$$

Next determine open sets N_i and O_i in B^2 such that

$$T_i \subset N_i \subset f_i^{-1}(W), \quad X_i^* \subset O_i \subset \mathrm{Cl}\, O_i \subset Z_i - Y_i,$$

$$N_i \cap O_i = \varnothing \qquad (i = 1, 2)$$

and

$$f_1'(\mathrm{Cl}\, N_1) \cap f_2'(\mathrm{Cl}\, N_2) = \varnothing = f_1'(\mathrm{Cl}\, O_1) \cap f_2'(\mathrm{Cl}\, O_2).$$

Since $C \cap f_1'(B^2) \cap f_2'(B^2) \subset f_1'(X_1^*) \cup f_2'(X_2^*) \cup (\bigcup_{j=k+1}^{r} V_j)$, it follows that

$$C \cap [f_1'(B^2) \cap f_2'(B^2 - O_2)] \cup [f_1'(B^2 - O_1) \cap f_2'(B^2)] \subset \bigcup_{j=k+1}^{r} V_j.$$

Let Q_1 and Q_2 denote disjoint sets in $\mathrm{Fr}\, U_k$, promised by the hypothesis that

Fr U_k satisfies the HMP, such that $U_k \cup Q_1$ and $(M/G - \text{Cl } U_k) \cup Q_2$ are 1-LC at points of Fr U_k. Then one can produce maps F_i': $\text{Cl}(N_i \cup O_i) \rightarrow M/G$ such that

$$F_i'(\text{Cl } N_i) \subset (W - \text{Cl } U_k) \cup Q_2,$$
$$F_i'(\text{Cl } O_i) \subset U_k \cup Q_1,$$

and

$$F_i' \mid \text{Fr } N_i \cup \text{Fr } O_i = f_i' \mid \text{Fr } N_i \cup \text{Fr } O_i \qquad (i = 1, 2)$$

with controls on the closeness of F_i' to f_i' ($i = 1, 2$) so that

$$F_1'(\text{Cl } O_1) \cap F_2'(\text{Cl } O_2) = \varnothing$$

and

$$[F_1'(\text{Cl } N_1) \cap f_2'(B^2 - O_2)] \cup [f_1'(B^2 - O_1) \cap F_2'(\text{Cl } N_2)] \subset \bigcup_{j=k+1}^{r} V_j.$$

(See Fig. 30-2.) Consequently, the maps F_i defined as F_i' on $N_i \cup O_i$ and as f_i' elsewhere fulfill the requirements for the reduction given in Step 2. ∎

Lemma 3. *Suppose G is a cell-like decomposition of an n-manifold M, $n \geq 5$, such that M/G is locally spherical and the $(n - 1)$-spheres Fr U satisfy the HMP; f_1 and f_2 maps of B^2 to M/G; $C \subset M/G$ a closed q-dimensional subset of M/G; W an open subset of M/G containing $f_1(B^2) \cap f_2(B^2) \cap C$; and $\varepsilon > 0$. Then there exists maps F_1, F_2: $B^2 \rightarrow M/G$ satisfying*

(a) $F_1(B^2) \cap F_2(B^2) \cap C = \varnothing$,
(b) $\rho(F_i, f_i) < \varepsilon$, *and*
(c) $F_i \mid B^2 - f_i^{-1}(W) = f_i \mid B^2 - f_i^{-1}(W)$ ($i = 1, 2$). ∎

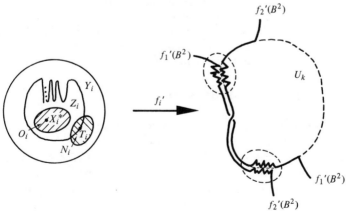

FIG. 30-2

Proof. The argument proceeds by induction on q, and Lemma 2 establishes the initial case $q = 0$. Assume that Lemma 3 holds for all $(q - 1)$-dimensional closed subsets of M/G. In the q-dimensional set C find closed subsets $A_1, A_2, \ldots$ of dimension $\leq q - 1$ such that $C - \bigcup A_j$ is 0-dimensional. Apply the inductive hypothesis repeatedly, imposing controls with an eye toward the limit, to obtain (limit) maps $f_i^*: B^2 \to M/G$ such that $f_1^*(B^2) \cap f_2^*(B^2) \cap (\bigcup A_j) = \varnothing$, $\rho(f_i^*, f_i) < \varepsilon/2$, and

$$f_i^* \mid B^2 - f_i^{-1}(W) = f_i \mid B^2 - f_i^{-1}(W) \qquad (i = 1, 2).$$

Then $C \cap F_1^*(B^2) \cap f_2^*(B^2)$ is 0-dimensional, since it is a subset of $C - \bigcup A_j$. Now apply Lemma 1 again, this time with open set W^* in W containing $C \cap f_1^*(B^2) \cap f_2^*(B^2)$ and having no component of diameter $\geq \varepsilon/2$, to obtain the required maps F_1 and F_2. ∎

When $q = n$ and $C = M/G$, Lemma 3 reveals that M/G satisfies the DDP. Theorem 1 follows from the cell-like approximation theorem.

Stripping the argument to its essentials, one discovers that it proves the following result.

Theorem 4. *Suppose G is a cell-like usc decomposition of an n-manifold, $n \geq 5$, such that each point x of $\pi(N_G)$ has arbitrarily small neighborhoods U_x such that* Fr U_x *is a simply connected $(n - 1)$-manifold or a simply connected ANR with the DDP and* Fr U_x *satisfies the HMP. Then G is shrinkable.*

Corollary 1B, mentioned near the beginning of this section, can be derived immediately from Theorem 4.

EXERCISES

1. Redo Step 4 of Lemma 2 to take care of the case $n = 5$.
2. Suppose G is a cell-like decomposition of an n-manifold M, $n \geq 5$, such that M/G has a closed subset S for which $(M/G) - S$ is an n-manifold and each point $x \in S$ has arbitrarily small neighborhoods U_x whose frontiers Fr U_x are $(n - 1)$-spheres, with $S \cap$ Fr U_x 0-LCC and 1-LCC embedded in Fr U_x. Show that G is shrinkable.

31. SLICED DECOMPOSITIONS OF E^{n+1}

A decomposition G of $X \times E^1$ (or of $X \times S^1$) is said to be *sliced* if each decomposition element g of G is contained in some slice $X \times \{s\}$, where $s \in E^1$ (or $s \in S^1$). Product decompositions $G = G^* \times E^1$ of $X \times E^1$ arising from a decomposition G^* of X serve as a natural and important class of sliced decompositions. Furthermore, at least for $n \geq 4$, there are powerful results, described in Section 26, concerning the shrinkability of cell-like product decompositions of $E^n \times E^1$. This section develops a connection, based on relatively elementary techniques, between the shrinkability of such product decompositions and that of sliced decompositions. The main result shows that a sliced decomposition of E^{n+1} is shrinkable if, for every decomposition G^s of E^n induced by G from a slice $E^n \times \{s\}$, $s \in E^1$, the associated product decomposition $G^s \times E^1$ of E^{n+1} is shrinkable. Most of what is described here was done in conjunction with D. K. Preston (Daverman–Preston [1]).

With any sliced decomposition G of $X \times S^1$ (or of $X \times E^1$), for fixed $s \in S^1$ one can distinguish two new decompositions associated with the slice $X \times \{s\}$. The first is the decomposition G^s of X given by

$$G^s = \{g \subset X \mid g \times \{s\} \in G\},$$

and the second is the decomposition $G(s)$ of $X \times S^1$ consisting of all those $g \in G$ contained in the slice $X \times \{s\}$ together with the singletons from $X \times (S^1 - \{s\})$. The latter coincides with the trivial extension of $G^s \times \{s\}$ on $X \times \{s\}$ to all of $X \times S^1$.

Lemma 1. *Suppose that G is a sliced cell-like decomposition of $S^n \times S^1$ and that C is a compact 0-dimensional subset of S^1 satisfying*

(a) *N_G is contained in $S^n \times C$, and*
(b) *for each $c \in C$ the decomposition $G(c)$ is shrinkable.*

Then G itself is shrinkable.

Proof. Fix a metric ρ on $(S^n \times S^1)/G$, and determine $\delta \in (0, \varepsilon/3)$ so that if $F: S^n \times S^1 \to S^n \times S^1$ is a homeomorphism moving points less than δ, then $\rho(\pi, \pi F) < \varepsilon/3$.

According to hypothesis (b) above and Theorem 13.1, for each $c \in C$ the decomposition $G(c)$ is strongly shrinkable. Consequently, there exists a homeomorphism h_c of $S^n \times S^1$ to itself shrinking each $g \in G(c)$ to diameter less than $\varepsilon/3$, satisfying $\rho(\pi, \pi h_c) < \varepsilon/3$, and fixing points outside $S^n \times U_c$, where U_c denotes the δ-neighborhood of c in S^1.

Corresponding to each $c \in C$ there is an open interval J_c such that $c \in J_c \subset U_c$, $\mathrm{Bd}\, J_c \cap C = \varnothing$, and, for each $g \in G \cap (S^n \times J_c)$, the diameter of $h_c(g)$ is less than $\varepsilon/3$. From the open cover $\{J_c \mid c \in C\}$ we extract a finite subcover $\{J_{c(i)} \mid i = 1, \ldots, N\}$, and we cut back these intervals slightly so that the collection consists of pairwise disjoint intervals.

For $i = 1, \ldots, N$ we name a homeomorphism f_i of $\mathrm{Cl}\, U_{c(i)}$ onto $\mathrm{Cl}\, J_{c(i)}$ that keeps $J_{c(i)} \cap C$ pointwise fixed and then name the product homeomorphism

$$F_i = \mathrm{Id} \times f_i: S^n \times S^1 \to S^n \times S^1.$$

Finally, we produce the required shrinking homeomorphism h as the one equal to $F_i h_{c(i)} F_i^{-1}$ on $S^n \times J_{c(i)}$ ($i = 1, \ldots, N$) and equal to the identity elsewhere. ∎

Lemma 2. *Let G be a decomposition of S^n such that $G \times S^1$ is a shrinkable decomposition of $S^n \times S^1$, C a compact subset of S^1, and θ a map of $S^n \times S^1$ to itself realizing the decomposition $G \times C$, trivially extended over $S^n \times S^1$. Then for each point c of C, $\theta(S^n \times \{c\})$ is bicollared in $S^n \times S^1$.*

Proof. Let π denote the decomposition map $S^n \times S^1 \to (S^n \times S^1)/(G \times S^1)$. It follows from Corollary 13.2B and Theorem 13.1 that the decomposition $\theta(G \times S^1)$ is strongly shrinkable. Thus, there exists a map θ' of $S^n \times S^1$ to itself fixing $\theta(S^n \times C)$ and realizing $\theta(G \times S^1)$. This yields

$$
\begin{array}{ccc}
S^n \times S^1 & \xrightarrow{\ \pi\ } & (S^n \times S^1)/(G \times S^1) = (S^n/G) \times S^1 \\
\Big\downarrow{\scriptstyle \theta'\theta} & \swarrow{\scriptstyle \theta'\theta\pi^{-1}} & \\
S^n \times S^1 & &
\end{array}
$$

with $\theta'\theta\pi^{-1}$ being a homeomorphism. Clearly $(S^n/G) \times \{c\}$ is bicollared in $(S^n/G) \times S^1$. As a result, the homeomorphism above transports $(S^n/G) \times \{c\}$ to the bicollared set

$$
\theta'\theta\pi^{-1}((S^n/G) \times \{c\}) = \theta'\theta(S^n \times \{c\})
$$

$$
= \theta(S^n \times \{c\}). \quad \blacksquare
$$

Lemma 3. *Let G be a shrinkable sliced decomposition of $S^n \times S^1$, θ a map of $S^n \times S^1$ to itself realizing the decomposition G, and s a point of S^1 such that $G^s \times S^1$ is a shrinkable decomposition of $S^n \times S^1$. Then $\theta(S^n \times \{s\})$ is bicollared in $S^n \times S^1$.*

Proof. Because $G^s \times S^1$ is shrinkable, the decomposition $G(s)$ of $S^n \times S^1$ must be shrinkable as well (Theorem 13.2). Hence, there exists a map f of $S^n \times S^1$ to itself realizing the decomposition $G(s)$. As in the proof of Lemma 2, the decomposition $f(G)$ is strongly shrinkable, and thus there exists a map f' of $S^n \times S^1$ to itself fixing $f(S^n \times \{s\})$ and realizing $f(G)$. At this point we have a diagram

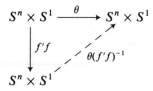

with the induced function $\theta(f'f)^{-1}$ a homeomorphism. By Lemma 2, the set $f'f(S^n \times \{s\}) = f(S^n \times \{s\})$ is bicollared in $S^n \times S^1$, and then its image $\theta(S^n \times \{s\})$ under $\theta(f'f)^{-1}$ must also be bicollared. $\quad \blacksquare$

Lemma 4. *Suppose that for $j = 1, 2, \ldots$ there exists an embedding h_j of a compact space $X_j \times [-1, 1]$ in $S^n \times S^1$ satisfying*

(a)   *$h_j(X_j \times \{\pm 1\})$ is bicollared,*

(b)   *$h_j(X_j \times [-1, 1]) \cap h_k(X_k \times [-1, 1]) = \varnothing$ whenever $j \neq k$, and*

(c)   *The maximal diameter λ_j of a fiber arc $h_j(p \times [-1, 1])$, $p \in X_j$, approaches 0 as $j \to \infty$.*

Then the decomposition G of $S^n \times S^1$ into points and the arcs $h_j(p \times [-1, 1])$, where $p \in X_j$ and $j = 1, 2, \ldots$, is usc and shrinkable.

    Proof. Exercise. (*Hint:* See the proof of Theorem 8.6.)

Theorem 5. *Suppose G is a sliced cell-like decomposition of $S^n \times S^1$ satisfying*

(a)   *for each $s \in S^1$ the decomposition $G(s)$ of $S^n \times S^1$ is shrinkable, and*

(b)   *S^1 contains a countable dense set $D = \{d(i)\}$ for which the decompositions $G^{d(i)}$ of S^n yield an $(n + 1)$-manifold factor (that is, $G^{d(i)} \times E^1$ is shrinkable).*

Then G itself is shrinkable.

    Proof. *Step 1. Stretching the original decomposition.* By reproducing the model of a monotone decomposition of S^1 with nondegenerate elements dense in S^1, we construct a cell-like map f of S^1 to itself such that $f^{-1}(d(i))$ is an interval for each $d(i) \in D$ and that otherwise $f^{-1}(s)$ is a point. (Recall that any two countable dense subsets of S^1 are equivalently embedded there.) We name the product map $F = \text{Id} \times f$ of $S^n \times S^1$ to itself and consider the induced decomposition

$$G_F = \{F^{-1}(g) \mid g \in G\},$$

which closely resembles G except that elements from the D-levels have been stretched out along the vertical, or the S^1, direction.

    Our intention is to prove that G_F is shrinkable. Before proceeding with that, we point out how to establish the shrinkability of G, assuming the shrinkability of G_F, based on the diagram below:

$$
\begin{array}{ccc}
S^n \times S^1 & \xrightarrow{\ \ F\ \ } & S^n \times S^1 \\
\Big\downarrow{\scriptstyle \pi_F} & \raisebox{1ex}{$\scriptstyle F^* = \pi_F F^{-1}$}\nearrow & \Big\downarrow{\scriptstyle \pi} \\
(S^n \times S^1)/G_F & \dashrightarrow{H} & (S^n \times S^1)/G
\end{array}
$$

The natural function $H = \pi F(\pi_F)^{-1}$ clearly is a homeomorphism. Both π_F and F clearly are approximable by homeomorphisms. As relatively easy consequences, one can show, in order, that $F^* = \pi_F F^{-1}$ and $\pi = HF^*$ are approximable by homeomorphisms as well.

Step 2. Slicing the stretched decomposition. There are two related decompositions instrumental to the shrinking of G_F. The first of these is the sliced decomposition, a kind of level stratification, defined as

$$G_1 = \{g_F \cap (S^n \times \{s\}) \mid g_F \in G_F \text{ and } s \in S^1\}.$$

Step 3. Opening the sliced decomposition. The second related decomposition is a fenestration of G_1 obtained by eliminating nondegenerate elements from a certain dense and open subset of levels. Specifically, for the union N_f of the nondegenerate elements of $\{f^{-1}(s) \mid s \in S^1\}$, G_2 consists of the singletons from $S^n \times \text{Int } N_f$ together with

$$\{g_F \cap (S^n \times \{s\}) \mid g_F \in G_F \text{ and } s \in \text{Cl}(S^1 - N_f)\}.$$

In other words, G_2 has for its nondegenerate elements those (nondegenerate) elements of G_1 from levels not interior to any nondegenerate element of f.

Step 4. Shrinking G_F. According to Lemma 1, the fenestrated decomposition G_2 is shrinkable. Hence, there exists a map θ_2 of $S^n \times S^1$ to itself that realizes this decomposition, in the sense that

$$G_2 = \{\theta_2^{-1}(x) \mid x \in S^n \times S^1\},$$

and where θ_2 is the end of a pseudo-isotopy ψ_t^2 defined on $S^n \times S^1$ such that $\rho(\pi_F, \pi_F \psi_t^2) < \varepsilon/3$ (see Theorem 13.3).

Naturally, we next look at the modified decomposition $\theta_2(G_1)$, whose nondegenerate elements are partitioned into countably many (curvilinear) product decompositions. Explicitly, for each inverse $A_i = f^{-1}(d(i))$ in S^1, $\theta_2(G_1 \cap (S^n \times \text{Int } A_i))$ is topologically equivalent with $G^{d(i)} \times E^1$, which is shrinkable by hypothesis. In particular, $\theta_2(G_1 \cap (S^n \times A_i))$, which corresponds to the image under θ_2 of that decomposition induced from $G^{d(i)} \times S^1$ over $(S^n/G^{d(i)}) \times A_i$, can be shrunk to arbitrarily small size by means of a homeomorphism fixed outside $\theta_2(S^n \times A_i)$. Since no more than a finite number of indices i give rise to a decomposition element of $\theta_2(G_1 \cap (S^n \times A_i))$ having diameter larger than any preassigned positive number, we can assemble a finite number of such homeomorphisms to show that the decomposition $\theta_2(G_1)$ is shrinkable, and again we can produce a map θ_1 of $S^n \times S^1$ to itself that realizes $\theta_2(G_1)$, where θ_1 is the end of a pseudo-isotopy ψ_t^1 such that $\rho(\pi, \pi \psi_t^1) < \varepsilon/3$.

Finally, we turn to the resultant decompositions $\theta_1 \theta_2(G_F)$, from which all the strata of G_1 have been crushed to points, leaving only products of an arc with the decomposition spaces $X_i = S^n/G^{d(i)}$ associated with the dense set D of special levels, where the fiber arcs of these products form the nondegenerate decomposition elements. By Lemma 4 (refer to Lemma 3 concerning the bicollarability hypothesis in Lemma 4), $\theta_1\theta_2(G_F)$ is shrinkable, and there exists a map θ_F realizing this decomposition, where θ_F is the end of a pseudo-isotopy ψ_t^F such that $\rho(\pi_F, \pi_F\psi_t^F) < \varepsilon/3$.

Now we see that $\theta_F\theta_1\theta_2$ realizes the decomposition G_F. Choosing $\gamma \in [0, 1)$ very close to 1, we determine a homeomorphism $\psi_\gamma^F\psi_\gamma^1\psi_\gamma^2$ of $S^n \times S^1$ to itself fulfilling the conditions required to verify that G_F is shrinkable. ∎

Remark. The compactness of the domain $S^n \times S^1$ considerably simplifies the proofs, but is not necessary for them. In the applications that follow we transfer the setting from the compact $S^n \times S^1$ to the more natural $E^n \times E^1$.

Another Remark. Those who regularly meditate on Q-manifolds will perceive that similar methods establish the result analogous to Theorem 5 pertaining to sliced decompositions of $Q \times I$.

Corollary 5A. *If G is a sliced decomposition of E^{n+1} such that, for each $s \in E^1$, $G^s \times E^1$ is shrinkable, then G is shrinkable.*

Corollary 5B. *If G is a sliced decomposition of E^{n+1} such that, for each $s \in E^1$, G^s is a shrinkable decomposition of E^n, then G is shrinkable.*

From this one can obtain a decomposition-theoretic proof for a result of Dyer and Hamstrom [1].

Corollary 5C. *Every sliced cell-like decomposition of E^3 is shrinkable.*

This follows from Corollary 5B and the classical Moore theorem (Theorem 25.1).

We also obtain another proof for a result of D. L. Everett [1].

Corollary 5D. *If G is a 0-dimensional cell-like decomposition of E^n, considered as $E^n \times \{0\}$ in $E^n \times E^1$, then the trivial extension G^T [consisting of elements of G and singletons from $E^n \times (E^1 - \{0\})$] to all of E^{n+1} is shrinkable.*

Proof. By elementary methods like those found in Hurewicz–Wallman [1] there exists a map $f: E^n/G \to E^1$ such that $f \mid \pi(N_G)$ is an embedding. Let θ denote the E^n-coordinate preserving homeomorphism of E^{n+1} to itself defined by

$$\langle x, s \rangle \to \langle x, s + f\pi(x) \rangle.$$

Then $\theta(G)$ is a sliced cell-like decomposition of E^{n+1} such that each slice $E^n \times \{s\}$ contains at most one nondegenerate element $\theta(g)$. By Corollary

18.5A, each such $\theta(g)$ is itself cellular in E^{n+1}, so each $(\theta(G))(s)$ is shrinkable. Furthermore, for s from a dense subset of E^1, $(\theta(G))^s$ contains only singletons, trivially implying that its product with E^1 is shrinkable. ■

Corollary 5E. *Suppose G is a sliced cell-like decomposition of E^{n+1} ($n \geq 4$) such that (E^{n+1}/G) is a finite-dimensional space and E^1 contains a dense subset D for which the decompositions*

$$G^d = \{g \subset E^n \,|\, g \times d \in G\} \qquad (d \in D)$$

yield E^{n+1} factors (that is, $G^d \times E^1$ is shrinkable). Then G itself is shrinkable.

Proof. According to Theorem 25.10, for every $s \in E^1$ the decomposition $G(s)$ is shrinkable. ■

Corollary 5F. *If G is a sliced cell-like decomposition of E^{n+1} ($n \geq 4$) such that, for each $s \in E^1$, the decomposition G^s is $(n - 3)$-dimensional or closed-$(n - 2)$-dimensional, then G is shrinkable.*

Proof. It was established in Corollary 24.3C and Theorem 26.9 that under these conditions $G^s \times E^1$ is shrinkable. ■

Corollary 5G. *Every $(n - 3)$-dimensional or closed-$(n - 2)$-dimensional sliced cell-like decomposition of E^{n+1} ($n \geq 4$) is shrinkable.*

The next two corollaries are due to D. K. Preston [1].

Corollary 5H. *If G is a sliced cell-like decomposition of E^n such that E^n/G is finite-dimensional, then $(E^n/G) \times E^1$ is homeomorphic to E^{n+1}.*

Corollary 5I. *If G is a decomposition of E^{n+2} such that E^{n+2}/G is finite-dimensional and each $g \in G$ is contained in some hyperplane $E^n \times \{z_g\}$, where $z_g \in E^2$, then G is shrinkable.*

The proofs are exercises.

EXERCISES

1. Prove Lemma 4.
2. Prove Corollaries 5H and 5I.

NONSHRINKABLE DECOMPOSITIONS

Nonshrinkable cellular decompositions exist in dimensions greater than 4. Chapter VI presents a variety of examples. In so doing it lays out two construction techniques, the first a workable modification of the classical notion of a defining sequence (for closed-0-dimensional decompositions) and the second a more complicated but completely general method, which can be used to fabricate any cell-like decomposition whatsoever.

32. NONSHRINKABLE CELLULAR DECOMPOSITIONS OBTAINED BY MIXING

An efficient method for generating a nonshrinkable cell-like decomposition of S^n, when $n > 3$, is to multiply suspend a cell-like subset C of S^3 having nonsimply connected complement (Exercise 18.6). For generating nonshrinkable cellular ones, the spinning technique of Section 28 is equally efficient. Several other techniques are available as well. A foundational one, treated briefly in this section, macrocosmically entails the tubing together of two disjointly and wildly embedded Cantor sets, directed by a given homeomorphism h between them; microcosmically, the nucleus of the technique involves obtaining a homeomorphism h that mixes the admissible subsets of the embedded Cantor sets. This notion of mixing is a concept we previously encountered while examining 3-dimensional examples. Because no substantial changes occur in moving to the higher-dimensional cases, all we will do here is quickly review the construction and lightly sketch in the supporting devices.

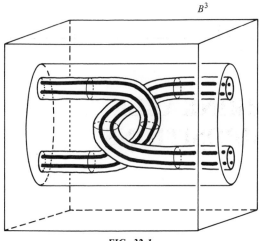

B^3

FIG. 32-1

Historically, the original nonshrinkable cellular decomposition of S^n, $n > 3$, was Eaton's dogbone space (Eaton [2]). We shall examine a modification of it suggested by D. G. Wright [2], just as we did in the 3-dimensional case, where our primary attention fell not on the original example, Bing's dogbone space, but on Example 9.5. At the hub of Wright's modification stands the ramified Bing Cantor set, suitably spun to put it in S^n.

Lemma 1. *For the decomposition K of B^3 depicted in Fig. 32-1, $\mathrm{Sp}^{n-3}(K)$ is a shrinkable decomposition of S^n ($n \geq 5$).*

Proof. The shrinkability of $\mathrm{Sp}^{n-3}(K)$ follows from Theorem 28.9. Details are left to the reader. ■

A Cantor set in S^3 having as its defining sequence the 0-spin of the structures shown in Fig. 32-1 is called a *ramified Bing Cantor set*. Two features of the defining sequence deserve retrospection: first, how the 0-spin compares with the defining sequence for the (unramified) Bing Cantor set of Example 9.1, and second, that it is ambiently equivalent to both the upper half and lower half of what is shown in Fig. 9-5.

Theorem 2. *There exists a nonshrinkable cellular decomposition G of S^n, $n \geq 5$, into points and flat arcs.*

Proof. The decomposition G is determined by a defining sequence $\{T_j \mid j = 0, 1, 2, \ldots\}$ schematically identical to that shown in Fig. 9-5. Each component of each T_j consists of three parts, two being thickened $(n - 2)$-spheres and the third a tube joining them.

Lemma 1 promises a defining sequence $\{K_j\}$ for a spun, ramified Bing Cantor set X in S^n. Hence, it also gives a defining sequence $\{K_j'\}$ for another spun, ramified, Bing Cantor set X' there, arranged so that K_0 and K_0' are disjoint. Each component of a K_j (K_j') is a thickened $(n-2)$-sphere.

Connect K_0 and K_0' by a tube (equivalent to $I \times B^{n-1}$), forming the initial stage T_0. Repeat at even-numbered stages, tying components of K_{2j} to those of K_{2j}' to form stage T_j, while following the algorithm of Example 9.5 in assigning the component of K_{2j+2}' which is to be joined in T_j to a given component of K_{2j+2}. Use thinner and thinner tubes as j increases, each running straight through its predecessors, in order to ensure that the components of $\bigcap T_j$ are arcs locally flatly embedded at their interior points. By the proof of Proposition 12.2, each component of $\bigcap T_j$ is then a flat arc.

The argument that G is nonshrinkable coincides with the one given for the nonshrinkability of Example 9.5, because the K_j's have similar properties with respect to mappings of 2-cells into the n-sphere as do the elements of a standard defining sequence for the (unspun) ramified Bing Cantor set in S^3 (see Exercise 3) and because the identical rule is employed for tubing together the thickened $(n-2)$-spheres. Due to this rule, the nondegenerate elements expose an implicit homeomorphism S from X to X' such that, for every pair of admissible subsets A of X and A' of X', $S(A) \cap A' \neq \emptyset$ (see the definition of admissibility given prior to Lemma 9.12). ■

Remark.   The preceding construction also works when $n = 4$, the main difference then being the extra effort it takes to show $\mathrm{Sp}^1(K)$ is shrinkable, which can be done either by brute force or by applying (Neuzil [1]). Altogether avoiding the question of whether $\mathrm{Sp}^1(K)$ is shrinkable, one still can resort to this construction for producing a nonshrinkable cellular decomposition of S^4; the only loss will be the certainty that the nondegenerate elements consist of flat arcs.

EXERCISES

1. Prove Lemma 1.
2. Show that $\mathrm{Sp}^0(K)$ is a shrinkable decomposition of S^3.
3. Suppose H is a compact 2-manifold with boundary in a 2-cell B and $f: H \to K_0$ is a virtually I-essential map. Show that $f(H)$ contains an admissible subset of $\bigcap K_j$.

33. NONSHRINKABLE NULL SEQUENCE CELLULAR DECOMPOSITIONS OBTAINED BY AMALGAMATING

In this section the objective is to explain why there is a nonshrinkable decomposition of E^n $(n > 3)$ determined by a null sequence of cellular sets. Bing did this for $n = 3$ with his minimal example (Example 9.5). Unlike the graphic representation he gave when describing his, our rendering here just establishes an existence result.

A more explicit representation of a nonshrinkable decomposition in which the nondegenerate elements form a null sequence of cellular arcs has been described by J. J. Walsh and the author [2]. Generated from a specific source decomposition in B^3, it is spun to sweep out a relatively specific decomposition of S^n. A complete analysis of the source decomposition would demand tortuous manipulation of 3-dimensional embedding techniques; providing full detail would necessitate some exacting and highly specialized methods, beyond the purview of this text.

Theorem 1. *There exists a cellular, nonshrinkable usc decomposition $\mathcal{K}$ of E^n ($n \geq 5$) whose nondegenerate elements form a null sequence.*

Proof. The guide leading up to the example $\mathcal{K}$ is the nonshrinkable decomposition G of E^n into points and a Cantor set of flat arcs constructed in the preceding section. At the heart of that construction is an embedding λ of $C \times [-1, 1]$ in E^n, where C denotes a Cantor set, such that

$$N_G = \lambda(C \times [-1, 1]),$$

$$\lambda(C \times [-1, 1]) \quad \text{intersects} \quad E^{n-1} \times \{0\} \quad \text{at} \quad \lambda(C \times \{0\}),$$

$$\lambda(C \times [-1, 0]) \subset E^n_- = E^{n-1} \times (-\infty, 0],$$

$$\lambda(C \times [0, 1]) \subset E^n_+ = E^{n-1} \times [0, \infty),$$

and

for $0 < \delta < 1$, $\lambda(C \times [-1 + \delta, 1 - \delta])$ has embedding dimension 1.

If it seems desirable, one can insist $\lambda(C \times \{0\})$ be standardly embedded in $E^{n-1} \times \{0\}$.

To streamline notation, let $Q = E^n/G$, $Q_+ = \pi(E^n_+)$, $Q_- = \pi(E^n_-)$, and $X = Q_+ \cap Q_- = \pi(E^{n-1} \times \{0\})$. Also, set $Z = \pi\lambda(C \times [-1, 1])$. Clearly Q is an n-manifold except at points of Z.

The plan is to locate a special F_σ set F in Z such that $Q - X$ is 1-ULC in $F \cup (Q - X)$ [that is, very small loops in $Q - X$ will be contractible in small subsets of $F \cup (Q - X)$] and then to produce a cell-like approximation $f \colon E^n \to Q$ to π that is 1-1 over $F \cup (Q - X)$. How this is to be secured will be described later on, in Lemma 2.

For the moment assume such a set F and map f have been obtained. Then the decomposition G_f induced by f, which has the nonmanifold Q as decomposition space, cannot be shrinkable. By Theorem 20.1, there exists $\varepsilon > 0$ such that no ε-amalgamation of G_f is shrinkable.

This sets the stage for another appearance of Edwards's amalgamation trick, first exposed in Theorem 20.5, backed up this time by special devices to bring about cellular elements. With N_f denoting the nondegeneracy set of G_f, which is contained in $f^{-1}(Z - F)$, we express N_f as the union of compact sets $H_1, \ldots, H_k, \ldots$ such that any compact component of H_k is a point inverse of f having diameter at least $1/k$. Then we cover $f(N_f)$ by a null sequence $\{Y_j\}$ of pairwise disjoint cell-like sets in $X - F$ such that diam $Y_j < \varepsilon$ for each j and $f(H_k)$ is covered by a finite number of sets from $\{Y_j\}$; the amalgamation techniques developed in Theorem 20.5 accomplish exactly this end, and the extra requirement here, that each Y_j misses the 0-dimensional F_σ set F, presents little difficulty.

Since the sequence $\{Y_j\}$ is null and only a finite number of the sets $f^{-1}(Y_j)$ contain big elements from G_f, the sequence $\{f^{-1}(Y_j)\}$ is also null. The latter determines the desired decomposition $\mathcal{K}$. Certainly each $f^{-1}(Y_j)$ is cell-like (Exercise 17.1), and certainly $\mathcal{K}$ is nonshrinkable, being an ε-amalgamation of G_f.

What yet remains is a proof that $\mathcal{K}$ is cellular. It depends on $Q - X$ being 1-ULC in $F \cup (Q - X)$. As a consequence, each neighborhood U of a given Y_j contains a smaller neighborhood V of Y_j, with $V \cap X$ connected, such that every loop in $V - X$ is null homotopic in $(U \cap F) \cup (U - X) \subset U - Y_j$. Due to the connectedness of $V \cap X$, each loop in $V - Y_j$ is freely homotopic there to the composition of loops in $V - X$. By the Seifert-van Kampen Theorem, Y_j satisfies the cellularity criterion in Q, implying $f^{-1}(Y_j)$ satisfies it in E^n (Corollary 18.4A). Therefore $f^{-1}(Y_j)$ is cellular (Theorem 18.5). ∎

Lemma 2. *In the notation introduced in Theorem 1, there exists an F_σ set F in Z such that $Q - X$ is 1-ULC in $F \cup (Q - X)$ and there exists a cell-like map $f: E^n \to Q$ approximating π that is 1-1 over $F \cup (Q - X)$.*

Proof. The required F_σ set is extracted later, based on geometric properties of the embeddings of $\lambda(C \times I) \cap E_\pm^n$ in $E_\pm^n$. Select PL triangulations T_{2j+1} of E_-^n, with mesh $T_{2j+1} \to 0$. Adjust them, using the hypothesized embedding dimension properties of $\lambda(C \times [-1, 1])$, so that $T_{2j+1}^{(2)} \cap \lambda(C \times [-1, 0]) \subset \lambda(C \times \{-1\})$ (to avoid unnecessary trouble, it is permissible to presume $T_{2j+1}^{(2)} \cap \lambda(C \times \{0\}) = \varnothing$). Set $F_{2j+1} = T_{2j+1}^{(2)} \cap \lambda(C \times [-1, 0])$, and note that dem $F_{2j+1} \leq 0$ [in E^n or in $E^{n-1} \times (-\infty, 0)$]; moreover, $E_-^n - \lambda(C \times [-1, 0])$ is 1-ULC in $(\bigcup_j F_{2j+1}) \cup (E_-^n - \lambda(C \times [-1, 0]))$. Similarly, obtain compact subsets F_{2j} of $\lambda(C \times \{1\})$, with dem $F_{2j} \leq 0$, such that $E_+^n - \lambda(C \times [0, 1])$ is 1-ULC in $(\bigcup_j F_{2j}) \cup (E_+^n - \lambda(C \times [0, 1]))$.

Define F as $\bigcup_j \pi(F_j)$. It should be obvious that $Q - X$ is 1-ULC in $F \cup (Q - X)$.

The map f will be the limit of cell-like maps $f_0 = \pi, f_1, ..., f_k, ...$ such that f_k is 1-1 over $\pi(F_1 \cup \cdots \cup F_k)$. In addition, f will be made 1-1 over $F \cup (Q - Z)$ by expressing Z as an intersection of a decreasing collection of open sets W_k and by requiring f_k to be 1-1 over $\pi(F_k) \cup (Q - W_k)$, later maps f_{k+j} to be in agreement with f_k over this set, and the convergence throughout to be rapid enough that, for each k, the restrictions of f and f_k agree on

$$f_k^{-1}(\pi(F_k) \cup (Q - W_k)).$$

The controls are the familiar ones of Proposition 23.4.

In order to procure such maps f_k ($k > 0$), one must show that the successive decompositions G_k of E^n induced by the maps f_{k-1} over $\pi(F_k)$ are shrinkable. To explain why G_k is shrinkable in general, it suffices to explain why, in particular, G_1 is shrinkable. The nondegenerate elements of G_1 are those components of $\lambda(C \times [-1, 1])$ meeting F_1. One can show that E^n/G_1 has the DDP by approximately lifting given maps $\mu_1, \mu_2 : B^2 \to E^n/G_1$ to disjoint PL embeddings $m_1, m_2 : B^2 \to E^n$, adjusting so neither image meets F_1 (recall: dem $F_1 = 0$), and then modifying further so that neither image meets $\lambda(C \times [-1, 1)) \cap N_{G_1}$ and the two new images are still disjoint. Projecting back to E^n/G_1 produces disjoint disks there, close to $\mu_1(B^2)$ and $\mu_2(B^2)$. Hence, G_1 is shrinkable by Theorem 24.3 Alternatively, one can exploit what was established implicitly above, namely

$$\text{dem}[\lambda(C \times [-1, 1)) \cap N_{G_1}] \le 1,$$

to physically compress the nondegenerate elements of G_1 near $\lambda(C \times \{1\})$, while abiding by the usual shrinkability regulations.

Following standard practice, one can obtain f_1 from $f_0 = \pi$, by defining $f_1 = f_0 \theta_1^{-1}$, where θ_1 is a map of E^n to itself realizing G_1 and where f_1 is close to f_0. Successive maps can be obtained similarly, subject to the convergence controls alluded to above. ∎

In [2] D. G. Wright particularized these methods a bit; he also established the existence of nonshrinkable, null, cellular decompositions of E^n, where $n \ge 3$. In so doing, he imposed enough regimentation on the process to detect cellularity geometrically, without reference to the cellularity criterion.

A diversion of sorts, not entirely frivolous, as we shall see, can be used to summarize more precisely the effort just expended. Let $k \ge -1$ denote an integer. Say that a cell-like decomposition G of an n-manifold M is *secretly k-dimensional* (a term coined by F. D. Ancel and W. T. Eaton) provided there exists a cell-like map f of M onto M/G for which $\dim f(N_f) \le k$. For example, each cellular decomposition G of M with $\dim M/G < \infty$ is secretly $(n - 1)$-dimensional (Exercise 24.5). Moreover, asking whether G is shrinkable amounts to asking whether it is secretly (-1)-dimensional.

In the opposite vein, say that a cell-like decomposition G of M is *intrinsically k-dimensional* (where $k \geq -1$) if it is secretly k-dimensional but not secretly $(k - 1)$-dimensional. At this juncture we have encountered some intrinsically 0-dimensional cell-like (cellular) decompositions of E^n ($n \geq 3$) but no intrinsically positive-dimensional ones. That will be changed before long.

EXERCISES

1. Suppose G is a cell-like decomposition of an n-manifold M, $n \geq 3$, such that dim $\pi(N_G) = k$ and no $g \in H_G$ satisfies the cellularity criterion. Show that G is intrinsically k-dimensional.
2. Let G be a secretly $(n - 1)$-dimensional cell-like decomposition of E^n, $n \geq 5$. Prove that E^n/G contains a Cantor set C such that the decomposition $G(C)$ induced over C is shrinkable.
3. Suppose G is a finite-dimensional, cell-like decomposition of a compact n-manifold M, $n \geq 5$, such that for each $\varepsilon > 0$ there exists a secretly k-dimensional ε-amalgamation K of G. Show that then G is secretly k-dimensional.
4. If G is a secretly k-dimensional cell-like decomposition of an n-manifold M, $n \geq 5$, show that any two maps μ_1, μ_2: $B^2 \to M/G$ can be approximated, arbitrarily closely, by maps μ_1', μ_2' such that

$$\dim[\mu_1'(B^2) \cap \mu_2'(B^2)] \leq k.$$

34. NESTED DEFINING SEQUENCES FOR DECOMPOSITIONS

Repeatedly, notable and peculiar cell-like decompositions have been specified by means of a sequence of nested manifolds-with-boundary. That kind of defining sequence, effective merely for generating closed-0-dimensional decompositions, is too constricted to function as a useful systematizing device. Here the concept is enlarged to speed the construction of decompositions whose nondegeneracy set has positive-dimensional image. The techniques were developed originally in Cannon–Daverman [1].

It is not hard to produce examples where this occurs. Given an embedding e of $B^k \times I$ in S^n, one can consider the decomposition G of S^n whose nondegenerate elements are the arcs $e(\{b\} \times I)$, $b \in B^k$; then $\pi(N_G)$ is another k-cell. It is more difficult, and more in the spirit of what we intend, to produce a cellular decomposition of S^n ($n > 1$), every element of which is nondegenerate. Once the methodology is in place we suggest how to build these totally nondegenerate cellular decompositions.

The same methodology will be mobilized in the ensuing section to form a cell-like decomposition of S^n ($n \geq 3$) with no cellular elements (in particular, with no singletons). As a warm-up, it is also used at the end of this section to generate an easier cell-like decomposition of E^3 having a large collection of nondegenerate (and noncellular) elements.

Let X be a space and $\mathfrak{M}$ a collection of subsets of X, not necessarily covering X. Given an arbitrary set Z in X, define its *star in* $\mathfrak{M}$ as

$$\text{St}(Z, \mathfrak{M}) = Z \cup (\bigcup\{M \in \mathfrak{M} \mid M \cap Z \neq \varnothing\}),$$

also written as $\text{St}^1(Z, \mathfrak{M})$, and, recursively for any integer $k > 1$, its kth *star in* $\mathfrak{M}$ as

$$\text{St}^k(Z, \mathfrak{M}) = \text{St}(\text{St}^{k-1}(Z, \mathfrak{M}), \mathfrak{M}).$$

When $Z = \{x\}$, we write $\text{St}^k(\{x\}, \mathfrak{M})$ simply as $\text{St}^k(x, \mathfrak{M})$.

Now suppose X is a PL n-manifold. A *nested PL defining sequence* (in X) is a sequence $\mathcal{S} = \{\mathfrak{M}_1, \mathfrak{M}_2, \ldots\}$ satisfying the following three conditions:

Disjointness criterion. For each index i, the set $\mathfrak{M}_i$ is a locally finite collection $\{A_j\}$ of PL n-manifolds-with-boundary (embedded in X as subpolyhedra), whose interiors are pairwise disjoint;

Nesting criterion. For each index $i > 1$, each $A \in \mathfrak{M}_i$ has a unique predecessor, $\text{Pre}\,A$, in $\mathfrak{M}_{i-1}$ that properly contains A.

Boundary size criterion. For each index i, each $A \in \mathfrak{M}_i$, and each pair of distinct points x, y from ∂A, there is an integer $s > i$ such that no element of $\mathfrak{M}_s$ contains both x and y.

Nothing in these criteria forces the elements of $\mathfrak{M}_i$ to become small as i grows large, for the resultant decomposition then would be trivial. Instead, the size control pertains only to the various pseudo-$(n-1)$-skeleta, $\bigcup\{\partial A \mid A \in \mathfrak{M}_k\}$, taken with fixed index k; the implicit "triangulations" of such sets induced by successive $\mathfrak{M}_i$'s have meshes tending toward zero with increasing i (at least, when restricted to subcompacta).

The *decomposition G of X associated with a nested defining sequence* $\mathcal{S} = \{\mathfrak{M}_1, \mathfrak{M}_2, \ldots\}$ is the relation prescribed by the rule: distinct points $x, y \in X$ belong to the same element of G provided there exists an integer k, depending simply on x and y, such that each $\mathfrak{M}_i$ has a chain $A_1, \ldots, A_k$ of elements of length k (or less) connecting x to y (that is, $x \in A_1$, $y \in A_k$, and $A_j \cap A_{j+1} \neq \varnothing$ for $j \in \{1, 2, \ldots, k-1\}$). Clearly G is a decomposition (partition) of X. Let $\pi: X \to X/G$ denote the decomposition map.

As an aid to thoroughly understanding G, it is advantageous to have a more explicit, set-theoretic description of G.

Lemma 1.   (a)   *For each $x \in X$, $\pi^{-1}\pi(x) = \bigcap_i \text{St}^2(x, \mathfrak{M}_i)$.*

(b)   *No element $g \in G$ contains more than one point of the set $\partial\mathcal{S}$ defined as $\bigcup\{\partial A \mid A \in \bigcup_i \mathfrak{M}_i\}$.*

(c) *If $x \in g \in G$ and if either $x \in \partial S$ or $g \cap \partial S = \varnothing$, then $\pi^{-1}\pi(x) = \bigcap_i \mathrm{St}(x, \mathfrak{M}_i)$.*

Proof. Certainly $\bigcap_i \mathrm{St}^2(x, \mathfrak{M}_i) \subset \pi^{-1}\pi(x)$. To examine the reverse inclusion, consider distinct points x, y in the same decomposition element $\pi^{-1}\pi(x)$ and determine the minimal positive integer k such that each $\mathfrak{M}_i$ has a chain $\{A_{i1}, \ldots, A_{ik}\}$ of length k from x to y. Since a particular $\mathfrak{M}_i$ contains at most finitely many such chains, an easy counting argument substantiates that these can be arranged with $A_{ij} = \mathrm{Pre}\, A_{i+1,j}$ for all i and all $j \in \{1, \ldots, k\}$. When i is sufficiently large, the sets $A_{i1}, \ldots, A_{ik}$ must be distinct, for otherwise k would not be minimal.

Now we prove that $k \leq 2$. If not, the sets $Y_j = A_{j1} \cap A_{j2}$ and $Z_j = A_{j2} \cap A_{j3}$ ($j \geq i$) are disjoint, nonempty, compact subsets of ∂A_{j2}. Furthermore, $Y_j \supset Y_{j+1}$ and $Z_j \supset Z_{j+1}$, because the contrary property would cause k to increase. As a result, there exist points $p \in \bigcap_{j \geq i} Y_j$ and $q \in \bigcap_{j \geq i} Z_j$; these (distinct) points from ∂A_{i2} contradict the boundary size criterion. Hence, $k = 2$ and $\pi^{-1}\pi(x) = \bigcap_i \mathrm{St}^2(x, \mathfrak{M}_i)$.

If $x \in \partial A_{i1}$ for some i and if the minimal chain length from x to y equaled 2, we could define sets $Y_j = \{x\}$ and $Z_j = A_{j1} \cap A_{j2}$ and could obtain points $p = x$ and $q \in \bigcap Z_j$ as before, reaching a contradiction. Consequently, when $x \in \partial S$, $\pi^{-1}\pi(x) = \bigcap_i \mathrm{St}(x, \mathfrak{M}_i)$. The boundary size criterion ensures that $\pi^{-1}\pi(x) \cap \partial S$ is then equal to $\{x\}$.

Finally, if $g = \pi^{-1}\pi(x)$ contains no element of ∂S, the minimal chain length k is 1, for otherwise g would contain a point $p \in \bigcap_{j \geq i}(A_{j1} \cap A_{j2}) \subset g \cap \partial S$. ∎

Theorem 2. *The decomposition G associated with a nested PL defining sequence $S = \{\mathfrak{M}_1, \mathfrak{M}_2, \ldots\}$ is usc.*

Proof. To see this directly from the definition, start with a neighborhood U of some $g \in G$, where $g = \bigcap \mathrm{St}^2(x, \mathfrak{M}_i)$. Lemma 1 and compactness considerations imply the existence of an integer $r > 0$ such that $\mathrm{St}^4(x, \mathfrak{M}_r) \subset U$. Define a neighborhood V of g as

$$V = U - \bigcup\{A \in \mathfrak{M}_r \,|\, A \cap \mathrm{St}^2(x, \mathfrak{M}_r) = \varnothing\}.$$

Thus, for each $A \in \mathfrak{M}_r$ satisfying $A \cap V \neq \varnothing$, $A \subset \mathrm{St}^3(x, \mathfrak{M}_r)$, and

$$\mathrm{St}^2(V, \mathfrak{M}_r) = V \cup \mathrm{St}^4(x, \mathfrak{M}_r) \subset U.$$

By Lemma 1, any $g' \in G$ intersecting V is contained in $\mathrm{St}^2(V, \mathfrak{M}_r) \subset U$. ∎

Due to the obvious property, recorded below, that decompositions associated with nested PL defining sequences cannot raise dimension, not all usc decompositions can emanate from such defining sequences.

Proposition 3. *If G is a usc decomposition associated with some nested PL defining sequence $\{\mathfrak{M}_1, \mathfrak{M}_2, \ldots\}$ in X, then $\dim X/G \leq \dim X$.*

Proof. Given a neighborhood W of a point z in X/G, one can select a point $x \in \pi^{-1}(z)$ and an index r such that $\mathrm{St}^3(x, \mathfrak{M}_r) \subset \pi^{-1}(W)$. Let Σ denote the $(n-1)$-complex equal to the frontier of $\mathrm{St}^3(x, \mathfrak{M}_r)$. Then $\pi(\mathrm{St}^3(x, \mathfrak{M}_r))$ contains a neighborhood of z having frontier $\pi(\Sigma)$, and $\dim \pi(\Sigma) = \dim \Sigma$ by Lemma 1. ∎

When do nested PL defining sequences generate cell-like decompositions? The next result sets forth a readily identified sufficient condition.

Proposition 4. *The usc decomposition G of X associated with a nested PL defining sequence $\mathfrak{S} = \{\mathfrak{M}_1, \mathfrak{M}_2, \ldots\}$ is cell-like if $\mathfrak{S}$ satisfies the null homotopy criterion: For each index $i > 1$ and each $A \in \mathfrak{M}_i$, the inclusion mapping $A \to \mathrm{Pre}\, A$ is null-homotopic.*

Proof. There are two cases.

Case 1. Suppose $x \in g \in G$ and $x \in \partial A \subset A \in \mathfrak{M}_i$. Fix a neighborhood U of g. By Lemma 1, there exists an integer $r \geq i$ such that

$$g \subset \mathrm{St}(x, \mathfrak{M}_r) \subset U.$$

Enumerate the elements $A_1, \ldots, A_k$ of $\mathfrak{M}_{r+1}$ containing x. According to the null homotopy criterion, each A_j can be contracted to x, keeping x fixed, in $\mathrm{Pre}\, A_j \subset \mathrm{St}(x, \mathfrak{M}_r)$. These individual contractions, restricted to $g \cap A_j$, can be assembled into a contraction of g in U, which is well-defined because

$$g \cap (\bigcup\{\partial A_j \mid j = 1, \ldots, k\}) = \{x\},$$

by Lemma 1.

Case 2. Suppose $x \in g$ and $g \cap \partial \mathfrak{S} = \varnothing$. Then $g = \bigcap_i \mathrm{St}(x, \mathfrak{M}_i)$ and either $\mathrm{St}(x, \mathfrak{M}_i)$ ultimately equals $\{x\}$ or $\mathrm{St}(x, \mathfrak{M}_i)$ always equals a single element of $\mathfrak{M}_i$, in which situation $\mathrm{St}(x, \mathfrak{M}_{i+1})$ contracts in $\mathrm{St}(x, \mathfrak{M}_i)$ by hypothesis. In either case, g is obviously cell-like. ∎

If G is a decomposition of a PL n-manifold X endowed with a nested PL defining sequence $\mathfrak{S}$, not only is $\dim X/G \leq n$ but also X/G contains a multitude of embedded 2-cells, in $\pi(\partial \mathfrak{S})$. Certain cell-like decomposition spaces obtained from S^n, which were mentioned in Section 26, contain no 2-cells whatever, so not all cell-like decompositions have such defining sequences.

Then why bother with them? Having mentioned some of their limitations, we submit two justifications for their inclusion. First, the nesting aspect is both a natural and desirable feature: useful decompositions can be easily described by means of nested PL defining sequences, and the elements from those decompositions that can seem less amorphous than the ones from those that cannot. Second, with such defining sequences the problem of shrinkability for the stabilized decomposition is solvable ($n \geq 4$). In a way the latter remark also serves as a justification for the rather lengthy final result of Section 26.

Theorem 5. *Suppose G is a cell-like decomposition of a PL n-manifold X, $n \geq 4$, associated with a nested PL defining sequence $\mathcal{S}$. Then $G \times E^1$ is a shrinkable decomposition of $X \times E^1$.*

Proof. The decomposition map $\pi\colon X \to X/G$ is 1-1 on $\partial \mathcal{S} \subset X$. Consequently, X/G is locally encompassed by manifolds at enough points (perhaps not all points) to permit application of Corollary 26.13A. ∎

Example 1. A totally nondegenerate cellular decomposition G_1 of S^n ($n > 1$). A nested PL defining sequence $\{\mathfrak{M}_1, \mathfrak{M}_2, \ldots\}$ for G_1 is obtained by modifying a sequence of triangulations $\{T_1, T_2, \ldots\}$, where T_{i+1} subdivides T_i and mesh T_i approaches 0. One can regard these T_i's as a defining sequence for the trivial decomposition of S^n.

Choose nonvoid open sets W_1 and W_2 in S^n having disjoint closures. These are to function as a crude pair of calipers, gauging that the elements of G_1 are indeed nondegenerate, because every one will touch both $\overline{W}_1$ and $\overline{W}_2$.

The idea simply is to perturb T_i so that that all of its elements meet both W_1 and W_2. This is easy enough with T_1. Generally, assuming T_k has been modified to give $\mathfrak{M}_k$ ($k \geq 1$), where each $A \in \mathfrak{M}_k$ intersects both W_1 and W_2, one constructs a PL homeomorphism of S^n fixing each ∂A, $A \in \mathfrak{M}_k$, and causing each image of an n-simplex $\sigma \in T_{k+1}$ to touch both W_1 and W_2, by manually pushing a point x_i from the adjusted σ into W_i ($i = 1, 2$) with a motion supported inside the unique $A \in \mathfrak{M}_k$ containing the adjusted σ. Define $\mathfrak{M}_{k+1}$ as the resulting image of T_{k+1}.

Note. J. H. Roberts [1] devised the first example of a totally nondegenerate, cell-like decomposition of E^2, back in 1929. Related examples in E^n ($n > 2$) spring up immediately on crossing with E^{n-2}.

It is an exercise that the elements of G_1 are cellular. The next example supplies an extensive stock of noncellular ones.

Example 2. A cell-like decomposition G_2 of E^3 whose set of noncellular elements has 2-dimensional image. In particular, the image will be a 2-cell, and it will stem from a planar 2-cell B in E^3 touching the individual elements $g \in H_{G_2}$, each of which is noncellular, in exactly one point.

The source for G_2 is another nested PL defining sequence $\{\mathfrak{M}_1, \mathfrak{M}_2, \ldots\}$. Each A in each $\mathfrak{M}_i$ will be a solid torus and will include precisely four elements from $\mathfrak{M}_{i+1}$. The first stage $\mathfrak{M}_1$ will consist of one torus, and the containment

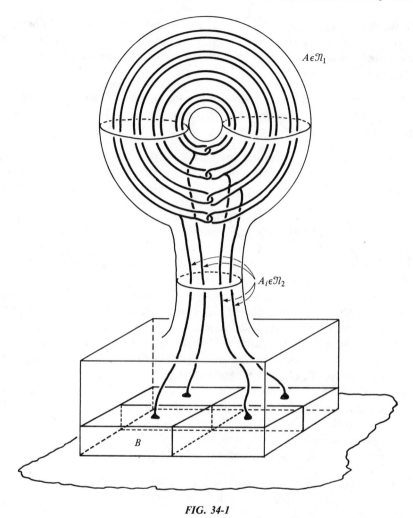

FIG. 34-1

pattern, essentially unvarying under passage from stage to stage, will be fixed by the manner in which the four tori $A_i \in \mathfrak{N}_2$ lie in $A \in \mathfrak{N}_1$. That is depicted in Fig. 34-1.

Clearly G_2 is cell-like. Thicken the first stage solid torus $A \in \mathfrak{N}_1$ slightly to A^* (with $A \subset \text{Int } A^*$), and identify a meridional simple closed curve J on ∂A^*. No nondegenerate element g can be cellular in E^3 because J then would be contractible in $A^* - g$, but g must contain at least one copy Wh of the Whitehead continuum, standardly embedded in A^*, and J is not contractible in $A^* - \text{Wh}$ (see Section 9).

EXERCISES

1. Prove that every monotone usc decomposition of E^1 or S^1 is associated with a nested PL defining sequence.
2. Show that there exists a totally nondegenerate cellular usc decomposition of E^n $(n > 1)$.
3. Construct a cell-like decomposition of S^n $(n > 3)$ whose set of noncellular elements has $(n - 1)$-dimensional image.

35. CELL-LIKE BUT TOTALLY NONCELLULAR DECOMPOSITIONS

This section reproduces the construction due to Cannon and Daverman [1] of a nested PL defining sequence $\mathcal{S}$ for a cell-like, totally noncellular decomposition G of an arbitrary, compact PL n-manifold M, $n \geq 3$. The strategy is to specify $\mathcal{S}$ in such a way that each $g \in G$ is 1-dimensional and contains a wild Cantor set.

The sequence $\mathcal{S}$ is compiled inductively.

Description of $\mathfrak{M}_1$. Let $\mathfrak{M}_1 = \{M\}$ and set $S(M) = \partial\sigma$, for some $(n - 1)$-simplex σ in M.

Observe that $S(M)$ has a PL-product neighborhood $S(M) \times B^2$ in M. Choose a simple closed curve J in M corresponding to $\{p\} \times \partial B^2$, $p \in S(M)$, in $S(M) \times B^2$. This curve J is a useful object of reference because it cannot be contracted in $M - S(M)$.

Inductive Hypothesis j. Suppose $\mathfrak{M}_1, \mathfrak{M}_2, \ldots, \mathfrak{M}_j$ and compact, orientable $(n - 2)$-manifolds $S(A)$, $A \in \bigcup_{i=1}^{j} \mathfrak{M}_i$, are given satisfying the following properties:

(a) For $i \in \{1, \ldots, j\}$, $\mathfrak{M}_i$ is a cover of M;

(b) $\mathfrak{M}_1, \ldots, \mathfrak{M}_j$ satisfy the disjointness and nesting criteria in the definition of nested PL defining sequence;

The remaining properties all pertain to arbitrary $A \in \mathfrak{M}_i$, where $1 < i \leq j$.

(c) diam $[\partial A \cap \partial \operatorname{Pre} A] < 1/i$;
(d) the inclusion map $A \to \operatorname{Pre} A$ is null homotopic;
(e) there exists a $(1/i)$-map of A to a 1-complex;
(f) $S(A)$ is a finite union of pairwise disjoint $(n - 2)$-spheres and $S(A)$ has a PL-product neighborhood $S(A) \times B^2$ in

$$\operatorname{Int} A \cap \operatorname{Int}(S(\operatorname{Pre} A) \times B^2)$$

with $S(A)$ corresponding to $S(A) \times \{0\}$ in $S(A) \times B^2$, such that the curve J cannot be contracted in $M - S(A)$.

Description of $\mathfrak{M}_{j+1}$. Assuming inductive hypothesis j, we specify $\mathfrak{M}_{j+1}$ and the $(n-2)$-manifolds $S(A)$, $A \in \mathfrak{M}_{j+1}$, so that $\mathfrak{M}_1, \mathfrak{M}_2, \ldots, \mathfrak{M}_{j+1}$, together with the associated $(n-2)$-manifolds, fulfill the parallel inductive hypothesis $(j+1)$. We fix $A \in \mathfrak{M}_j$ and $S = S(A)$. Clearly it suffices to describe the elements of $\mathfrak{M}_{j+1}$ contained in A (that is, those elements of $\mathfrak{M}_{j+1}$ having A as predecessor). We choose a collar neighborhood $\partial A \times [0, 1]$ on ∂A in A (which is empty in case $j = 1$) disjoint from $S \times B^2$, with ∂A corresponding to $\partial A \times \{0\}$. We also choose ε, $0 < \varepsilon < 1/j$, so small that 2ε-subsets of A are contractible in A. We will construct n-manifolds $A_1, \ldots, A_s$ filling up A and $(n-2)$-manifolds $S_1 = S(A_1), \ldots, S_s = S(A_s)$ in Int $A_1, \ldots$, Int A_s [as well as in $S(A) \times B^2$] in five steps.

Step 1. Splitting the surface $S = S(A)$. Rather than dealing with this step, and thus with Property (f), in truly recursive fashion, we announce that our global intention is to develop the various surfaces $S(A)$ as follows: given elements $A_i \in \mathfrak{M}_i$ ($i = 1, 2, \ldots$), with $A_1 \supset A_2 \supset \cdots$, we want the collection $\{S(A_i) \times B^2\}$ to intersect in a Cantor set embedded in a Euclidean piece of M exactly like the spun Bing Cantor set of Corollary 28.9B. For $M \in \mathfrak{M}_1$, $S(M) \times B^2$ initiates this project correctly. Assuming that for $A_1, \ldots, A_j = A$, nested as above, with $A_i \in \mathfrak{M}_i$, the associated sets $S(A_i) \times B^2$ begin to describe a spun Bing Cantor set, we find a finite collection $R_1, \ldots, R_r$ of $(n-2)$-spheres in Int $S(A) \times B^2$ such that

    (i) these spheres have pairwise disjoint PL-product neighborhoods $R_1 \times B^2, \ldots, R_r \times B^2$ in Int $S(A) \times B^2$, each of diameter less than $1/(j+1)$; and

    (ii) the collection $\{S(A_1) \times B^2, \ldots, S(A_j) \times B^2, \bigcup(R_i \times B^2)\}$ is topologically equivalent to an initial set of stages (not necessarily consecutive) from some preassigned representation of the spun Bing Cantor set.

(Anyone more familiar with other wild Cantor sets in E^n, such as those of Antoine [1] or Blankinship [1], or with other constructions of Cantor sets defined by sets having PL-product neighborhoods $R_i \times B^2$, may mentally substitute the appropriate structures from the construction patterns of their favorite examples.) Of course, since "J" cannot be shrunk missing the spun Bing Cantor set, it cannot be shrunk missing any stage of it. See Fig. 35-1 for a specific 3-dimensional representation of this pattern.

Step 2. Decomposing A into cells. Let T_∂ denote a triangulation of ∂A. Then $T_\partial \times [0, 1]$ defines a cell-decomposition of the collar $\partial A \times [0, 1]$, and $T_\partial \times [0, 1]$ extends to a PL cell-decomposition T_1 of A. Shortening the collar and subdividing both T_∂ and A minus its collar, if necessary, we may assume that each of the n-cells $C_1, \ldots, C_s$ of T_1 has diameter less than ε. Furthermore,

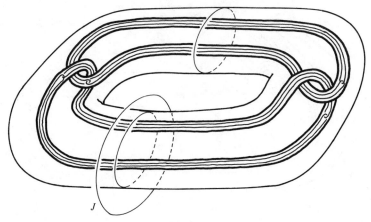

FIG. 35-1

we may assume each C_i to be so small that there exists a point p_i in Int B^2 such that C_i misses

$$(R_1 \cup \cdots \cup R_r) \times \{p_i\} \subset (R_1 \cup \cdots \cup R_r) \times B^2 = (R_1 \times B^2) \cup \cdots \cup (R_r \times B^2).$$

Step 3. Ramifying the surfaces $R_1 \cup \cdots \cup R_r$. Pick distinct points $p_1, \ldots, p_s$, one for each n-cell C_i of T_1 as in the preceding step, and let $S_i = (R_1 \cup \cdots \cup R_r) \times \{p_i\}, i = 1, \ldots, s$. See Fig. 35-2. Note that S_i looks like the core of a deeper stage than $S(A) \times B^2$ from the defining structure of a spun Bing Cantor set, so J cannot be contracted in $M - S_i$. We reemphasize that S_i misses C_i.

Step 4. Connecting C_i *to* s_i. Let $\alpha_1, \ldots, \alpha_s$ denote pairwise disjoint compact sets, each the union of r disjoint PL arcs in Int A such that, for each $i = 1, \ldots, s$, the r arcs of α_i irreducibly join the r components of S_i to C_i.

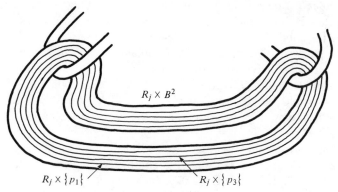

FIG. 35-2

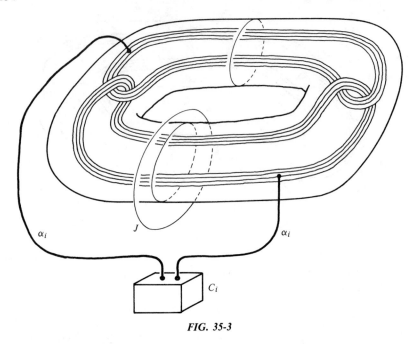

FIG. 35-3

The map collapsing C_i as well as each component of S_i individually to (distinct) points defines an ε-map from the compact connected set $C_i \cup \alpha_i \cup S_i$ onto the endpoint wedge of r arcs. In addition, the size restrictions on C_i and on the components of S_i imply that $C_i \cup \alpha_i \cup S_i$ is contractible in A. See Fig. 35-3.

Step 5. Defining $A_1, ..., A_s$. After slight general position adjustments, we may assume that $\alpha_i \cap S_k = \varnothing$ whenever $i \neq k$ and that each α_i and S_i meet each ∂C_k transversely. We choose a finite simplicial subdivision T_2 of T_1 such that each C_i, each α_i, and each S_i is covered by a full subcomplex of T_2. If the subdivision T_2 is sufficiently fine, the sets $\alpha_1 \cup S_1, ..., \alpha_s \cup S_s$ will have pairwise disjoint simplicial (regular) neighborhoods $N_1, ..., N_s$ in the first derived subdivision T_3 of T_2, and the sets $A_1, ..., A_s$ defined below, like their predecessors $C_1 \cup \alpha_1 \cup S_1, ..., C_s \cup \alpha_s \cup S_s$, will admit an ε-map to a finite graph. We define A_i as $(C_i \cup N_i) - \bigcup_{k \neq i} \text{Int } N_k$ and $S(A_i) \times B^2$ as a PL-thickening of S_i in both A_i and the previously mentioned product $(R_1 \cup \cdots \cup R_r) \times B^2$.

This finishes the inductive description of the defining sequence $\mathcal{S}$. With the specification of $A_1, ..., A_s$, we have completed the construction of the elements from $\mathfrak{M}_{j+1}$ found in $A \in \mathfrak{M}_j$, and in the course of that construction we also have designated the associated $(n - 2)$-manifolds $S(A_i)$, $i \in \{1, ..., s\}$.

The six properties of inductive hypothesis $(j + 1)$ are identifiable features of the construction.

Determination of the defining sequence prompts the following:

Theorem 1. *The sequence* $S = \{\mathfrak{M}_1, \mathfrak{M}_2, \ldots\}$ *defined above is a nested PL defining sequence for a cell-like decomposition* G *of* M *such that no* $g \in G$ *is cellular.*

Proof. The three criteria required in order that S be a nested PL defining sequence drop out of properties (b) and (c) of the inductive hypothesis. By Theorem 34.2, G is usc. Furthermore, because of property (d), S satisfies the null homotopy criterion, and Proposition 34.4 attests that G is cell-like.

Fix $g \in G$. First, we claim that $\dim g \leq 1$. The proof of this fact resembles the proof of Proposition 34.4, showing g to be cell-like. In the notation of that argument (Case 1), instead of using that $A_j \cap g$ contracts to x in Pre A_j, one uses that $A_j \cap g$ ε-maps to some 1-complex and that, as a consequence, g itself ε-maps to a finite wedge of such 1-complexes, all wedged together at the appropriate images of x. In the other case of Proposition 34.4, the same conclusion is more easily obtained. Thus, $\dim g \leq 1$.

Second, we note that g contains a Cantor set C_g in $\text{Int}(S(M) \times B^2)$ such that J is not contractible in $M - C_g$. To see this, choose sets $A_j \in \mathfrak{M}_j$ such that $A_1 \supset A_2 \supset \cdots$ and each meets g. Then $\bigcap_j (S(A_j) \times B^2)$ is the desired Cantor set.

Third, because $\dim g \leq 1$, the curve J is homotopic in $M - C_g$ to a curve J' in $M - g$. If g were cellular, J' would be contractible, not only in $M - g$, but then also in $M - S(A_k)$ for some large index k, which is ruled out by property (f) of the inductive hypothesis. ∎

Corollary 1A. *There exists a compact, generalized n-manifold X, $n \geq 3$, such that $X - \{x\}$ fails to be locally simply connected at x, for every $x \in X$.*

Corollary 1B. *Let M denote a compact PL n-manifold, $n \geq 3$. There exists a cell-like decomposition G of M and there exists an embedding $\lambda: \partial B^2 \to M^n/G$ such that every map $F: B^2 \to M/G$ extending λ satisfies $F(B^2) = M/G$.*

Proof. Require during the construction delineated for Theorem 1 that the special curve J lie in ∂S. Then the decomposition map π will inspire a homeomorphism λ of ∂B^2 onto $\pi(J)$.

Consider any map $F: B^2 \to M/G$ extending λ. Should $F(B^2)$ avoid some point $\pi(g) \in M/G$, F could be lifted to a map contracting J in $M - g$ (Theorem 16.7), which was proved impossible in Theorem 1. ∎

Corollary 1C. *There exists a cell-like decomposition G of S^n ($n \geq 3$) such that no $g \in G$ is cellular but $G \times E^1$ is shrinkable.*

Proof.   For $n \geq 4$ see Theorem 34.5. Alternately, in all cases $n \geq 3$ the construction of Theorem 1 can be varied slightly to impose some other, relatively minor, technical controls, with which one can bring about a shrinking by bare hands. See (Cannon–Daverman [1]).   ∎

Rephrased, Exercise 33.1 shows that the decomposition of Theorem 1 is intrinsically n-dimensional. Cellular decompositions cannot achieve the same degree of pathology, however; Exercise 24.5 indicates that all finite-dimensional cellular decompositions of S^n are secretly $(n - 1)$-dimensional. Exploiting a more subtle intertwining of the various surfaces $S(A)$ arising in the proof of Theorem 1, D. J. Garity and the author [1, 2] have constructed nested defining sequences for intrinsically k-dimensional $(0 < k < n)$ cellular decompositions of S^n.

EXERCISES

1. What modifications must be incorporated to produce a totally noncellular, cell-like decomposition of E^n $(n \geq 3)$?
2. Let S be a nested PL defining sequence for a cell-like decomposition G of E^n such that no $g \in G$ is cellular (satisfies the cellularity criterion), where $G \times E^1$ is shrinkable. Show that no line of the form $\{x\} \times E^1$ is standardly embedded in $(E^n/G) \times E^1 = E^{n+1}$.
3. If G is the cell-like decomposition of S^n promised in Corollary 1B, prove that there is no cell-like map of S^n/G onto an n-manifold.

36. MEASURES OF COMPLEXITY IN DECOMPOSITION SPACES

Due to their geometric simplicity, manifolds represent the quintessential objects of our attention; the other generalized manifolds, by contrast, riddled with baroque complexities, are less ideal. We will be better able to topologically distinguish the various generalized manifolds if we can weigh, or quantify, such complexities. What are the effective measures?

Several methods have been exposed already. A crude one is just the dimension of the set $S(X)$ of nonmanifold points in a given generalized manifold X. Another is the dimension of the set

$$Q(X) = \{x \in X \,|\, X - \{x\} \text{ is not 1-LC at } x\}.$$

When X is the cell-like image of an n-manifold M $(n > 3)$, with G the associated cell-like decomposition, $Q(X)$ coincides with the image of the noncellular elements from G; furthermore, for $n > 4$ dim $S(X) = k$ iff G is intrinsically closed-k-dimensional.

The notion of intrinsic dimension, discussed briefly at the end of Section 33, provides a more refined measure. Its primary advantage materializes from the direct connection it provides to the nondegeneracy set N_G of some decomposition G efficiently giving rise to X, instead of to $\operatorname{Cl} N_G$, which is more closely aligned with $S(X)$. Of course, $\dim[S(X) = S(M/G)]$ serves as an upper bound to the intrinsic dimension of G, but the latter can assume (for varying G) any integral value between -1 and $\dim S(M/G)$.

D. J. Garity [2, 3] has discovered a still more discriminating measure, involving the following generalizations of the disjoint disks property. A metric space X is said to satisfy the *disjoint k-tuples property*, abbreviated as DD_k, if every collection of k maps $\mu_i: B^2 \to X$ ($i = 1, 2, ..., k$) can be approximated, arbitrarily closely, by maps $\mu_i': B^2 \to X$ such that $\bigcap_i \mu_i'(B^2) = \varnothing$; similarly, X is said to satisfy DD_ω if, for each $\varepsilon > 0$, and each countable collection of maps $\mu_i: B^2 \to X$, there exist maps $\mu_i': B^2 \to X$ such that $\rho(\mu_i', \mu_i) < \varepsilon$ and $\bigcap_i \mu_i'(B^2) = \varnothing$.

The countable property deserves mention because cellular decompositions are precisely the cell-like decompositions satisfying DD_ω.

Theorem 1. *Let G be a cell-like decomposition of an n-manifold M, $n > 3$. Then G is cellular if and only if M/G satisfies DD_ω.*

Proof. Assume M/G satisfies DD_ω. The goal is to show that each $g \in G$ satisfies the cellularity criterion in M; by Proposition 18.4 it suffices to show that each point x of M/G satisfies the cellularity criterion there. Given a map $f: B^2 \to M/G$, one can set $\mu_i = f$ for $i \in \{1, 2, ...\}$ and apply DD_ω to find some close approximation μ_j' to f avoiding x, where $\mu_j' | \partial B^2$ is homotopic to $f | \partial B^2$ under a short homotopy in $M/G - \{x\}$. This essentially verifies the cellularity criterion.

Conversely, assume G is cellular. Fix $\varepsilon > 0$ and maps $\mu_1, \mu_2, ...$ of B^2 to M/G. One can find n-cells $C_1, C_2, ...$ in M such that (1) M/G has an open cover $\{V_1, V_2, ...\}$ where $\pi^{-1}(V_j) \subset C_j$ and (2) $\operatorname{diam} \pi(C_j) < \varepsilon$ for $j \in \{1, 2, ...\}$. Then the maps μ_j can be adjusted to μ_j' with $\rho(\mu_j', \mu_j) < \varepsilon$ and $V_j \cap \mu_j'(B^2) = \varnothing$. Consequently, $\bigcap_j \mu_j'(B^2) = \varnothing$, and DD_ω holds. ∎

Theorem 1 is also valid in case $n = 3$, provided each $g \in G$ is known to have a neighborhood embeddable in E^3.

Before taking up property DD_k, we characterize the secretly s-dimensional cell-like decompositions in terms of the dimension of intersecting pairs of 2-cell images in the decomposition space.

Proposition 2. *Let G be a finite-dimensional cell-like decomposition of an n-manifold M, $n > 4$. Then G is secretly s-dimensional if and only if each pair of maps $\mu_1, \mu_2: B^2 \to M/G$ is approximable by maps $\mu_1', \mu_2': B^2 \to M/G$ such that*

$$\dim[\mu_1'(B^2) \cap \mu_2'(B^2)] \le s.$$

Proof. Suppose the map approximation property holds. Select a countable dense subset $\{\langle \alpha_i, \beta_i \rangle\}$ from the space of all pairs of maps from B^2 to M/G, where, from this hypothesis,

$$\dim[\alpha_i(B^2) \cap \beta_i(B^2)] \leq s, \qquad \text{for each } i.$$

Then $Q = \bigcup_i [\alpha_i(B^2) \cap \beta_i(B^2)]$ is an s-dimensional F_σ subset of M/G, and it is included in some s-dimensional G_δ subset R of M/G. By Proposition 24.4, decompositions induced over arbitrary compact subsets of $(M/G) - R$ are shrinkable. Hence, according to Proposition 23.4, the given decomposition map is approximable by a cell-like map $p \colon M \to M/G$ that is 1-1 over $(M/G) - R$. As a result,

$$\dim p(N_p) \leq \dim R \leq s.$$

The converse is Exercise 33.4.   ■

The ensuing result supplies the technical core for the forthcoming analysis. Its proof requires a minor generalization of the one given for Proposition 24.1; details are left as an exercise.

Proposition 3. *Let X denote a locally compact, separable ANR and $k > 1$ an integer. Then X satisfies DD_k if and only if each map $f \colon B^2 \to X$ can be approximated, arbitrarily closely, by a map $F \colon B^2 \to X$ such that* card $F^{-1}(x) < k$ *for each $x \in X$.*

With Proposition 3 one can make a rough estimate about the possible dimension of 2-cell images in X.

Corollary 3A. *If X as above satisfies DD_k, then each map $f \colon B^2 \to X$ can be approximated by a map F for which* $\dim F(B^2) \leq k$.

Proof. This follows from the classical dimension theory result, discovered by W. Hurewicz [2], that if $h \colon Y \to Z$ is a closed surjective map between separable metric spaces and card $h^{-1}(z) \leq m$ for all $z \in Z$, then $\dim Z \leq \dim Y + (m - 1)$. See also (Engelking [1, p. 134]).

Next comes the central result, revealing the pertinence of DD_k to secret dimension.

Theorem 4. *If G is a finite-dimensional cell-like decomposition of an n-manifold M, $n > 4$, and M/G satisfies DD_k, then G is secretly $(k - 3)$-dimensional.*

Proof. For $k = 2$ this is just a restatement of Edwards's cell-like approximation theorem; for $k > n + 1$ it is a triviality because of Theorem 1 and the fact that finite-dimensional cellular decompositions of M are secretly $(n - 1)$-dimensional (see Exercise 24.5). The argument given below works when $2 < k < n - 1$. The case $k = n - 1$ is dealt with in the exercises, and the remaining two cases (as well as the others) are proved by Garity [2].

Consider maps $\mu_1, \mu_2 \colon B^2 \to M/G$. Corollary 3A ensures that maps of B^2 to M/G can be adjusted so as to have $(n - 2)$-dimensional images (assuming $k < n - 1$). Then an old technique, modifying a map on a infinite 1-skeleton to avoid $(n - 2)$-dimensional compacta, combines with Proposition 3 to furnish approximations $\tilde{\mu}_1, \tilde{\mu}_2 \colon B^2 \to M/G$ where

$$A = \tilde{\mu}_1^{-1}(\tilde{\mu}_1(B^2) \cap \tilde{\mu}_2(B^2)) \quad \text{is} \quad \text{0-dimensional}$$

and no point of M/G has more than $(k - 1)$ preimages under $\tilde{\mu}_1 \cup \tilde{\mu}_2$. This means that $\tilde{\mu}_1(B^2) \cap \tilde{\mu}_2(B^2)$ is the (at most) $(k - 2)$-to-1 image of A under $\tilde{\mu}_1$, so, for dimension theory reasons (Engelking [1, p. 134]),

$$\dim[\tilde{\mu}_1(B^2) \cap \tilde{\mu}_2(B^2)] = \dim \tilde{\mu}_1(A) \le k - 3.$$

Proposition 2 secures the desired conclusion. ∎

This automatically upgrades a result given as Exercise 26.10.

Corollary 4A. *If G is a finite-dimensional, cell-like decomposition of an n-manifold, $n > 3$, then $G \times E^1$ is secretly 0-dimensional.*

Proof. The fact that the decomposition space satisfies DD_3 is Exercise 25.3. ∎

Corollary 4A leads to an alternative proof that $G \times E^2$ is shrinkable, without reference to the intermediary DADP of Section 26, for now we see that $[(M \times E^1)/(G \times E^1)] \times E^1$ has the DDP by the proof of Corollary 24.3C (Exercise 24.7).

The converse of Theorem 4 is false. It fails in part because the decomposition space determined by any cell-like, noncellular decomposition does not satisfy DD_ω, let alone DD_3. More important, when attention is restricted to the cellular case, it fails because of decompositions like an analogue of Example 28.1, an $(n - 3)$-spin of a modified Example 9.3, which is a closed-0-dimensional cellular decomposition of S^n whose associated decomposition space satisfies DD_4 but not DD_3. Garity [2, 3] has described other cellular decomposition spaces satisfying DD_{k+1} but not DD_3, built with mixing and tubing methods like those used in Section 32.

The failure of the converse to Theorem 4 upholds the contention that the properties DD_k do not measure the same complexities as intrinsic dimension does. Reinforcing that contention, we now can pinpoint differences between the secretly countable and some secretly 0-dimensional cellular decompositions, such as the spin of a modified Example 9.3: the secretly countable ones all have DD_3 (Exercise 1).

EXERCISES

1. If G is a countable cellular usc decomposition of E^n, show that E^n/G has DD_3.
2. Prove Proposition 3.
3. Let G be a cellular usc decomposition of an n-manifold M, $n > 2$. Show that every pair of maps $\mu_1, \mu_2 : B^2 \to M/G$ can be approximated by maps $\tilde{\mu}_1, \tilde{\mu}_2$ such that

$$\dim \tilde{\mu}_i^{-1}[\tilde{\mu}_1(B^2) \cap \tilde{\mu}_2(B^2)] \leq 1.$$

4. Let G denote a finite-dimensional, cell-like decomposition of an n-manifold M, $n > 4$, such that M/G satisfies DD_k.
   (a) Use Exercise 3 to prove G is secretly $(k - 2)$-dimensional when $k > n - 2$;
   (b) For $k = n - 1$ use the consequence of (a) that M/G satisfies DADP (Proposition 25.9) to prove G is secretly $(k - 3)$-dimensional.
5. Show that the decomposition of Example 28.1 satisfies DD_3.
6. For $k > 3$ construct a closed-0-dimensional decomposition G of S^n ($n \geq 3$) such that S^n/G satisfies DD_{k+1} but not DD_k.

37. DEFINING SEQUENCES FOR DECOMPOSITIONS

Nested PL defining sequences can be put to effective use, evidenced in Section 35, for manufacturing nontrivial generalized manifolds. Nevertheless, several limitations prevent this type of defining sequence from being the paradigm. Another type is available, somewhat more cumbersome but all-encompassing in the sense that every usc decomposition of a locally compact metric space is associated with this more general variety of defining sequence.

A paraphrase of (Daverman–Walsh [1]), this section lays the foundation for a broad notion of defining sequence, introduces a criterion for cell-likeness of the resulting decomposition, and treats the naturality of each.

Throughout this part X will denote (at the least) a locally compact metric space. A *defining sequence* (in X) is a sequence $\mathcal{S} = \{\mathfrak{M}_1, \mathfrak{M}_2, \ldots\}$ satisfying two criteria:

Disjointness criterion. For each index i, $\mathfrak{M}_i$ is a locally finite collection of compact subsets of X having pairwise disjoint interiors.

Star nesting criterion. For each i and $x \in X$

$$\text{St}^3(x, \mathfrak{M}_{i+1}) \subset \text{Int } \text{St}^2(x, \mathfrak{M}_i).$$

The *decomposition G of X associated with a defining sequence* $\mathcal{S} = \{\mathfrak{M}_1, \mathfrak{M}_2, \ldots\}$ is the relation prescribed by the rule: for $x \in X$, $G(x)$ is the subset of X consisting of all $y \in X$ such that $y \in \text{St}^2(x, \mathfrak{M}_i)$ for every integer $i > 0$.

Such a relation G obviously is reflexive and symmetric.

Lemma 1. *The relation G associated with a defining sequence $\{\mathfrak{M}_1, \mathfrak{M}_2, \ldots\}$ is transitive.*

Proof. Suppose $y \in G(x)$ and $z \in G(y)$. It follows that $z \in St^4(x, \mathfrak{M}_i)$ for all $i > 0$. The heart of the proof is the claim that

$$St^4(x, \mathfrak{M}_{k+1}) \subset St^3(x, \mathfrak{M}_k)$$

for each $x \in X$ and each integer $k > 0$. Then by the claim and the star nesting criterion,

$$z \in St^4(x, \mathfrak{M}_{i+2}) \subset St^3(x, \mathfrak{M}_{i+1}) \subset St^2(x, \mathfrak{M}_i),$$

implying $z \in G(x)$.

In order to establish the claim, note that each $x' \in St^4(x, \mathfrak{M}_{k+1})$ satisfies $x' \in St(x^*, \mathfrak{M}_{k+1})$ for some $x^* \in St^3(x, \mathfrak{M}_{k+1})$. By the star nesting criterion,

$$x^* \in St^3(x, \mathfrak{M}_{k+1}) \subset Int\, St^2(x, \mathfrak{M}_k).$$

Choose $A^* \in \mathfrak{M}_{k+1}$ such that $x^* \in A^* \subset St^3(x, \mathfrak{M}_{k+1})$. According to the disjointness criterion, A^* has nonempty interior, and the local finiteness of $\mathfrak{M}_k$ yields a point $x^{**} \in A^*$ such that some $A \in \mathfrak{M}_k$ contains x^{**} in its interior. The uniqueness of this A containing x^{**}, guaranteed again by disjointness, indicates that $A = St(x^{**}, \mathfrak{M}_k)$. Then the fact that $x^{**} \in A^* \subset St^2(x, \mathfrak{M}_k)$ implies $A \subset St^2(x, \mathfrak{M}_k)$, and the observation that $x' \in St^2(x^{**}, \mathfrak{M}_{k+1})$ reveals

$$x' \in St^2(x^{**}, \mathfrak{M}_{k+1}) \subset St^2(x^{**}, \mathfrak{M}_k) = St(A, \mathfrak{M}_k) \subset St^3(x, \mathfrak{M}_k),$$

as required. ∎

Therefore, the equivalence relation G induces a decomposition (partition) of X.

Theorem 2. *The decomposition G associated with a defining sequence $\mathbb{S}$ is usc.*

The proof coincides with that of Theorem 34.2.

With this we have an entirely satisfactory notion of defining sequence for usc decompositions. Although the star nesting criterion may seem awkward or artificial at first, the superficially more natural requirement of

$$St^2(x, \mathfrak{M}_{k+1}) \subset Int\, St(x, \mathfrak{M}_i)$$

simply does not work. It cannot be attained even when $\mathbb{S} = \{\mathfrak{M}_1, \mathfrak{M}_2, \ldots\}$ represents a nested PL defining sequence on a PL manifold. The next result makes plain the comprehensiveness of the notion at hand.

Theorem 3. *If G is an arbitrary usc decomposition of a locally compact metric space X, then G is the decomposition associated with some defining sequence.*

Proof. Let X/G denote the associated decomposition space and $\pi: X \to X/G$ the induced map. For the moment, assume the existence of a defining sequence $\{\mathcal{P}_1, \mathcal{P}_2, \ldots\}$ in X/G for the trivial decomposition into singletons; furthermore, assume each $\mathcal{P}_k$ is a cover of X/G and $\mathrm{Int}\, \pi^{-1}(\mathrm{Fr}\, A) = \varnothing$ for all $A \in \mathcal{P}_k$. Then it is an easy matter to verify that $\{\mathfrak{M}_1, \mathfrak{M}_2, \ldots\}$ is a defining sequence, where $\mathfrak{M}_k = \{\pi^{-1}(A)\,|\,A \in \mathcal{P}_k\}$, and that G is the associated decomposition.

What remains is the production of the defining sequence $\{\mathcal{P}_1, \mathcal{P}_2, \ldots\}$. The forthcoming construction can also be adapted to obtain $\mathcal{P}_1$. Generally, presuming $\mathcal{P}_{k-1}$ has already been obtained, we find a locally finite open cover $\mathcal{U}$ of X/G such that:

(1) each element of $\mathcal{U}$ has compact closure and diameter less than $1/k$;
(2) $\mathrm{Int}\, \pi^{-1}(\mathrm{Fr}\, U) = \varnothing$ for each $U \in \mathcal{U}$; and
(3) for each $z \in X/G$, $\mathrm{St}^3(z, \bar{\mathcal{U}}) \subset \mathrm{St}^2(z, \mathcal{P}_{k-1})$, where $\bar{\mathcal{U}} = \{\bar{U}\,|\,U \in \mathcal{U}\}$.

If an initial choice of $\mathcal{U}$ satisfying conditions (1) and (3) does not satisfy (2) as well, perform the following modification. Let $\{V(U)\,|\,U \in \mathcal{U}\}$ be an open cover such that $V(U) \subset \mathrm{Cl}\, V(U) \subset U$ for all $U \in \mathcal{U}$ (see Dugundji [1, p. 152]). For each $U \in \mathcal{U}$ determine a map $f: \mathrm{Cl}\, U \to [0, 1]$ with $f(\mathrm{Fr}\, U) = 0$ and $f(\mathrm{Cl}\, V(U)) = 1$. Since the compact set $\pi^{-1}(U)$ is second countable, $U_t = f^{-1}((t, 1])$ has the property that $\mathrm{Int}\, \pi^{-1}(\mathrm{Fr}\, U_t) = \varnothing$ for all but countably many $t \in (0, 1)$. Alter the given cover $\mathcal{U}$ by replacing each $U \in \mathcal{U}$ with such a U_t.

Let $U_1, U_2, \ldots, U_\alpha, \ldots$ be a well-ordering of the elements of $\mathcal{U}$. For each α define P_α as $\mathrm{Cl}(U_\alpha - \bigcup_{\beta < \alpha} \mathrm{Cl}\, U_\beta)$. Finally, let $\mathcal{P}_k$ be the collection consisting of the nonempty P_α's. The local finiteness of $\mathcal{U}$ implies $\pi^{-1}(\mathrm{Fr}\, P_\alpha)$ meets only a finite subcollection of $\{\pi^{-1}(\mathrm{Fr}\, U_\beta)\,|\,\beta \le \alpha\}$, so $\mathrm{Int}\, \pi^{-1}(\mathrm{Fr}\, P_\alpha) = \varnothing$. ∎

A criterion that is vitally instrumental for analyzing the cell-likeness of a decomposition G associated with a defining sequence $\{\mathfrak{M}_1, \mathfrak{M}_2, \ldots\}$ is the

Star null homotopy criterion. For each index i and each $x \in X$ there exists an integer $k > i$ such that $\mathrm{St}^3(x, \mathfrak{M}_k)$ is contractible in $\mathrm{St}^2(x, \mathfrak{M}_i)$.

Given another defining sequence $\mathcal{S}' = \{\mathfrak{M}'_1, \mathfrak{M}'_2, \ldots\}$ for G, for every index k and every $x \in X$ one can find some integer j with $\mathrm{St}^3(x, \mathfrak{M}'_j) \subset \mathrm{St}^3(x, \mathfrak{M}_k)$. Consequently, when this criterion is satisfied by some defining sequence for G, it is satisfied by every one.

Proposition 4. *Let X denote a locally compact ANR. A necessary and sufficient condition for a decomposition G of X associated with a defining sequence $S = \{\mathfrak{M}_1, \mathfrak{M}_2, \ldots\}$ to be cell-like is that S satisfy the star null homotopy criterion.*

The proof is elementary; one implication, in a more general setting, is an exercise.

With an infinite recycling of the techniques from Section 35, one can grind out a defining sequence S for a cell-like decomposition G of S^n ($n > 2$) more convoluted than that secured from Theorem 35.1: if J is any simple closed curve embedded in S^n/G, every contraction of J contains an open subset of S^n/G, and the decomposition space includes neither any ANR nor any cell-like set having dimension strictly between 1 and n (Daverman–Walsh [1]).

EXERCISES

1. If S is a defining sequence for a decomposition G of a locally compact metric space X and S satisfies the star null homotopy criterion, show that each $g \in G$ is cell-like in X.
2. If X is a locally compact ANR containing no cell-like set of dimension strictly between 1 and n ($n > 2$), show that there is no cell-like map of X onto an n-manifold, with or without boundary.

VII

APPLICATIONS TO MANIFOLDS

The study of cell-like decompositions is an indispensable component of the fabric of geometric topology. It is interwoven with several topics: analysis of manifolds, embedding theory, simple-homotopy theory. Chapter VII highlights several strands from this intricate texture.

At the beginning this chapter assembles some formidable machinery involving gropes, due to Cannon. Ultimately gropes aid in procuring resolutions of generalized n-manifolds. Although Quinn has given a very general result about the existence of resolutions, explicit blueprints outlining the construction of this machinery remain necessary for deriving some of the envisioned applications, primarily the ones showing how arbitrary embeddings of compact objects in manifolds can be approximated by nice embeddings. This fact exemplifies once again the strong magnetic attraction between decomposition theory and embedding theory.

Quinn's important result combines with Edwards's cell-like approximation theorem to (essentially) provide the crowning touch: a criterion for recognizing n-manifolds.

Last of all, Section 41 puts to work other methods from embedding theory, suggested by penetrating insights of Miller, and makes close inspection of the resulting cell-like decompositions to show how finite-dimensional ANRs can be positioned in manifolds so as to have manifold mapping cylinder neighborhoods.

Throughout this part one should keep in mind the vague warning from the Introduction that, for the kind of far-reaching applications being sought, there will be a new price for progress. In the strict confines of the text, we will become wholesale dealers of mathematics, who feel no reluctance to trade on results from other topics in geometric topology with little attempt at motivation or justification, whenever it helps the cause.

38. GROPES AND CLOSED *n*-CELL-COMPLEMENTS

Gropes appear to have originated, buried implicitly deep in the construction, in the marvelous work of M. A. Štan'ko [3] showing how to approximate embeddings of codimension 3 compacta in manifolds by 1-LCC embeddings. When he extended Štan'ko's work, particularly while delivering certain formal addresses on the subject, R. D. Edwards [2] assigned them a more conspicuous role, and later on, perhaps more significantly, he reiterated their benefits in his initial unpublished work on the double suspension theorem [3]. Since then, it is J. W. Cannon who has been the primary manipulator and champion of gropes, under various designations, through his work with F. D. Ancel (Ancel–Cannon [1]) on the locally flat approximation theorem for codimension one embeddings, his own results (Cannon [6]) about normal forms for decompositions, which led to the final solution of the double suspension problem, and his work with J. L. Bryant and R. C. Lacher (Cannon–Bryant–Lacher [1]) about resolutions of certain generalized manifolds. Cannon also has described the connections of gropes with wildness and decomposition problems in a revelatory survey paper [3]. The preceding list of prominent applications should furnish motivation enough for studying gropes now.

Gropes unite homology and homotopy properties in a special way. That conjunction will be partially sketched in this section, but other noteworthy aspects will be brought forward in the sequel. Here the resoundingly geometric central issue will be the manner in which thickened gropes can be attained from cell-like decompositions of the *n*-cell.

The indispensable building block for concocting a grope is the disk-with-handles, specifically, a compact, connected, orientable 2-manifold D having just one boundary component. In the typical case where it is is not a disk, D contains simple closed curves A_1 and B_1 in Int D meeting each other transversely in a single point, and the decomposition whose only nondegenerate element is $A_1 \cup B_1$ yields a disk-with-handles having one less handle than D. Iteration provides disjoint pairs $(A_1, B_1), ..., (A_k, B_k)$ of simple closed curves in Int D such that A_i meets B_i transversely in a single point and the decomposition whose nondegenerate elements are the sets $A_i \cup B_i$, $i \in \{1, ..., k\}$, yields a disk. The union $J = A_1 \cup B_1 \cup \cdots \cup A_k \cup B_k$ is called a *complete handle curve* for D.

A grope is composed of infinitely many disks-with-handles, attached together in the following manner. Start with a disk-with-handles D_0 on which is prescribed a complete handle curve $J_0 = A_1 \cup B_1 \cup \cdots \cup A_k \cup B_k$. For the next stage D_1, name a collection $D(A_1)$, $D(B_1)$, ..., $D(A_k)$, $D(B_k)$ of mutually exclusive disks-with-handles, one for each of the simple closed curves of J_0, and identify $\partial D(A_i)$ with A_i and $\partial D(B_i)$ with B_i. Also specify a prescribed complete handle curve J_1 for D_1 (meaning a complete handle curve for each component of D_1). Repeat for the next stage D_2, naming a collection of mutually exclusive disks-with-handles, one for each of the simple closed curves in J_1, identifying each boundary in D_2 with the corresponding curve in J_1 (and permitting no other intersections of D_2 with $D_0 \cup D_1$), and also specifying a prescribed complete handle curve J_2 for D_2. Repeating this infinitely often, one obtains an infinite 2-dimensional complex $D = D_0 \cup D_1 \cup D_2 \cup \cdots$ called a *grope*, or sometimes, an *infinite commutator*, in which $D_{i+1} \cap (D_0 \cup D_1 \cup \cdots \cup D_i) = J_i = \partial D_{i+1}$. We call D_i the *ith stage of D* and $\partial D = \partial D_0$ the *boundary of D*.

Abstractly, a *compactified grope* D^+ (occasionally called a *generalized 2-disk* in the literature, a term avoided here because D^+ is *not* a generalized 2-manifold) is the Freudenthal, or endpoint, compactification of a grope D. More concretely, the compactified grope can be realized as a subset of E^3 by controlling the embedding of a grope shown in Fig. 38-1. View each of the handle curves of J_0 as being spanned by thickened disks, or pillboxes, two of which intersect only along their boundaries and do so only if their associated handle curves intersect, and put the next stage disks-with-handles in the appropriate pillboxes. Make the pillboxes for this new stage lie interior to some pillbox at the initial stage, and pictured in Fig. 38-2. Iterate. If P_i denotes the pillboxes used at stage i and if sizes are limited by requiring the pillboxes at stage i to have diameter less than $1/i$, then the compactified grope D^+ corresponds to

$$\bigcap (D_0 \cup \cdots D_i \cup P_i).$$

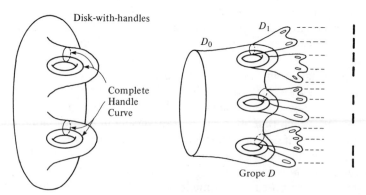

FIG. 38-1

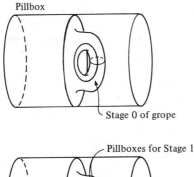

Pillbox

Stage 0 of grope

Pillboxes for Stage 1

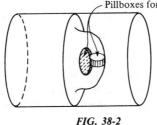

FIG. 38-2

The resulting intersection is what we call a *standard realization* of D^+ in E^3.

Gropes serve as topological representation of elements in, as well as certain subgroups of, perfect groups. Recall that a group π is said to be *perfect* if π equals its commutator subgroup; equivalently, π is perfect if and only if its Abelianization is trivial. The fundamental group of any grope is perfect. Taking the opposite tack, let π denote the fundamental grope of some space X and let $[f]$ be an element of some perfect subgroup P of π, where $f: S^1 \to X$ is a (based) loop. Then $[f]$ is homotopic to a product of commutators in P, so f bounds a singular disk-with-handles D_0 in X, where images of the simple closed curves from a complete handle curve for D_0 represent conjugates of elements of P. Accordingly, the image handle curves bound disks-with-handles, whose handle curves then bound disks-with-handles, whose handle curves bound, and so on. The infinite process gives a map of a grope D into X agreeing with f on the boundary.

The applications envisioned will cause us to operate on embedded gropes, not just those in E^3 as shown in Fig. 38-2, but also those in arbitrary manifolds of dimension $n \geq 5$. For this we will need the following deep result of W. B. R. Lickorish and L. C. Siebenmann [1].

Theorem 1 (Lickorish–Siebenmann). *Fix an integer $n \geq 5$ and a locally finite k-complex K, where $2k + 1 \leq n$, and consider the closed PL embeddings $F: K \to M$ of K into arbitrary PL n-manifolds M and the regular neighborhoods $U(F(K))$ of $F(K)$ in M. Then the PL homeomorphism types* *of the pairs* $(U(F(K)), F(K))$ *are classified by the tangent PL microbundle of the open PL manifolds* Int $U(F(K))$. *Equivalently, the homeomorphism types are in 1–1 correspondence with the homotopy set* $[K, BPL]$, *where BPL is the classifying space for stable tangent PL microbundles.*

Corollary 1A. *Let D be a grope and* $F: D \to M$ *any closed PL embedding of D in a PL n-manifold M, $n \geq 5$, and let U be a regular neighborhood of $F(D)$ in M. Then the PL homeomorphism type of $(U, F(D))$ is uniquely determined by D and n.*

   Proof. In (Cannon–Bryant–Lacher [1]) it is shown that the homotopy set $[D, BPL]$ consists of a single element, so Theorem 1 applies. ∎

   Consequently, in order to understand all regular neighborhoods of all PL embeddings of a given grope in n-manifolds (for fixed $n \geq 5$), it suffices to completely understand just one. The most convenient embedding to inspect is the one given by the standard realization of D^+ in

$$E^3 = E^3 \times \{0\} \subset E^3 \times E^{n-3} = E^n.$$

Thickening this copy of D in E^n precipitates a familiar sort of wild object in S^n called a crumpled n-cube, meaning the solid in S^n bounded by a wild $(n-1)$-sphere. (Precisely, a *crumpled n-cube C* is a space homeomorphic to the closure of one of the components of $S^n - S$, where S denotes any $(n-1)$-sphere topologically embedded in S^n; its *boundary*, denoted Bd C, is the set corresponding to S, or the set at which C fails to be a boundaryless n-manifold.) A special class of crumpled n-cubes stands out: the closed n-cell-complements. A crumpled n-cube C is a *closed n-cell-complement* if C can be embedded in S^n as the closure of the complement of an n-cell. Impending is a proof (Corollary 40.1D) that all crumpled n-cubes $(n \geq 5)$ are closed n-cell-complements, but the approach to be taken requires early confirmation that thickened gropes belong to this special class.
   What makes closed n-cell-complements advantageous is the fact that they crop up from well-regulated, admissible, cell-like decompositions of the n-cell.

Lemma 2. *If C is any closed n-cell-complement, then there exists a collar $c: S^{n-1} \times I \to B^n$ on $S^{n-1} = \partial B^n$ and there exists a map f of B^n onto C whose nondegenerate point inverses are the fiber arcs $c(\{s\} \times I)$, $s \in S^{n-1}$, of this collar.*

   Proof. Assume C is embedded in S^n so that $\mathrm{Cl}(S^n - C)$ is an n-cell B. There is an embedding $w: \partial B \times I \to B$ with $w(\langle b, 0 \rangle) = b$ for all $b \in \partial B$. Thus, $C \cup w(\partial B \times [0, \tfrac{1}{2}])$ is homeomorphic to B^n, and $f: B^n \to C$

corresponds to the retraction of $C \cup w(\partial B \times [0, \frac{1}{2}])$ to C sending $w(\{b\} \times [0, \frac{1}{2}])$ to $b \in \partial B = \text{Bd } C$.   ∎

Proposition 3. *Let* $D^+ \subset E^3 \times \{0\} \subset E^3 \times E^{n-3} = E^n$, $n \geq 3$, *be the* (*extended*) *standard realization of a compactified grope* D^+, *and let* U *be a regular neighborhood of* D *in* $E^n - (D^+ - D)$. *Then* $C^n = U \cup D^+$ *is a closed n-cell-complement that strong deformation retracts to* D^+.

Showing C^n strong deformation retracts to D^+ is straightforward; the chief difficulty in Proposition 3 is showing C^n is a closed *n*-cell-complement. Several preliminary matters must be discussed before looking at Proposition 3 itself.

By way of introduction to the analysis, focus on the case $n = 3$, in the situation where the grope D^+ is what has been called the fundamental grope, which at every stage consists of disks-with-a-single-handle. See Fig. 38-3. Almost by visual comparison one can see that a familiar admissible cell-like decomposition G_3 of B^3, shown in Fig. 38-4, yields a quotient space equivalent to a natural pinched thickening C^3 of D^+; simply compress the tubes comprising the components of the defining sequence down into small pillboxes spanning these tubes near their centers. Here $C^3 = B^3/G_3$ coincides with the closed 3-cell-complement bounded by the famous Alexander horned sphere.

The case $n = 4$ (still dealing with the fundamental grope D^+) exposes a big clue about the relationship between G_3 and the higher-dimensional

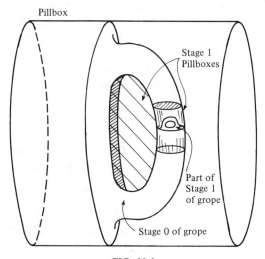

FIG. 38-3

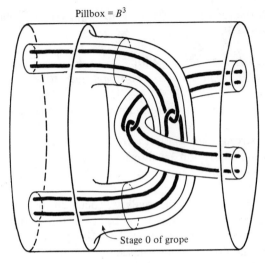

Pillbox = B^3

Stage 0 of grope

FIG. 38-4

thickenings. The enlargement to form C^4 can be regarded as a pinched thickening of C^3 modulo its frontier. Thus,

$$C^4 = \{\langle c, t \rangle \in E^3 \times E^1 \mid c \in C \text{ and } |t| \leq d(c)\},$$

where $d: C \to [0, 1]$ is a map such that $d^{-1}(0) = \text{Bd } C$. The frontier of C^4 in E^4 consists of two copies of C^3, corresponding to the graphs of d and $-d$, which copies are attached together via the identity homeomorphism along their frontiers. By the foregoing analysis, this frontier is equivalent to the space associated with the 0-spin of the decomposition on B^3 of Corollary 28.9B, which yields S^3. Hence, C^4 is a crumpled 4-cube.

In order to explain why it is a closed 4-cell complement and also to study C^n when $n > 4$, it will pay to have some further terminology. For any crumpled n-cube C, define the *inflation* Infl C *of* C as

$$\text{Infl } C = \{\langle c, t \rangle \in C \times E^1 \mid c \in C \text{ and } |t| \leq d(c)\},$$

where $d: C \to [0, 1]$ is a map such that $d^{-1}(0) = \text{Bd } C$. Generally, let C be a space, S a closed subset of C, and $d: C \to [0, 1]$ a map such that $S = d^{-1}(0)$. By the *inflation of C relative to S* we mean the space

$$\text{Infl}(C, S) = \{\langle c, t \rangle \in C \times E^1 \mid c \in C \text{ and } |t| \leq d(c)\}.$$

[The topological type of Infl C or Infl(C, S) does not depend on the choice of map d.] Somewhat similarly, given an admissible cell-like decomposition G of the n-cell B^n, define the *inflated decomposition* $I(G)$ of B^{n+1} and a *doubled decomposition* $2G$ of $S^n = \partial B^{n+1}$ as the decompositions whose

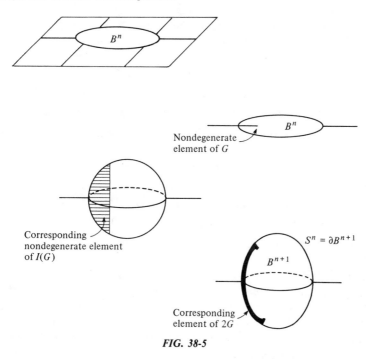

FIG. 38-5

nondegenerate elements are, respectively, the sets $(g \times E^1) \cap B^{n+1}$ and $(g \times E^1) \cap S^n$, where g is a nondegenerate element of G and $E^{n+1} = E^n \times E^1$. See Fig. 38-5.

The next two results set forth relationships among pinched thickenings, inflations, and inflated decompositions. Details are left as exercises.

Lemma 4. *Suppose D^+ is a compactified grope in E^n having a pinched thickening C^n, as in Proposition 3, that is a crumpled n-cube. Then, considered as a subset of E^{n+1}, D^+ has a pinched thickening C^{n+1} homeomorphic to Infl C^n. Moreover, if there exists an admissible cell-like decomposition G_n of B^n for which B^n/G_n is equivalent to C^n, then C^{n+1} is homeomorphic to $B^{n+1}/I(G_n)$.*

Lemma 5. *Suppose G is an admissible cell-like decomposition of B^n, $C = B^n/G$, and S represents the image of S^{n-1} in C. Then $I(G)$ is an admissible cell-like decomposition of B^{n+1}, $B^{n+1}/I(G)$ is homeomorphic to $\mathrm{Infl}(C, S)$, and $S^n/2G$ is homeomorphic to the decomposition space $S^n/\mathrm{Sp}^0(G)$ associated with the 0-spin of G.*

Let G denote an admissible, cell-like decomposition of B^n. An embedding $q: B^n/G \to E^n$ is said to be *collared* if there exists an embedding

$\lambda: S \times I \to E^n$, where S represents the image of S^{n-1} in B^n/G, such that $\lambda(S \times I) \cap q(B^n/G) = \lambda(S \times \{0\})$ and $\lambda(\langle s, 0 \rangle) = q(s)$ for all $s \in S$. When B^n/G is a crumpled n-cube, it admits a collared embedding in E^n precisely when it is a closed n-cell-complement.

Lemma 6. *In the setting of Lemma 5, there is a collared embedding of $B^{n+1}/I(G)$ in E^{n+1} provided the following two conditions are satisfied:*

(a) *the decomposition $I(G)$, extended trivially over E^{n+1}, is a shrinkable decomposition of E^{n+1}, and*

(b) *the decomposition $2G \times E^1$ is a shrinkable decomposition of $S^n \times E^1$.*

Proof. Condition (a) gives an embedding $B^{n+1}/I(G) \to E^{n+1}$, which is collared by condition (b) and the proof of Lemma 31.2. ∎

At this point it is instructive to return our attention to the special crumpled 4-cube C^4. As shown previously, $C^3 = B^3/G_3$ (G_3 being an admissible cell-like decomposition of B^3). By Lemma 4, $C^4 = \text{Infl } C^3$, so $C^4 = B^4/I(G_3)$. Certainly the trivial extension G_3^T of G_3 over E^3 is shrinkable, so $I(G_3)^T$ is shrinkable (see Exercise 6); furthermore, $2G_3 \times E^1$ is a shrinkable decomposition of $S^3 \times E^1$, by Exercise 9.1 or by Theorem 27.1. Hence, Lemma 6 ensures that C^4 has a collared embedding in E^4, implying this crumpled 4-cube is a closed 4-cell-complement.

One should extrapolate from this argument to assimilate the philosophy that the crucial question asks whether the thickened object C^n is a crumpled n-cube, for once that question is decided affirmatively, the refinements needed for establishing C^n is a closed n-cell-complement become routine.

Proof of Proposition 3. Exactly as in the case of the fundamental grope, the standard realization of an arbitrary D^+ in E^3 has a pinched thickening C^3 there that is equivalent to B^3/G_3, where G_3 is a closed-0-dimensional, admissible, cell-like decomposition of B^3 whose trivial extension G_3^T over E^3 is shrinkable. In particular, one can see that the decomposition $G_3 \cap \partial B^3$ induced by G_3 on ∂B^3 is the trivial one, from which it follows directly that $C^3 = B^3/G_3$ is a closed 3-cell-complement.

Consider $n > 3$ and a pinched thickening C^n of D^+ in E^n. The main problem now is to show that C^n is a crumpled n-cube. Once that is done, we can verify it is indeed a closed n-cell-complement by observing $C^n = B^n/G_n$, where G_n is a closed-0-dimensional admissible decomposition of B^n resulting from successive inflations, beginning with G_3. Just as with the fundamental grope, G_n^T is a shrinkable decomposition of E^n (Exercise 6) and, due to the closed-0-dimensional nature of $2G_{n-1}$, $2G_{n-1} \times E^1$ is a shrinkable decomposition of $S^{n-1} \times E^1$ (Corollary 24.3C or Theorem 27.1), so Lemma 6 will attest that C^n is a closed n-cell-complement.

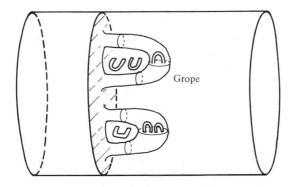

Defining sequence for the decomposition

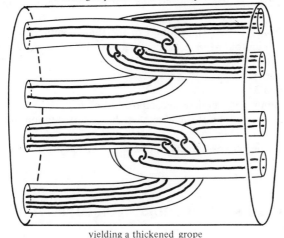

yielding a thickened grope

FIG. 38-6

To show C^n is a crumpled *n*-cube, we start out knowing C^n embeds in E^n, which means we need only show its frontier there is an $(n - 1)$-sphere. The frontier S corresponds to the image of S^{n-1} in $E^n = E^n/G_n^T$ and, hence, S is equivalent to $S^{n-1}/2G_{n-1}$. To complete the argument, we must take into account the differences between the decompositions G_3 of B^3 associated with the fundamental grope and with an arbitrary grope. The latter, exemplified by the decomposition of Lemma 32.1, are just ramified versions of the former. See Fig. 38-6. The other fact to check (another exercise) is that $2G_{n-1}$ equals $\mathrm{Sp}^{n-4}(G_3)$. Then, for the case $n = 4$, $\mathrm{Sp}^0(G_3)$, being just a ramified version of Example 9.1, is shrinkable by the classical techniques employed in Section 9. Consequently, when $n > 5$ the shrinkability of $\mathrm{Sp}^{n-4}(G^3)$ follows from Corollary 28.9A and when $n = 5$ from (Neuzil[1]), or barehanded methods alluded to at the close of Section 32.

Corollary 3A.   *Suppose D is a grope embedded in a PL n-manifold M, n $\geq$ 5, as a closed and PL subset, and suppose U is a regular neighborhood of D in M. Then the Freudenthal compactification C^n ($\approx U \cup D^+$, properly topologized) of U is a closed n-cell-complement.*

Proof.   See also Corollary 1A.   ∎

EXERCISES

1. Show directly that every compactified grope D^+ is contractible.
2. Let f_1 and f_2 be 1-LCC embeddings of a compactified grope D^+ in E^n, $n \geq 6$, such that $f_1|D^+ - D = f_2|D^+ - D$ and $f_i|D$ is PL $(i = 1, 2)$. Then there exists a homeomorphism H of E^n to itself such that $Hf_1 = f_2$ and $H|f_1(D^+ - D)$ is the identity.
3. If F_1 and F_2 are 1-LCC embeddings of the same compactified grope D^+ in a connected n-manifold M, $n \geq 6$, then there exists a homeomorphism H of M to itself such that $HF_1 = F_2$.
4. Prove Lemma 4.
5. Prove Lemma 5.
6. If G is an admissible cell-like decomposition of $B^n \subset E^n \subset E^{n+1}$ whose trivial extension G^T over E^{n+1} is shrinkable, then the trivial extension $I(G)^T$ of $I(G)$ over E^{n+1} is also shrinkable.
7. Show that the decomposition $2G_{n-1}$ of S^{n-1} arising in the proof of Proposition 3 equals $\text{Sp}^{n-4}(G^3)$.

39. REPLACEMENT PROCEDURES FOR IMPROVING GENERALIZED MANIFOLDS

Supported by the analysis of thickened gropes in Section 38, this section conscripts gropes into service as surveyors, as markers of the optimal spots for making improvements to the nonmanifold set in certain generalized manifolds. Bringing about such improvements is likely to entail changing the homeomorphism types of the objects studied. Such change engenders an advance similar to the output of the Štan'ko reembedding theorem in that, given a preassigned subset Z of the source space containing the nonmanifold set, one obtains another embedding of Z in the resultant space, better by virtue of being 1-LCC, whose image also contains the nonmanifold set there.

The work is divided into two separate portions, the first of which deals with the replacement process *per se* and the second, with the blueprint for constructing gropes and thickenings which, when replaced, consummate the desired improvements. The ideas and methods are taken from (Cannon–Bryant–Lacher [1]).

The first portion, the replacement process, applies to a given generalized n-manifold Y containing a specified null sequence of pinched closed n-cell-complements. In that setting we carefully describe a process of replacing the crumpled objects by pinched, real n-cells. The kind of pinching licensed in this section does not conform to the pinched thickening of gropes done in the preceding one but, instead, involves the banding together of collections of boundary points from the n-cells and n-cell-complements. Perhaps the subtlest aspect of the process is the determination of the appropriate topology to impose on the newly constructed space; to make this choice of topology as straightforward as possible, we embed Y as a closed subset of some high-dimensional Euclidean space E^k (it is known that dim $Y \leq n$, implying that Y embeds as a closed subset of E^{2n+1}, but the particular value of k is irrelevant for our purposes).

Formally, we consider the following setting:

Y, a generalized n-manifold embedded as a closed subset of E^k ($k \geq 2n$);

$f_i: B_i \rightarrow C_i$ ($i = 1, 2, 3, \ldots$), continuous surjective functions from n-cells B_i to closed n-cell complements C_i, as in Lemma 38.2, whose nondegenerate point preimages are the arc fibers of some possibly wild collars $(\partial B_i) \times I$ on ∂B_i; and

$g_i: C_i \rightarrow Y$ ($i = 1, 2, 3, \ldots$), continuous functions that embed Int C_i in Y in such a way that $\{g_i(C_i)\}$ forms a null sequence of sets having pairwise disjoint interiors.

In this setting we have the following construction:

Regard Y as a subset of

$$E^k = E^k \times \{0\} \subset E^k \times E^2 = E^{k+2}.$$

Let $q_1, q_2, q_3, \ldots$ denote distinct points from the unit circle in E^2, and let $H_1, H_2, H_3, \ldots$ be the half-spaces in E^{k+2} isometric to E^{k+1}_+, with $q_i \in H_i$ and $\partial H_i = E^k$. Note that any two such H_i intersect only along E^k. For fixed i, $g_i f_i: B_i \rightarrow Y$ takes B_i into $E^k = \partial H_i$, so (because $k + 1 \geq 2n + 1$) there exists a map $h_i: B_i \rightarrow H_i$ satisfying

$$h_i | \text{Int } B_i \text{ is an embedding into Int } H_i = H_i - E^k,$$

$$h_i | \partial B_i = g_i f_i | \partial B_i,$$

and

$$\text{diam } h_i(B_i) < 2 \text{ diam } g_i f_i(B_i) = 2 \text{ diam } g_i(C_i).$$

Define X to be the subspace of E^{k+2} given as

$$X = (Y - \bigcup g_i(C_i)) \cup (\bigcup h_i(B_i)).$$

Proposition 1. *The space X of the Construction is a generalized n-manifold embedded as a closed subset of E^{k+2}, and there exists a cell-like map $p: X \to Y$ that reduces to the identity on $X \cap Y$.*

Proof. *Part 1. X is closed in E^{k+2}.* Since $\{h_i(B_i)\}$ forms a null sequence of compact sets, this should be obvious.

Part 2. X is an ANR. According to Theorem 14.8, it suffices to show the finite-dimensional space X is locally contractible. Before doing that, we describe a controlled homotopy equivalence $X \to Y$. Recalling that f_i takes ∂B_i homeomorphically onto Bd C_i, we can produce another map $e_i: C_i \to B_i$ for which $e_i f_i | \partial B_i$ is the identity. Thus, $e_i f_i \simeq \mathrm{Id}$ (rel ∂B_i) and $f_i e_i \simeq \mathrm{Id}$ (rel Bd C_i), since the ranges in every case are ARs. For $B_i' = h_i(B_i)$ and $C_i' = g_i(C_i)$, the maps f_i and e_i induce maps $f_i': B_i' \to C_i'$ and $e_i': C_i' \to B_i'$ such that $e_i' f_i' \simeq \mathrm{Id}$ (rel $h_i(\partial B_i)$) and $f_i' e_i' \simeq \mathrm{Id}$ (rel $g_i(\mathrm{Bd}\ C_i)$). Piecing these maps e_i' and f_i' together, we obtain homotopy equivalences $e: Y \to X$ and $p: X \to Y$ defined by the rules

$$p(x) = \begin{cases} x & \text{if } x \in X \cap Y \\ f_i'(x) & \text{if } x \in B_i' \end{cases}$$

$$e(y) = \begin{cases} y & \text{if } y \in Y \cap X \\ e_i'(y) & \text{if } y \in C_i'. \end{cases}$$

Then $ep \simeq \mathrm{Id}_X$ (rel $X \cap Y$) via an induced homotopy sending each B_i' into itself and $pe \simeq \mathrm{Id}_Y$ (rel $X \cap Y$) via a similar homotopy sending each C_i' into itself.

Now we can show X is locally contractible at $x \in X$. Since the individual sets $h_i(\mathrm{Int}\ B_i)$ are open in X and locally contractible, it suffices to consider only $x \in X \cap Y$. Given a neighborhood U of x in X, we choose successively (1) a neighborhood T of $x = p(x)$ in Y for which $e(T) \subset U$, (2) a smaller neighborhood T' of x in Y that contracts, via a contraction c_t, in T, (3) another neighborhood $V \subset U$ of x in X for which $p(V) \subset T'$, and (4) a still smaller neighborhood $V' \subset V$ of x such that $ep | V': V' \to U$ is homotopic in U to the inclusion $V' \to U$ (because of the homotopy between ep and Id_X fixing $X \cap Y$). In addition, $ec_t p$ provides another homotopy in $e(T) \subset U$ between $ep | V'$ and a constant map, and the composition of these two homotopies gives a contraction of V' in U, as required.

Part 3. p is a cell-like map. Clearly $p: X \to Y$ is a closed surjection reducing to the identity on $X \cap Y$. In order to study its nondegenerate point preimages, examine the map $f_i: B_i \to C_i$ for some fixed i. By hypothesis the nondegenerate point preimages of the latter are fibers of an interior collar

$(\partial B_i) \times I$ on ∂B_i in B_i. If $\{s\} \times I$ is one of the fibers crushed out by f_i, then $h_i(\{s\} \times I)$ is an arc crushed out by p. There exists a deformation of $\bigcup h_i((\partial B_i) \times I)$ starting with the identity and contracting each fiber $h_i(\{s\} \times I)$ to the endpoint $h_i(\langle s, 0 \rangle)$. Restrictions of this contraction reveal the point inverses of p to be contractible in themselves, showing them to be cell-like subsets of X.

Part 4. *X is a generalized n-manifold.* It suffices to verify this for $x \in X \cap Y$.

Claim 1. $p_*: H_*(X, X - p^{-1}p(x)) \to H_*(Y, Y - \{p(x)\})$ *is an isomorphism.*

By Theorem 17.1, the maps $p: X \to Y$ and $p|: X - p^{-1}p(x) \to Y - \{p(x)\}$ are homotopy equivalences. Thus, this claim follows from the algebraic five lemma, applied to the homology ladder

$$\cdots H_k(X - p^{-1}p(x)) \to H_k(X) \to H_k(X, X - p^{-1}p(x)) \to H_{k-1}(X - p^{-1}p(x)) \to H_{k-1}(X) \to \cdots$$

$$\cdots H_k(Y - \{p(x)\}) \to H_k(Y) \to H_k(Y, Y - \{p(x)\}) \to H_{k-1}(Y - \{p(x)\}) \to H_{k-1}(Y) \to \cdots$$

with vertical maps $(p|)_* \cong$, $p_* \cong$, p_*, $(p|)_* \cong$, $p_* \cong$.

Claim 2. *The inclusion-induced* $\psi_*: H_*(X, X - p^{-1}p(x)) \to H_*(X, X - \{x\})$ *is an isomorphism.*

Examine the commutative diagram

$$\cdots \bar{H}_k(X - \{x\}) \to \bar{H}_k(X) \to H_k(X, X - \{x\}) \to \bar{H}_{k-1}(X - \{x\}) \to \bar{H}_{k-1}(X) \to \cdots$$

$$\cdots \bar{H}_k(X - p^{-1}p(x)) \to \bar{H}_k(X) \to H_k(X, X - p^{-1}p(x)) \to \bar{H}_{k-1}(X - p^{-1}p(x)) \to \bar{H}_{k-1}(X) \to \cdots$$

with vertical maps γ_*, $\cong$, ψ_*, γ_*, $\cong$.

We shall prove the central vertical arrow ψ_* is an isomorphism by verifying the first and fourth arrows γ_* are isomorphisms.

Step 1. γ_* *is monic.* Let z denote a reduced k-cycle in $X - p^{-1}p(x)$ that bounds a $(k + 1)$-chain b in $X - \{x\}$. Only finitely many of $\{h_i(B_i)\}$ can hit both x and b. Let $h_i(B_i)$ be one of them, let $(\partial B_i) \times I$ be the specified interior collar on ∂B_i, whose fiber arcs are the nondegenerate point preimages of $f_i: B_i \to C_i$, and let X_i denote the compact subset of ∂B_i mapped by h_i to x. Without changing z and without changing $b - h_i(\text{Int } B_i)$, we shall alter b in $h_i(\text{Int } B_i)$ so that the altered b misses $h_i(X_i \times I)$. Finite iteration then will yield that z bounds in $X - p^{-1}p(x)$.

Subdividing b, if necessary, we express b as the sum of two chains b' and b'' such that b' is disjoint from z and lies in $h_i(\text{Int } B_i)$ while b'' misses $h_i(X_i \times I)$. Thus $\partial b'$ is a reduced k-cycle in $h_i(\text{Int } B_i - X_i \times (0, 1))$. Regard

B_i as the standard n-cell in S^n. Then $(S^n - \text{Int } B_i) \cup (X_i \times I)$ is contractible, so by Lefschetz duality (Spanier [1, p. 298])

$$\tilde{H}_k(h_i(\text{Int } B_i - X_i \times (0, 1])) \cong \tilde{H}_k(S^n - [(S^n - \text{Int } B_i) \cup (X_i \times I)])$$

$$\cong 0.$$

This implies the existence of a $(k + 1)$-chain c in $h_i(\text{Int } B_i - X_i \times (0, 1])$ such that $\partial b' = \partial c$, and the $(k + 1)$-chain $c + b''$ is the desired alteration of b showing z bounds in $X - h_i(X_i \times I)$.

 Step 2. γ^* *is surjective.* This time let z denote a reduced k-cycle in $X - \{x\}$, and again suppose $h_i(B_i)$ meets both x and z. As in Step 1, $z = b' + b''$ with b' in $h_i(B_i)$ and b'' missing $h_i(X_i \times I)$. Also as in Step 1, $\partial b'$ bounds a k-chain c in $h_i(\text{Int } B_i - X_i \times (0, 1])$. Since $\tilde{H}_k(h_i(\text{Int } B_i))$ is trivial, the cycle $b' - c$ is the boundary of a $(k + 1)$-chain d in $h(\text{Int } B_i) \subset X - \{x\}$. Thus,

$$z - (b'' + c) = (b' + b'') - (b'' + c) = b' - c = \partial d,$$

so z is homologous in $X - \{x\}$ to the cycle $b'' + c$, which misses $h_i(X_i \times I)$. Again, finite iteration moves z off all of $p^{-1}p(x)$.

 Steps 1 and 2 complete the proof of Claim 2. Claims 1 and 2 combine to give that

$$H_*(X, X - \{x\}) \cong H_*(Y, Y - \{p(x)\}),$$

and, since Y is a generalized n-manifold, X is one too. ∎

Remark.   It would be helpful to have a simple topological proof that, if the space Y of the construction is a manifold, so also is X.

 Before selecting specific thickened gropes and then engaging Proposition 1, we derive some technical results needed for the main theorems. The first pair sheds more light on the way gropes unite homology and homotopy properties.

 To be explicit, in a generalized n-manifold X the *singular set*, or the *nonmanifold set*, denoted $S(X)$, consists of all points at which X fails to be a genuine n-manifold.

Lemma 2.   *Suppose Y is a generalized n-manifold, Z is a closed $(n - 3)$-dimensional subset of Y, $\alpha: B^2 \to Y$ is a map, and $\varepsilon > 0$. Then there exist a compactified grope D^+ and a map $\beta: D^+ \to Y$ satisfying*

   (a) $\beta(D^+) \subset N(\alpha(B^2); \varepsilon)$,
   (b) $\beta | \partial D^+$ *is a loop within ε of $\alpha | \partial B^2$, and*
   (c) $\beta^{-1}(Z) = D^+ - D$.

Proof. It follows from Lemma 26.2 that Z, being $(n - 3)$-dimensional, is 0-LCC and 1-lcc (each neighborhood U of $z \in Z$ contains a smaller neighborhood V of z such that every loop in $V - Z$ is null homologous in $U - Z$).

Stage 0. Choose an approximation $\alpha_0: B^2 \rightarrow Y$ to α such that $\alpha_0^{-1}(Z)$ is 0-dimensional and lies in Int B^2. Identify pairwise disjoint subdisks $\Delta_1, \Delta_2, ..., \Delta_m$ of Int B^2 whose interiors cover $\alpha_0^{-1}(Z)$ and whose images are tiny. In particular, assume $\alpha_0(\Delta_i)$ to be so small that $\alpha_0 \mid \partial \Delta_i$ bounds a small singular disk-with-handles $\beta_{0i}: D_{0i} \rightarrow Y - Z$. Let

$$D_0 = B^2 - \bigcup(\text{Int } \Delta_i) \cup (\bigcup D_{0i}),$$

appropriately conjoined, and let $\beta_0: D_0 \rightarrow Y - Z$ be the map defined as α_0 on $B^2 - \bigcup \text{Int } \Delta_i$ and as β_{0i} on D_{0i}. Specify a complete handle curve J_0 for D_0 with $J_0 \subset \bigcup D_{0i}$.

Stage i. Assume D_{i-1}, J_{i-1}, and $\beta_{i-1}: D_{i-1} \rightarrow Y - Z$ already have been defined, with the image of each component of J_{i-1} very small. For each simple closed curve J in J_{i-1}, let $B(J)$ be an abstract 2-disk bounded by J and let $\alpha_J: B(J) \rightarrow Y$ be a small singular 2-disk bounded by $\beta_{i-1} \mid J_i$ with $\alpha_J^{-1}(Z)$ a 0-dimensional subset of Int $B(J)$. Exactly as in Stage 0, replace $B(J)$ by a small disk-with-handles $D(J)$ and α_J by a map $\beta_J: D(J) \rightarrow Y - Z$ near the image of α_J. Set $D_i = \bigcup D(J)$ and $\beta_i = \bigcup \beta_J: D_i \rightarrow Y - Z$. Specify a complete handle curve J_i for D_i, each component of J_i having very small image under β_i.

Stage ω. By installing suitable size controls while constructing D_0, $D_1, ...,$ and $\beta_0, \beta_1, ...,$ we obtain a natural extension $\beta: D^+ \rightarrow Y$ of the map $\bigcup \beta_i: D = \bigcup D_i \rightarrow Y - Z$. Such a map obviously satisfies the conclusions of lemma. ∎

Lemma 2 has a variation in a slightly different setting, the proof of which is left as an exercise.

Lemma 2'. *Suppose Y is a generalized n-manifold, Z is a connected generalized $(n - 1)$-manifold embedded in Y as a closed subset that separates Y into disjoint open sets W and W', $\alpha: B^2 \rightarrow \text{Cl } W$ is a map, and $\varepsilon > 0$. Then there exists a compactified grope D^+ and a map $\beta: D^+ \rightarrow Y$ satisfying*

(a) $\beta(D^+) \subset N(\alpha(B^2); \varepsilon) \cap \text{Cl } W$,
(b) $\beta \mid \partial D^+$ *is a loop within ε of $\alpha \mid \partial B^2$, and*
(c) $\beta^{-1}(Z) = D^+ - D$.

Below is another lemma detailing the existence of partial PL structures and allowing extensive application of Corollary 38.3A in generalized n-manifolds, which, due to their nonmanifold sets, tend not to admit global PL structures.

Lemma 3. *Suppose D is a grope, M is an n-manifold, $n \geq 5$, and $F: D \to M$ is a closed 1-LCC embedding. Then some neighborhood W of $F(D)$ in M admits a PL triangulation for which $F(D)$ is a subpolyhedron.*

Proof. According to (Bryant-Seebeck [1, 2]), $F(D)$ is locally tame. Hence, it has a neighborhood W that strong deformation retracts to $F(D)$. The structure theorem of (Kirby-Siebenmann [1]) attests that the obstruction to imposing a PL structure on W lies in $H^4(W; Z_2) \cong H^4(D; Z_2)$, which is trivial. In the resulting PL structure $F(D)$ is ambiently equivalent to a PL embedding, again by (Bryant-Seebeck [1, 2]), and the inverse of said equivalence transforms the PL structure on W to one in which $F(D)$ is a PL subcomplex. ∎

Next we have the first version of the main result. For its proof the tactics are to identify a dense collection of gropes in Y, to reorganize them into a null sequence of pairwise disjoint gropes, still sufficiently dense, and to replace thickenings of the reorganized collection, as in Proposition 1, thereby forming the desired generalized n-manifold X. It will be clear that the natural copy of Z in X incorporates $S(X)$. While it may be intuitively clear that Z is 1-LCC embedded, a large share of the argument is devoted to detailing why this 1-LCC property holds.

Theorem 4. *Suppose Y is a generalized n-manifold, $n \geq 5$, embedded as a closed subset of some Euclidean space E^k, $k \geq 2n$, and Z is a closed, $(n-3)$-dimensional subset of Y containing $S(Y)$.*

Then there exist a generalized n-manifold X, a 1-LCC embedding $\lambda: Z \to X$ such that $S(X) \subset \lambda(Z)$, and a proper cell-like mapping $p: X \to Y$ such that $p\lambda = \text{Id}_Z$.

Proof. Let $\alpha_1, \alpha_2, \alpha_3, \ldots: B^2 \to Y$ be a dense subset of the space of all maps $B^2 \to Y$. By Lemma 2, there exist compactified gropes ${}^1D^+, {}^2D^+, {}^3D^+, \ldots$ and maps $\beta_i: {}^iD^+ \to Y$ satisfying

(a) $\beta_i({}^iD^+) \subset N(\alpha_i(B^2); 1/i)$,
(b) $\beta_i | \partial^iD^+$ is a loop within $1/i$ of $\alpha_i | \partial B^2$, and
(c) $\beta_i^{-1}(Z) = {}^iD^+ - {}^iD$.

Since the nonmanifold part $S(Y)$ of Y lies in Z, it follows from (c) that each $\beta_i({}^iD)$ is a closed subset of the manifold $Y - Z$. Consequently, by the proof of Proposition 24.1 we may assume that the maps $\beta_i | {}^iD: {}^iD \to Y - Z$ are closed 1-LCC embeddings having pairwise disjoint images.

By Lemma 3 each of the sets $\beta_i({}^iD)$ has a neighborhood W_i in $Y - Z$ admitting a PL triangulation for which $\beta_i({}^iD)$ is a subpolyhedron. We could apply Corollary 38.3A to choose pinched closed n-cell complements C_i such that

$$\beta_i({}^iD^+) \subset C_i \subset \text{Cl } W_i,$$

but that action would be premature, because we also need to have the collection $\{C_i\}$ form a null sequence, so we must vary the construction somewhat.

First crumpled cubes. Choose a stage ^1D_k of ^1D such that images of components $^1D - (^1D_1 \cup \cdots \cup {}^1D_k)$ under β_1 have diameter less than 1. Let U_1 denote the interior of a regular neighborhood of $^1D_1 \cup \cdots \cup {}^1D_k$ in ^1D. Then $^1D - U_k$ is a finite union of pairwise disjoint gropes mapped by β_1 to sets of diameter less than 1. It is these gropes, rather than $\beta_1(^1D)$, that we shall thicken in W_1.

Determine a PL triangulation of W_1 such that $\beta_1(^1D - U_1)$ covers a subcomplex, and choose a second derived subdivision such that, if M_1 is the simplicial neighborhood of $\beta_i(^1D - U_1)$ in this second derived, then:

(1)   each component of M_1 has diameter less than 1, and
(2)   each simplex of M_1 meeting $\beta_1(^1D_j \cup {}^1D_{j+1} \cup \cdots)$ has diameter less than $1/j$ and misses $\beta_j(^jD)$ ($j > 1$).

Let $M_{11}, \ldots, M_{1,k(1)}$ be the components of M_1 whose closures touch Z. Their Freudenthal compactifications $C_{11}, \ldots, C_{1,k(1)}$ are closed n-cell-complements by Corollary 38.3A, and there is a unique map

$$g_{1j}: C_{1j} \to \operatorname{Cl} M_{1j} \subset Y$$

that reduces to the identity on M_{1j}. In addition, there exist n-cells B_{1j} and cell-like maps $f_{1j}: B_{1j} \to C_{1j}$ satisfying the conclusions of Lemma 38.2.

Later crumpled cubes. Assume all $f_{ij}: B_{ij} \to C_{ij}$ and $g_{ij}: C_{ij} \to \operatorname{Cl} M_{ij} \subset Y$ have been chosen for $i < m$, with the $C_{i,j}$ pairwise disjoint objects missing $\beta_r(^rD)$, $r > i$, and having diameter $< 1/i$. Proceed as with the first one, choosing a stage ^mD_s of ^mD such that images of components $^mD - (^mD_1 \cup \cdots \cup {}^mD_s)$ under β_m have diameter less than $1/m$ and miss all of the M_{ij}, $i < m$. Pick U_m as U_1 was picked, triangulate W_m as W_1 was triangulated so that $\beta_m(^mD - U_m)$ covers a subcomplex, and choose the second derived so that the simplicial neighborhood M_m of $\beta_m(^mD - U_m)$ satisfies

(1_m) each component of M_m has diameter less than $1/m$ and misses all previous M_{ij}'s and
(2_m) each simplex of M_m hitting $\beta_m(^mD_j \cup {}^mD_{j+1} \cup \cdots)$ has diameter less than $1/j$ and misses $\beta_j(^jD)$ ($j > m$).

Again let $M_{m1}, \ldots, M_{m,k(m)}$ be the components of M_m whose closures touch Z, and let $C_{m1}, \ldots, C_{m,k(m)}$ denote their Freudenthal compactifications, with $g_{mj}: C_{mj} \to \operatorname{Cl} M_{mj} \subset Y$ and $f_{mj}: B_{mj} \to C_{mj}$ the associated maps.

Repeating the above, one eventually obtains countably many maps $f_{ij}: B_{ij} \to C_{ij}$ and $g_{ij}: C_{ij} \to \operatorname{Cl} M_{ij} \subset Y$ producing the setting previously described. Let X be the generalized n-manifold, $p: X \to Y$ the proper cell-like map, and $\lambda: Z \to X$ the identity embedding promised in Proposition 1.

Clearly λ is a closed embedding and $p\lambda = \text{Id}_Z$. Moreover, $X - \lambda(Z)$ is a manifold because only finitely many changes are made near any point of $Y - Z$, due to the fact that each of the sets $g_{ij}(C_{ij})$ intersects Z and $\{g_{ij}(C_{ij})\}$ forms a null sequence.

Because $X - p^{-1}(Z)$ and $Y - Z$ are manifolds, we can apply the cell-like approximation theorem to modify p in a manner that introduces no changes over Z and provides a new cell-like surjection $p: X \to Y$ such that $p \mid p^{-1}(Y - Z): p^{-1}(Y - Z) \to Y - Z$ is 1-1. The nondegeneracy set of this new map p then corresponds to those collar arcs from B_{ij} whose images under $g_{ij}f_{ij}$ belong to Z. Since the set of points in any C_{ij} mapped by g_{ij} to Z is 0-dimensional, consisting solely of those points associated with the tip set $D^+ - D$ of that compactified grope giving rise to C_{ij}, the preimage of Z in B_{ij} under $g_{ij}f_{ij}$ is 1-dimensional. As a consequence, $p^{-1}(Z) - \lambda(Z)$ is also 1-dimensional.

The only thing left to do is to check that the embedding $\lambda: Z \to X$ is 1-LCC. Toward that end, consider a neighborhood U of $z_0 \in \lambda(Z)$. Find an open set U^* about $\lambda(Z) - U$ such that $p^{-1}p(z_0) \cap \text{Cl } U^* = \varnothing$, and set $O = U \cup U^*$.

Claim.   *There exist a neighborhood W of $p^{-1}(Z)$ in X and a map $R: W \to O$ fixing a neighborhood Q of $\lambda(Z)$, sending $W - \lambda(Z)$, into $O - \lambda(Z)$, and sending each $p^{-1}(z)$ into itself.*

Proof of the Claim.   Let $h: B_h \to X$ be any map from an n-cell B_h defining one of the pinched n-cells in X used in changing Y to X. Let $S_h = \partial B_h$ and let $S_h \times [0, 1]$ denote the collar on S_h in B_h (with $S_h = S_h \times \{0\}$) used to identify the nondegenerate point preimages of $B_h \to C_h$, where $g(C_h) = \text{Cl } M_h \subset Y$. Select a point b_h of $B_h - (S_h \times [0, 1])$ and name a retraction $r_h: B_h - \{b_h\} \to S_h \times [0, 1]$ for which $r_h(B_h - (S_h \times [0, 1])) = S_h \times \{1\}$. (If the existence of r_h is not obvious, just retract a neighborhood of $S_h \times [0, 1]$.) Pick $t_h \in (0, 1]$ and let $s_h: S_h \times [0, 1] \to S_h \times [0, t_h]$ be the retraction sending $\{s\} \times [t_h, 1]$ to $\langle s, t_h \rangle$. Then for some $t_h \in (0, 1]$ and some neighborhood W_h of $h(B_h \cap p^{-1}(Z))$ in $h(B_h)$, $h s_h r_h h^{-1}(W_h) \subset O$.

With $\mathfrak{F}$ representing the (finite) set of defining maps $h: B_h \to X$ for which $h(B_h)$ is not contained in O, define W as

$$W = (O - \bigcup\{h(B_h) \mid h \in F\}) \cup (\bigcup\{W_h \mid h \in \mathfrak{F}\}),$$

and define $R: W \to X$ by the formula

$$R(x) = \begin{cases} x & \text{if } x \in O - \bigcup\{h(B_h) \mid h \in \mathfrak{F}\} \\ h s_h r_h h^{-1}(x) & \text{if } x \in W_h \quad (h \in \mathfrak{F}). \end{cases}$$

This map R has the properties called for in the claim.   ∎

Apply the claim to locate a neighborhood V of $p^{-1}p(z_0)$ in W such that $R(V) \subset U$. Choose neighborhoods T and T' of $p(z_0)$, with $T' \subset T$, $p^{-1}(T) \subset V$, and, for some integer K, every $\mathrm{Cl}\, M_{ij}$ having diameter less than $1/K$ and meeting T' lies in T. Then pick neighborhoods V' and V'' of z_0 in $Q \cap V \cap U$ such that $p(V') \subset T'$ and every loop in V'' is null homotopic in V'.

Having made all the necessary constructions, we assert that every loop L in $V'' - \lambda(Z)$ is null homotopic in $U - \lambda(Z)$. Since $p^{-1}(Z) - \lambda(Z)$ is 1-dimensional, L can be adjusted in $V'' - \lambda(Z)$ to a loop $L^*: S^1 \to V'' - p^{-1}(Z)$. By construction, L^* extends to a map $\alpha: B^2 \to V'$. The singular disk $p\alpha: B^2 \to T'$ then is the uniform limit of some subsequence of α_1, $\alpha_2, \ldots: B^2 \to Y$. Choose $m > K$ so that $\beta_m({}^m D^+) \subset T'$, $\beta_m(\partial^m D) \subset T' - Z$, and $p^{-1}\beta_m | \partial^m D$ is homotopic to L^* in $V'' - p^{-1}(Z)$. To complete the argument, it suffices to verify $p^{-1}\beta_m | \partial^m D$ is null homotopic in $U - \lambda(Z)$.

Recall the manner in which portions of $\beta_m({}^m D^+)$ were thickened in $Y - Z$ to form M_m. With $M_{m1}, \ldots, M_{mk}$ denoting the components of M_m whose closures touch T', note that each M_{mj} is a pinched crumpled n-cube of diameter less than $1/K$, which implies it lies in T. Using the spaces and maps

$$
\begin{array}{ccc}
B_{mj} & \xrightarrow{\;h_{mj}\;} & B'_{mj} \subset X \\
{\scriptstyle f_{mj}}\big\downarrow & & \big\downarrow{\scriptstyle p} \\
C_{mj} & \xrightarrow{\;g_{mj}\;} & \mathrm{Cl}\, M_{mj} \subset Y
\end{array}
$$

associated with the sets C_{mj} in the construction, one finds that $p^{-1}\beta_m | \partial^m D^+$ contracts in

$$ p^{-1}(T' - Z) \cup \left(\bigcup h_{mj}(\operatorname{Int} B_{mj}) \right) \subset (X - \lambda(Z)) \cap p^{-1}(T). $$

Since R fixes $p^{-1}\beta_m(\partial^m D)$, $p^{-1}(T) \subset V$, and $R(V - \lambda(Z)) \subset U - \lambda(Z)$, it follows that $p^{-1}\beta_m | \partial^m D^+$ contracts in $(X - \lambda(Z)) \cap Rp^{-1}(T) \subset U - \lambda(Z)$, as was to be shown. ∎

Here is an alternative version of the main result.

Theorem 4'. *Suppose Y is a generalized n-manifold, $n \geq 5$, embedded as a closed subset of some Euclidean space E^k, $k \geq 2n$, and Z is a generalized $(n\text{-}1)$-manifold embedded as a closed subset of Y with $Z \supset S(Y)$.*

Then there exist a generalized n-manifold X, a 1-LCC embedding $\lambda: Z \to X$ such that $S(X) \subset \lambda(Z)$, and a proper cell-like mapping $p: X \to Y$ such that $p\lambda = \mathrm{Id}_Z$.

Proof. The only variation from the argument for Theorem 4 involves a splitting at the very beginning. To get started properly, assume Z to be connected and to separate Y into two components Y_1 and Y_2, properties which hold locally (by duality), and for $j = 1, 2$, let

$$\alpha_1^j, \alpha_2^j, \alpha_3^j, \ldots : B^2 \to \text{Cl } Y_j$$

be dense in the space of all maps $B^2 \to \text{Cl } Y_j$. Then proceed exactly as in Theorem 4, invoking Lemma 2' instead of Lemma 2, and replace pinched crumpled cubes in both $\text{Cl } Y_1$ and $\text{Cl } Y_2$ with pinched n-cells. ∎

EXERCISES

1. Prove Lemma 2'.

40. RESOLUTIONS AND APPLICATIONS

Throughout this text the preponderant stress has fallen upon cell-like decompositions and cell-like maps $f: M \to Y$ defined on manifolds. The bulk of the information given usually has pertained to M; based on that information, the overriding objective has been to deduce corresponding information about Y. What now looms up ahead concerns an important reverse problem: given a space Y, under what conditions can one obtain a cell-like surjection $f: M \to Y$ defined on an n-manifold M?

Succinctly, in the reigning terminology, the problem concerns the existence of resolutions. A *resolution for a space* Y is a pair (M, f) consisting of a manifold M and a proper cell-like surjection $f: M \to Y$. On a finite-dimensional space Y a necessary condition for the existence of a resolution is that Y be a generalized n-manifold (Corollary 26.1A). The condition is nearly sufficient; that fact, due to F. Quinn [3, 4], has profound implications, some of which will be presented in this section.

The exceedingly complex proof of Quinn's result involves controlled surgery techniques quite unlike anything used here, techniques far too lengthy and too complicated for us to adequately treat in this text. While the proof is beyond our grasp, nevertheless the result itself cannot be ignored; the most powerful unifying aspects and far-reaching consequences of cell-like decomposition theory elsewhere in geometric topology frequently depend on the existence of such resolutions. Hence, because of their overwhelming significance, developing resolutions by nonsurgical methods seems a worthwhile endeavor. That constitutes a partial purpose behind this section. In case the nonmanifold set lies in an $(n - 1)$-manifold we apply a result of Seebeck and Ferry and, using it, in case the singular set is low-dimensional we recite an argument of Cannon, Bryant, and Lacher, involving only relatively familiar techniques, both of which lead directly to the existence of resolutions and both of which also have fundamental applications. Then we take up Quinn's resolution theorem and study the unilateral applications.

The key result to be employed initially is stated below. Originally it was claimed by C. L. Seebeck and A. V. Černavskiĭ, independently, but eventually Černavskiĭ's claim was retracted. Later, better results were obtained by S. Ferry [1], who published the first proof. The theorem also was derived by Quinn, as a consequence of his end theorem [1].

Theorem 1 (Seebeck–Ferry). *Suppose M is a locally compact metric ANR containing a closed subset B such that $M - B$ is an n-manifold, $n \geq 5$, B is an $(n - 1)$-manifold 1-LCC embedded in M, and $H_*(M, M - \{b\}) \cong 0$ for all $b \in B$. Then M is an n-manifold with boundary and $\partial M = B$.*

Remark. In case M is known to be an n-manifold with boundary at some point b of B (connected), a situation arising in many applications, then a proof can be given based on traditional engulfing methods initiated by Price and Seebeck [1]. The local collar structure at b can be dragged along $(n - 1)$-cells in B to produce local collars everywhere throughout B. Theorem 1 obviously incorporates the 1-LCC characterization of flatness in codimension one (see Černavskiĭ [3] or Price-Seebeck [1] and Daverman [2]), which attests that every $(n - 1)$-manifold 1-LCC embedded in an n-manifold, $n \geq 5$, is locally bicollared.

Corollary 1A. *If X is a generalized n-manifold, $n \geq 5$, whose singular set $S(X)$ is contained in an $(n - 1)$-manifold Z embedded as a closed, 1-LCC subset of Y, then X is an n-manifold.*

Proof. Locally Z separates Y into two pieces and satisfies the hypotheses of Theorem 1 in the closure of each. ∎

Corollary 1B. *If Y is a generalized n-manifold, $n \geq 5$, whose singular set $S(Y)$ is contained in an $(n - 1)$-manifold Z embedded as a closed subset of Y, then Y has a resolution.*

Proof. Theorem 39.4′ provides another generalized manifold X, a copy of Z 1-LCC embedded in X and including $S(X)$, and a cell-like map $p: X \to Y$. By Corollary 1A, $p: X \to Y$ is a (manifold) resolution of Y. ∎

The existence of resolutions in the above setting brings about another proof of the locally flat approximation theorem for codimension one embeddings by F. D. Ancel and J. W. Cannon [1].

Corollary 1C (Ancel–Cannon). *Each closed embedding of an $(n - 1)$-manifold N in an n-manifold M, $n \geq 5$, can be approximated, arbitrarily closely, by a locally flat embedding.*

Proof. As in Corollary 1B there exists a cell-like map $p: X \to M$ defined on a generalized n-manifold X and a 1-LCC embedding $\lambda: N \to X$ such that $S(X) \subset \lambda(N)$ and $p\lambda = \text{Id}_N$. Corollary 1A ensures X is a manifold in which $\lambda(N)$ is locally bicollared and, therefore, Edwards's cell-like approximation theorem yields homeomorphic approximations $h: X \to M$ to p, which transform the local bicollars on $\lambda(N)$ to bicollared embeddings near N. ∎

In Corollary 1C as well as in several other applications to manifolds given in this section, the conclusions would be even more accessible if there were an elementary verification that when the replacement procedures of Section 39 are put into operation on an n-manifold Y, the resulting generalized manifold X is automatically an n-manifold.

Corollary 1B also furnishes a new proof that all crumpled n-cubes are closed n-cell-complements (Daverman [4]). In effect, this indicates that codimension one wildness can be studied one side at a time.

Corollary 1D. *Each crumpled n-cube C, $n \geq 5$, is a closed n-cell-complement. In fact, if C is embedded in S^n, then there exist embeddings $e: C \to S^n$, arbitrarily close to the inclusion, such that $S^n - e(\text{Int } C)$ is an n-cell.*

Proof. Applying the technique of Theorem 39.4′ only on the one side $S^n - \text{Int } C$, we find a generalized n-manifold Y, a cell-like map $p: Y \to S^n$, and an embedding $e: C \to Y$ such that $pe = \text{Id}_C$ and $Y - e(C)$ is 1-LCC. Theorem 1 implies that $K = Y - e(\text{Int } C)$ is an n-manifold with $(n - 1)$-sphere boundary. In addition, Y satisfies the DDP, a fact left as an exercise. Since Y has a resolution by Corollary 1B, the cell-like approximation theorem certifies Y is manifold. Another application, starting this time with $p: Y \to S^n$, yields homeomorphisms $\mu: Y \to S^n$ with μe as close as desired to the inclusion $C \to S^n$. Finally, by the generalized Schönflies theorem, K is an n-cell. ∎

Examined next are analogues to the above for generalized manifolds in which the dimension of the singular set falls in the strongly trivial range. Proposition 2 was derived by Cannon, Bryant, and Lacher [1], who through much extra exertion confirmed it for the slightly better range $2 \cdot (\dim S(Y)) + 2 \leq n$.

Proposition 2. *If X is a generalized n-manifold whose singular set $S(X)$ is 1-LCC embedded and has dimension k, where $2k + 3 \leq n$, then X is an n-manifold.*

Proof. First consider the case where $S(X)$ is compact 0-dimensional. Given a Cantor set K topologically embedded in a connected n-manifold M $(n \geq 3)$, J. W. Alexander [1] demonstrated how to build an n-cell B in M so that K lies in its boundary. Beginning with a small n-cell near K, one simply pushes out along branched feelers nearer and nearer to K. In particular, when $n > 3$, this can be organized so that, at the end, K is standardly embedded in ∂B, as a tame subset.

Precisely the same construction can be applied in X to show that $S(X)$ lies standardly in the boundary of some n-cell B in X. Here "standardly" means $S(X)$ is 1-LCC embedded in ∂B. If any reasonable sort of construction procedure is followed, $\partial B - S(X)$ will be locally flatly or 1-LCC embedded in $X - S(X)$, from which it will follow that B itself is 1-LCC embedded in X. Theorem 1 indicates $Cl(X - B)$ is an n-manifold with boundary, and X, being expressed as the union of two n-manifolds B and $Cl(X - B)$, which have common boundary, is an n-manifold without boundary.

The proof for the general case, where $2k + 3 \leq n$, depends on essentially the same idea. Focus on $s \in S(X)$. Since $S(X)$ is 1-LCC and low-dimensional, there exists a map (embedding) $m: S(X) \to X - S(X)$ homotopic in X to the inclusion. Find a closed n-cell B in $X - S(X)$ containing $m(s)$ in its interior. Then s has a closed neighborhood T in $S(X)$ for which $m(T) \subset \text{Int } B$. Due to the dimension restriction on $S(X)$, there exists a 1-LCC embedding $f: T \to \partial B$, which then is homotopic in X to the inclusion $T \to X$. It is easy to find a homotopy $F: T \times I \to X - \text{Int } B$ between f and the inclusion; since $S(X)$ and B are 1-LCC, it can be adjusted so that $F(T \times (0, 1)) \cap (S(X) \cup B) = \varnothing$, and then it can be adjusted further, using the manifold properties of $X - S(X)$, to be a 1-LCC embedding. Now the arcs $F(\{t\} \times I)$ can be squeezed down to T; specifically, one can find a map q of X with nondegeneracy set $\{F(\{t\} \times I) \mid t \in T\}$ and such that $q \mid S(X) = \text{Id}_{S(X)}$. Then $q \mid B$ is 1-1. Check that $q(B)$ is 1-LCC embedded. In an open subset U of s with $U \cap S(X) \subset T$, apply Theorem 1 to see that $U - q(\text{Int } B)$ is an n-manifold with boundary $U \cap q(\partial B)$ and, therefore, U is an n-manifold without boundary. Accordingly, every point of $S(X)$ has a manifold neighborhood in X, so X itself is a manifold. ∎

Proposition 2 has two antecedents, both about generalized manifolds having 0-dimensional singular sets. Bryant and Hollingsworth [1] proved that a generalized n-manifold X, $n \geq 5$, has a resolution provided $\dim S(X) = 0$ and $X \times E^1$ is a manifold. Bryant and Lacher [1] were able to do the same without hypothesizing that $X \times E^1$ is a manifold.

Corollary 2A. *If Y is a generalized n-manifold, $n \geq 5$, with k-dimensional singular set $S(Y)$, where $2k + 3 \leq n$, then Y has a resolution.*

See also Theorem 39.4.

Corollary 2B. *For $n \geq 4$ each homology n-sphere H^n bounds a contractible $(n + 1)$-manifold.*

Proof. The open cone Y on H^n is a generalized $(n + 1)$-manifold whose singular set $S(X)$ is the cone point. Theorem 39.4 produces another generalized $(n + 1)$-manifold X and a cell-like map $p: X \to Y$ carrying the potential nonmanifold set $S(X)$ onto $S(Y)$ in 1-1 fashion, such that $S(X)$ is 1-LCC. Since X is a manifold by Proposition 2, we can adjust p over $Y - S(Y)$ so that it carries $p^{-1}(Y - S(Y))$ homeomorphically onto $Y - S(Y)$. Then, for a natural copy of the (closed) cone C on H^n in Y, $p^{-1}(C)$ is an $(n + 1)$-manifold bounded by H^n, which is contractible because p is a fine homotopy equivalence (Theorem 17.1). ∎

Corollary 2C. *The double suspension of every homology n-sphere H^n is S^{n+2}.*

Proof. Assume $n \geq 3$. For the most part this was taken care of in Corollary 24.3D, except that in some cases extraneous results were invoked asserting that the double suspension of H^n has a resolution. Since $\Sigma^2 H^n$ is a generalized $(n + 2)$-manifold having 1-dimensional singular set, that step now follows from Corollary 2A. ∎

Finally, we will consider Quinn's full resolution theorem and its consequences.

Theorem 3 (Quinn [3, 4]). *A connected generalized n-manifold Y, $n \geq 5$, has a resolution if and only if a certain integer-valued local index $i(Y)$ equals 1.*

Remark. Whether the obstruction $i(Y)$ to the existence of a resolution ever is nonunitary is unknown. It is known to be locally defined and locally constant [4]. Hence, Y is resolvable whenever some nonempty open subset of Y is. According to Quinn [1] an object like Y has a resolution if its product with some E^k does, and consequently Y has a resolution if stably some nonempty open subset does.

Corollary 3A. *Each connected generalized n-manifold Y, $n \geq 5$, for which $S(Y) \neq Y$ has a resolution.*

As previously mentioned, Theorem 3 and Edwards's cell-like approximation theorem coalesce to provide a striking topological characterization of manifolds. The potentially unnecessary hypothesis for Theorem 3 brings about the solitary discordant note in the statement; elimination of the need for that one instantaneously would eliminate the need for the corresponding phrase in Theorem 4.

Theorem 4 (Edwards–Quinn). *A generalized n-manifold Y, $n \geq 5$, is an n-manifold if and only if it satisfies the disjoint disks property and each component of Y has a local index 1.*

Corollary 4A. *Let Y be a generalized n-manifold, $n \geq 4$, whose singular set $S(Y)$ is either $(n - 2)$-dimensional or lies in a generalized $(n - 1)$-manifold contained in Y as a closed subset. Then $Y \times E^1$ is an $(n + 1)$-manifold.*

Proof. By the proof of Corollary 24.3C or, respectively, of Theorem 26.12, $Y \times E^1$ has the DDP. ∎

Theorem 4 sustains an extension of Corollary 1C to cover the case where N is a generalized $(n - 1)$-manifold whose product with E^1 is a manifold.

Corollary 4B. *Let N be a generalized $(n - 1)$-manifold, $n \geq 5$, whose product with E^1 is a manifold. Each closed embedding of N in an n-manifold M can be approximated, arbitrarily closely, by a locally bicollared embedding.*

Proof. Theorem 39.4' gives a generalized n-manifold X in which a 1-LCC copy of N contains the singular set, plus a cell-like map $X \to M$. Corollary 3A promises a resolution for X. That resolution is used here, in combination with Lemma 26.11 (applied locally), to verify X is an n-manifold. With engulfing methods Cannon [2] has shown that locally N has product neighborhoods in X of the form $U \times (-1, 1)$ (U open in N), from which the conclusion follows as in Corollary 2B. ∎

The existence of resolutions for those generalized manifolds having codimension three singular sets leads to another proof of the very beautiful codimension three tame approximation theorem of M. A. Štan'ko [3], which inspired the development of gropes. His theorem attests that embeddings of codimension three compacta in manifolds can be approximated by 1-LCC embeddings. When the compacta themselves are manifolds, the 1-LCC embeddings are locally flat; when the compacta are polyhedra and the range manifold is PL, the 1-LCC approximations are ambiently equivalent to PL embeddings.

Corollary 4C (Štan'ko). *If Z is an $(n - 3)$-dimensional closed subset of an n-manifold M and $\varepsilon: Z \to (0, 1)$ is continuous, then there exists a 1-LCC embedding $\lambda: Z \to M$ satisfying $\rho(z, \lambda(z)) < \varepsilon(z)$ for all $z \in Z$.*

Proof. For $n < 5$ this stems from the classical techniques of Hurewicz and Wallman [1] showing that the embeddings are dense in the space of all maps from Z to M. For $n \geq 5$ this follows from a variant to the proof of Corollary 1C. ∎

There is also a uniqueness aspect to resolutions. The source manifold M appearing in any resolution of a given generalized n-manifold Y is, in most cases, known to be unique. When Y represents a genuine n-manifold ($n \geq 5$), for which the only resolutions are the near-homeomorphisms, this fact is obvious; with arbitrary generalized manifolds of this dimension, it is not quite so obvious but follows from the thin h-cobordism theorem of Quinn [1].

Theorem 5 (Quinn). *Let $f_1: M_1 \to Y$ and $f_2: M_2 \to Y$ be two resolutions of a generalized n-manifold Y, $n \geq 5$. Then, for each open cover $\mathcal{U}$ of Y, there exists a*

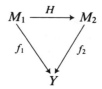

homeomorphism $H: M_1 \to M_2$ such that $f_2 H$ is $\mathcal{U}$-close to f_1.

The uniqueness of resolutions lends support to the following result suggested in (Cannon [6]) about the normal form for closed-$(n - 3)$-dimensional cell-like decompositions.

Theorem 6 (Normal form for decompositions). *Suppose G is a closed-$(n - 3)$-dimensional cell-like decomposition of an n-manifold M, $n \geq 5$. Set $Y = M/G$. Then there exist a closed 1-LCC embedding $\lambda: S(Y) \to M$ and a cell-like map $p: M \to Y$ satisfying:*

(a) $p\lambda = \mathrm{Id}_{S(Y)}$,
(b) *p is 1-1 over $Y - S(Y)$,*
(c) *$p^{-1}(S(Y)) - \lambda(S(Y))$ is 1-dimensional, and*
(d) *each nondegenerate preimage of p is a 1-dimensional, contractable set.*

Proof. Virtually all of this was established in the course of deriving previous results. In particular, in Section 39 (Proposition 39.1 and Theorem 39.4) it was shown how to obtain a cell-like map $p: X \to Y$ and 1-LCC embedding $\lambda: S(Y) \to X$ satisfying conclusions (a)–(c) and such that $S(X) \subset \lambda(S(Y))$. The proof actually reveals what the point inverses are like: for $s \in S(Y)$ $p^{-1}(s)$ is composed of a null sequence of cones over 0-dimensional compacta, all wedged together at the common cone point $\lambda(s)$. This verifies conclusion (d). By Theorem 4, X is an n-manifold.

In a sense this is almost enough: the study of $Y = M/G$ can be approached by studying the decomposition of X induced by p. Theorem 6 adds the final touch—the manifold X coincides with the given manifold M begetting Y. ∎

EXERCISES

1. Let C be any crumpled $(n - 1)$-cube in E^{n-1}, $n \geq 5$, such that $\mathrm{Sp}^0(C) = S^{n-1}$. Apply Theorem 1 to prove $\mathrm{Infl}(C)$ is a crumpled n-cube whose boundary is collared in $\mathrm{Cl}(E^n - \mathrm{Infl}(C))$.

2. Attach an n-cell B to an arbitrary crumpled n-cube C, $n \geq 5$, via a homeomorphism between their boundaries. Show directly that $Y = B \cup C$ is a generalized n-manifold.

3. Show that the space Y of Exercise 2 satisfies the DDP.

4. If B is an n-cell in a generalized n-manifold Y and $K \subset \partial B$ is a Cantor set satisfying (i) K is 1-LCC embedded in Y and (ii) $\partial B - K$ is 1-LCC embedded in $Y - \mathrm{Int}\, B$, then B is 1-LCC embedded in Y.

5. Show that the n-cell $q(B)$ identified in the proof of Proposition 2 is 1-LCC embedded.

6. Prove Corollary 4C.

7. Let $f_1: M_1 \to Y$ and $f_2: M_2 \to Y$ be two resolutions of a generalized n-manifold Y, $n \geq 4$. Identify the mapping cylinders of f_1 and f_2 together along Y to form a space W. Show that W is a compact $(n + 1)$-manifold having two boundary components M_1 and M_2, each of which is a strong deformation retract of W.

41. MAPPING CYLINDER NEIGHBORHOODS

Every finite-dimensional ANR Y can be embedded in some manifold of sufficiently high dimension. Taking a retrospective look at such ANRs Y, in this concluding section we seek optimal neighborhoods of the embedded objects. Given a closed subset X of a manifold M, we say W is a *manifold mapping cylinder neighborhood* of X in M (henceforth to be abbreviated as manifold MCN) provided (1) W is a manifold with boundary, (2) W is a closed neighborhood of X in M, and (3) there is a proper retraction $r: W \to X$ such that W is equivalent to the mapping cylinder, denoted $\mathrm{Map}(r')$, of the map $r' = r|\partial W$, under a homeomorphism preserving both X and ∂W. Certainly those compacta X having manifold MCNs must themselves be ANRs. The aim of this section is to establish a converse, that every ANR nicely embedded in a manifold M, subject to certain codimension restrictions, has a manifold MCN. The proof relies heavily on the approximation results previously developed and ultimately, in a crucial way, on a special case of Quinn's resolution theorem, Theorem 40.3.

This sort of result was first obtained in the PL category by J. H. C. Whitehead [2], who showed that a PL regular neighborhood of a subcomplex X of a PL manifold M is a (PL) manifold MCN of X. In the wider topological category, M. W. Hirsch [1] showed that stably a submanifold X of a manifold M has a disk-bundle neighborhood; that is, $X = X \times \{0\}$ has a disk-bundle neighborhood in $M \times E^k$, where k is large.

Work on the problem is closely connected to an old question first raised in 1954 by K. Borsuk [2], which asks whether every compact metric ANR Y has the homotopy type of a finite complex. In 1950 Whitehead [3] had proved that every such Y has the homotopy type of a countable CW complex and asked whether it would have the type of a finite-dimensional one. Affirmative answers to Borsuk's question were obtained for the 1-connected ANRs by C. B. deLyra [1] in 1957, for finite-dimensional manifolds by Kirby–Siebenmann [1] in 1969, and for Hilbert cube manifolds by T. A. Chapman [1] in 1973. The issue was completely settled in 1974 by J. E. West [1], inspired by work of R. T. Miller [1] at virtually the same time. Miller showed that if Y is a compact, k-dimensional ANR 1-LCC embedded in an n-manifold M, where $n - k \geq 4$ and $n \geq 5$, then $Y \times [0, 1)$ has a manifold MCN in $M \times (-1, 1)$. Transferring the problem to the infinite-dimensional world, West obtained the analogue for an arbitrary compact ANR Y nicely embedded (as a Z-set) in a Hilbert cube manifold M and employed infinite-dimensional techniques to prove that then Y itself has a Hilbert cube manifold MCN in $M \times (-1, 1)$; the final answer to Borsuk's question followed immediately from the work of Chapman. More recently, F. Quinn developed an alternative way to answer the question, based on his end theorem (Quinn [1]).

Our initial concern will be with the proof of Miller's result. Later, after locating a manifold MCN W of $Y \times [0, 1)$ in $M \times (-1, 1)$, we will reproduce an argument presented by R. D. Edwards in his 1978 lectures at the CBMS Regional Conference held in Oklahoma. In essence, his argument exploits decomposition methods to show the closure of W in $M \times E^1$, a kind of pinched neighborhood of $Y \times [0, 1]$ there, is a generalized manifold (in fact, is a manifold factor) MCN of Y itself. Application of Quinn's work on resolutions quickly will lead to a genuine manifold MCN.

For pinpointing precisely which embedded ANRs have manifold MCNs, Quinn's alternative argument mentioned above gives a better result than what is derived here. While he is able to treat any codimension 3 ANR Y 1-LCC embedded in an n-manifold M, we not only must restrict to codimension 4 ANRs Y but also must stabilize to $M \times E^1$ before successfully obtaining a manifold MCN of $Y = Y \times \{0\}$. On the other hand, for addressing the basic question of which abstract ANRs have manifold MCNs, the approach to be taken does offer two advantages. First, it involves a more constructive technique: given a specific embedded ANR Y, the procedures outlined can be retraced to recover a specific map determining the MCN. Second, it involves a simpler argument: without relying on anything more powerful than engulfing results, it displays the existence of manifold factor MCNs. Showing that manifold factors have resolutions, which requires ingredients used to prove Quinn's end theorem (see Quinn [1] or Chapman [3]), unquestionably is far less complicated than establishing the full-strength resolution theorem.

Theorem 1 (Miller). *Let Y be a compact, k-dimensional ANR 1-LCC embedded in an n-manifold M, where $n - k \geq 4$ and $n \geq 5$. Then $Y \times [0, 1)$ has a manifold MCN in $M \times (-1, 1)$.*

Following Miller, we say that a subset X of M is *constrictable* if for each $\varepsilon > 0$ there exists an ε-pseudo-isotopy ψ_t of M to itself, fixed on X and outside $N(X; \varepsilon)$, such that ψ_t retracts some neighborhood of X to X and is a homeomorphism over $M - X$. Given a constrictable subset X of M, which of necessity is an ANR, Miller devised an ingenious, captivating proof that $X \times [0, 1)$ has a manifold MCN in $M \times (-1, 1)$. Before looking at it, we spell out the link with Theorem 1.

Lemma 2. *If $Y \subset M$ is a compact ANR as in Theorem 1, then Y is constrictable in M.*

This depends on an engulfing argument. A sketch of the proof, for the case where M is a PL manifold, is given in an Appendix at the end of this section.
The following result furnishes the basic technical controls.

Lemma 3. *Suppose $Y \subset M$ is constrictable and $\varepsilon > 0$. Then there exists a proper, level-preserving, surjective map $F: M \times I \to M \times I$ such that, for all $t \in I$,*

(a) $F_t: M \to M$ is cell-like,
(b) $F_0 = \text{Id}_M$,
(c) $F_t | M - N(Y; \varepsilon) = \text{Id}$,
(d) F_t is 1-1 over $M - Y$, and
(e) for $s < t$, $F_s^{-1}(Y) \subset \text{Int}_M F_t^{-1}(Y)$.

Proof. Let ρ represent a metric on $M \times I$. By reparameterizing the pseudo-isotopies ψ_t stemming from the constrictability of Y, for every $a \in (0, 1]$ and $\varepsilon \in (0, a)$ we build a level-preserving surjective map $G[a, \varepsilon]: M \times I \to M \times I$ satisfying

  (i) $\rho(G[a, \varepsilon], \text{Id}) < \varepsilon$,
  (ii) $G[a, \varepsilon]$ is fixed off the ε-neighborhood of $Y \times I$ and on $(M \times [0, a - \varepsilon]) \cup (Y \times I)$,
  (iii) $G[a, \varepsilon]$ is 1-1 over $((M - Y) \times I) \cup (M \times [0, a))$,
  (iv) $G[a, \varepsilon]_a: M \to M$ retracts some neighborhood of Y to Y, and
  (v) $G[a, \varepsilon]_s = G[a, \varepsilon]_a$ for all $s \in [a, 1]$.

Of course, here

$$G[a, \varepsilon] = \begin{cases} \text{Id} & \text{for } t \in [0, a - \varepsilon] \\ \psi_1 & \text{for } t \in [a, 1] \\ \psi_u & \text{for } t \in [a - \varepsilon, a], \quad \text{where } u = u(t) \in [0, 1]. \end{cases}$$

The desired map $F: M \times I \to M \times I$ will be the limit of compositions of suitable $G[a, \varepsilon]$'s.

Fix $\varepsilon > 0$ and choose positive numbers $\varepsilon_0, \varepsilon_1, \ldots$ for which $\Sigma \varepsilon_i < \varepsilon/2$ and $\varepsilon_i < \frac{1}{2}^{i+1}$. Assume without loss of generality that $N(Y; \varepsilon)$ has compact closure.

Set $H_0 = G[1, \varepsilon_0]$.

Set $H_1 = G[1, \varepsilon_0] \circ G[\frac{1}{2}, \varepsilon_1']$, where among other properties to be mentioned when pertinent, $\varepsilon_1' \in (0, \varepsilon_1)$ is chosen small enough that $\rho(H_1, H_0) < \varepsilon_1$.

Set $H_2 = G[1, \varepsilon_0] \circ G[\frac{3}{4}, \varepsilon_2'] \circ G[\frac{1}{2}, \varepsilon_1'] \circ G[\frac{1}{4}, \varepsilon_2']$, with $\varepsilon_2' \in (0, \varepsilon_2)$ chosen to ensure (among other things) $\rho(H_2, H_1) < \varepsilon_2$.

Continue in this fashion, determining $H_3, \ldots, H_i, \ldots$ (see Fig. 41-1).

Define $F: M \times I \to M \times I$ to be the limit of these H_i's. Most of the properties required of F are standardly derived. It is a proper, level-preserving, surjective map, because $\{H_i\}$ forms a Cauchy sequence of such maps, all having common compact support. Clearly $F_0 = \text{Id}_M$ and $F_t | M - N(Y; \varepsilon) =$ inclusion. Each F_t, being the limit of cell-like maps, is itself cell-like (Theorem 17.4). One can impose additional bounds on $\varepsilon_j' \in (0, \varepsilon_j)$, as has been done countless times before, to force F_t to be 1-1 over $M - Y$.

The final property required of F, however, is less standard. To understand why it holds, we must understand $F^{-1}(Y \times I)$, which is approachable through studying the various $H_i^{-1}(Y \times I)$. Restricting the $\varepsilon_j' \in (0, \varepsilon_j)$ even

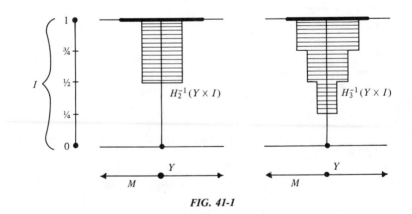

FIG. 41-1

more, for every dyadic rational d $= m/2^i \in (0, 1]$, we can force

$$F_d^{-1}(Y) = (H_i)_d^{-1}(Y)$$

and

$$F_s^{-1}|M - \text{Int}_M(F_d^{-1}(Y)) = F_d|M - \text{Int}_M(F_d^{-1}(Y)) \quad \text{whenever } s \in [0, d].$$

This implies $F_s^{-1}(Y) \subset F_d^{-1}(Y)$ and, more generally, $F_s^{-1}(Y) \subset F_t^{-1}(Y)$ for all $s, t \in I$ with $s < t$. Finally, given such $s, t \in I$ we can obtain dyadics $c = (m - 1)/2^i$ and $d = m/2^i$ for which $s < c < d < t$ and then see that

$$F_s^{-1}(Y) \subset F_c^{-1}(Y) = (H_i)_c^{-1}(Y)$$

$$\subset \text{Int}_M(H_i)_d^{-1}(Y) \subset F_d^{-1}(Y) \subset F_t^{-1}(Y). \quad \blacksquare$$

Proof of Theorem 1. Let $F: M \times I \to M \times I$ denote the map of Lemma 3. Restrict F to $M \times [0, 1)$ and extend via the identity on $M \times (-1, 0]$ to get a proper, level-preserving surjection $F: M \times (-1, 1) \to M \times (-1, 1)$.

According to Conclusion (e) of Lemma 3, the set

$$U = \bigcup\{F_t^{-1}(Y) \mid t \in [0, 1)\}$$

is open in M. Except for the unsatisfactory feature that $W^* = U \times [-\frac{1}{2}, 1)$ fails to be closed in $M \times (-1, 1)$, W^* supplies the precursor of the mapping cylinder structure. To overcome the nonclosedness of W^*, determine a first coordinate-preserving embedding h of W^* in itself such that $h(W^*)$ is closed in $M \times (-1, 1)$ and $h(U \times (-\frac{1}{2}, 1)) \supset F^{-1}(Y \times [0, 1))$. This can be accomplished by defining $u: U \to [0, 1)$ as

$$u(x) = \inf\{t \in [0, 1) \mid x \in F_t^{-1}(Y)\},$$

noting that u is continuous by conclusion (e) of Lemma 3, naming another continuous function $u^*: U \to [-\frac{1}{2}, 1)$ such that $u^*(x) < u(x)$ and $u^*(x) \to 1$ as $x \to M - U$, and defining h so $h(\langle x, -\frac{1}{2}\rangle) = \langle x, u^*(x)\rangle$ and $h(\langle x, t\rangle) = \langle x, t\rangle$ when $t > u(x)$. The effect is illustrated in Fig. 41-2.

Specify another first coordinate-preserving embedding $\Lambda: U \times I \to W^*$ with $\Lambda(\langle x, 0\rangle) = \langle x, u^*(x)\rangle$ and $\Lambda(\langle x, 1\rangle) = \langle x, u(x)\rangle$ for all $x \in U$. Note that

$$\Lambda(U \times [0, 1)) = h(W^*) - F^{-1}(Y \times [0, 1)).$$

Define W as $Fh(W^*)$. The rest should be transparent: W is a manifold with boundary, where $\partial W = Fh(U \times \{-\frac{1}{2}\})$, and W is naturally equivalent to the mapping cylinder of $F\Lambda i: U \to Y \times [0, 1)$, where $i: U \to U \times \{1\}$ denotes inclusion. Explicitly, the mapping cylinder structure lines in W, shown in Fig. 41-3, are the images under F of the vertical arcs, those of the form $\Lambda(\{x\} \times I)$, from $h(U \times \{-\frac{1}{2}\})$ up to $F^{-1}(Y \times [0, 1))$. $\blacksquare$

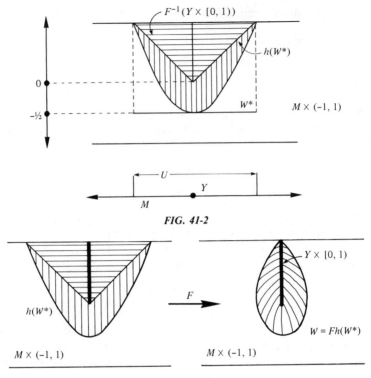

FIG. 41-2

FIG. 41-3

Having produced one manifold MCN of $Y \times [0, 1)$ in $M \times (-1, 1)$, next we improve it to produce another one on $Y \times \{0\}$. We proceed by examining

$$Z = W \cup (Y \times \{1\}) \subset M \times E^1,$$

a set to be regarded as a mapping cylinder neighborhood of $Y \times I$ pinched at $Y \times \{1\}$.

Several additional constructions help expose the mapping cylinder structure on Z. Let $R_t: Z \to Z$ denote the strong deformation retraction of Z to $Y \times I$ moving points at a constant rate along the mapping cylinder structure of W. Then R_t is cell-like (point inverses are contractable) and invariably $R_t^{-1}(Y \times \{1\}) = Y \times \{1\}$.

Define sets

$$A_t = F_t^{-1}(Y) - \bigcup \{F_s^{-1}(Y) \mid s \in [0, t)\}$$

and

$$Z_t = (Fh(F_t^{-1}(Y) \times \{-t/2, t\}) \cup (A \times [-t/2, t])).$$

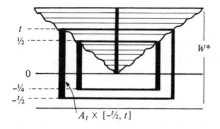

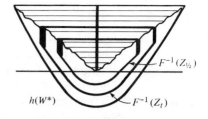

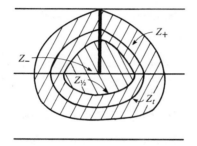

FIG. 41-4

Split Z apart along $Z_{1/2}$, to form two compacta Z_+ and Z_-, with $Z_+ = \bigcup\{Z_t \mid t \in [\frac{1}{2}, 1]\}$ and $Z_- = \bigcup\{Z_t \mid t \in [0, \frac{1}{2}]\}$ (see Fig. 41-4).

There is a map $s: Z_- \to Y$ defined as $s(z) = \text{proj} \circ R(z)$ (proj: $Y \times I \to Y$ denoting the projection). Geometrically $s(z)$ is determined by moving z along the mapping cylinder flow lines to some $\langle y_z, t_z \rangle \in Y \times [0, \frac{1}{2}]$ and setting $s(z) = y_z$. For $\langle y, t \rangle \in Y \times [0, \frac{1}{2}]$, $s(\langle y, t \rangle) = y$. As the composition of cell-like maps, s is cell-like.

Still to be defined is a map Φ of Z_+ to $X \times [\frac{1}{2}, 1]$, where X is some cell-like image of $Z_{1/2}$. This will be carried out componentwise. Let X be the decomposition space resulting from the usc decomposition G of $Z_{1/2}$ having non-degeneracy set $\{Fh(\{a\} \times [-\frac{1}{4}, \frac{1}{2}]) \mid a \in A_{1/2}\}$, and let $\pi: Z_{1/2} \to X$ be the decomposition map. Using coordinate projections p_1 and p_2 on $X \times [\frac{1}{2}, 1]$, define $p_2\Phi(z)$ to be the unique $t \in [\frac{1}{2}, 1]$ for which $z \in Z_t$. If $R_t(z) \cap Z_{1/2} \neq \emptyset$ for some t, define $p_1\Phi(z) = \pi(R_t(z))$ for the smallest such t; otherwise, represent $R_t(z)$ as $\langle y, s \rangle \in Y \times (\frac{1}{2}, 1]$ and define $p_1\Phi(z) = \pi(\langle y, \frac{1}{2} \rangle) \in Z_{1/2}/G$.

Lemma 4. $\Phi: Z \to X \times [\frac{1}{2}, 1]$ *is a cell-like map.*

Proof. That $\Phi | Z_{1/2}$ is cell-like is obvious, so concentrate on $\Phi | Z_s$, where $s > \frac{1}{2}$. Note that $\Phi | Y \times [\frac{1}{2}, 1]$ is 1-1.

By Exercise 17.1 it suffices to show ΦF is cell-like. Focus on $z \in X \times \{s\}$, $s > \frac{1}{2}$, the only points of interest being $z = \Phi(\langle y, s \rangle)$ for some $y \in Y$, and consider $F^{-1}\Phi^{-1}(z) = C_z^*$. Since $C_z = h^{-1}(C_z^*)$ is topologically equivalent to C_z^*, it is enough to prove that the more regularly parameterized C_z is cell-like.

There are three distinguishable parts to C_z:

the top $T_z = C_z \cap (M \times \{s\})$,
the bottom $B_z = C_z \cap (M \times \{-s/2\})$,
and the lateral sides $L_z = \text{Cl}(C_z - (T_z \cup B_z))$.

Here $T_z = F_s^{-1}(y) \times \{s\}$, which therefore is cell-like,

$$L_z = (F_s^{-1}(y) \cap A_s) \times [-s/2, s],$$

and

$$B_z = \bigcup \{A_t \cap F_t^{-1}(y) \mid t \in [\tfrac{1}{2}, s]\} \times \{-s/2\}.$$

Define Q_z as

$$Q_z = C_z \cup (B_z^* \times [-s/2, s]),$$

where $B_z = B_z^* \times \{-s/2\}$. Then Q_z is cell-like, due to the existence of the vertical cell-like map $Q_z \to T_z$ shown in Fig. 41-5.

We prove that C_z itself is cell-like by suggesting how to construct, for any neighborhood O of C_z, a contraction c_t of C_z in O. Given points $s(1) \in (0, \frac{1}{2})$ and $s(2) \in (\frac{1}{2}, s)$ we form an auxiliary cell-like decomposition $\mathcal{G}$ of $M \times (-1, 1)$ having

$$\{F_{s(1)}^{-1}(y) \times \{t\} \mid y \in Y, t \in (-1, -s(1))\}$$

$$\cup \{F_{s(2)}^{-1}(y) \times \{t\} \mid y \in Y, t \in [-s(1), s(2)]\}$$

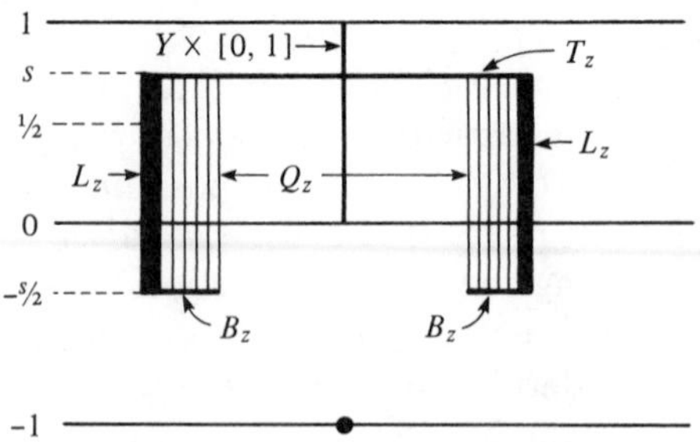

FIG. 41-5

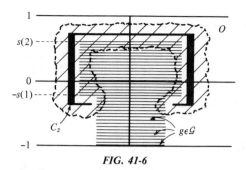

FIG. 41-6

as its nondegeneracy set. For some choice of $s(1)$, $s(2)$ and also some $s(3) \in (s, 1)$ one can find a neighborhood O^* of $y \in Y$, where $\Phi(\langle y, s \rangle) = z$, such that (see Fig. 41-6)

$$O \supset F_s^{-1}(O^*) \times (-s(3)/2, s(3)) - N_{\mathcal{G}}.$$

By the proof of Theorem 13.2, one can show the existence of a level-preserving map f of $M \times (-1, 1)$ onto itself realizing the decomposition $\mathcal{G}$, sending $N_{\mathcal{G}}$ onto $Y \times (-1, s(2)]$, and controlled to ensure that $f | C_z =$ inclusion and $f(O) \supset Q_z$. Define a contraction η_t of Q_z in $f(O) - (Y \times (-1, s(2)])$ and set $c_t = f^{-1} \eta_t | C_z$. ■

Corollary 4A. $X \times (\frac{1}{2}, 1)$ *is an* $(n + 1)$*-manifold.*

Proof. Observe that, under the natural map $\pi: Z_{1/2} \to X$, $\pi | Y \times \{\frac{1}{2}\}$ is 1-1 and $X - \pi(Y \times \{\frac{1}{2}\})$ is an n-manifold. This corollary follows from the variation to Corollary 24.3C given as Exercise 24.9. ■

Theorem 5 (Edwards). *There is a mapping cylinder neighborhood of* $Y \times \{0\}$ *in* $M \times E^1$ *determined by a cell-like map* $X \to Y \times \{0\}$.

Proof. The restriction $s | Z_{1/2}: Z_{1/2} \to Y$ induces a map $\tilde{s}: X \to Y$ and gives rise to a mapping cylinder indentification $\psi: X \times [\frac{1}{2}, 1] \to \text{Map}(\tilde{s})$ behaving like $\tilde{s}$ on $X \times \{\frac{1}{2}\}$. This has been arranged so that $\psi\Phi: Z \to \text{Map}(\tilde{s})$ extends to a well-defined $\Psi: Z \to \text{Map}(\tilde{s})$ for which

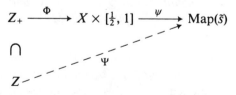

$\Psi | Z_- = s: Z_- \to Y \subset \text{Map}(\tilde{s})$. Consequently, Ψ is cell-like, for on $\Psi^{-1}(Y) = Z_-$, Ψ agrees with the cell-like map s, while on $\Psi^{-1}(\text{Map}(\tilde{s}) - Y) = Z - Z_-$, Ψ is also cell-like, essentially by Lemma 4.

Pushing upward again along the mapping cylinder structure lines, we devise a map d_1 of Z onto itself such that $d_1 | Z_- = R_1 | Z_-$ and d_1 is 1-1 over $Z - (Y \times [0, \frac{1}{2}])$. Compressing vertically downward (relative to the E^1 direction), we form a second map d_2 of Z onto itself such that $d_2(Y \times [0, \frac{1}{2}]) = Y \times \{0\}$ and d_2 is 1-1 over $Z - (Y \times \{0\})$. The composition $d = d_1 \circ d_2$ is 1-1 over $Z - (Y \times \{0\})$, satisfies $d^{-1}(Y \times \{0\}) = Z_-$, and induces the same decomposition on Z_- as s.

Define $\Psi': Z \to \mathrm{Map}(\tilde{s})$ as $\Psi' = \Psi d^{-1}$. Then Ψ' is cell-like, is 1-1 over Y, and sends each $\langle y, 0 \rangle \in Y \times \{0\}$ to y. Because $\psi(X \times (\frac{1}{2}, 1))$ is an $(n + 1)$-manifold by Corollary 4A, the cell-like approximation theorem indicates $\Psi' | \mathrm{Int}\, Z$ can be approximated by a homeomorphism $\Psi^*: \mathrm{Int}\, Z \to \psi(X \times [\frac{1}{2}, 1))$ for which $\Psi^* | Y \times \{0\} = \Psi' | Y \times \{0\}$. Finally, $\Psi^{*-1}(\psi(X \times [\frac{1}{2}, \frac{3}{4}]))$ is a mapping cylinder neighborhood of $Y \times \{0\}$.   ∎

Together Corollary 4A and Theorem 5 demonstrate that $Y \times \{0\}$ has a manifold factor mapping cylinder neighborhood in $M \times E^1$. The concluding step, which requires application of a weak form of Theorem 40.3, for generalized manifolds that are manifold factors (see Quinn [1]), promising a cell-like map $\mu: N \to X$ having a compact n-manifold N as its domain, is left as an exercise.

Theorem 6.   *Let Y be a compact ANR 1-LCC embedded in an n-manifold M, where $n \geq 5$ and $\dim Y \leq n - 4$. Then $Y \times \{0\}$ has a manifold MCN in $M \times E^1$ (naturally equivalent to $\mathrm{Map}(\tilde{s}\mu)$).*

APPENDIX

Here is an outline of the engulfing procedures needed for showing the constrictability of $Y \subset M$.

Lemma A1.   *Let Y be a compact ANR 1-LCC embedded in a (PL) n-manifold M, where $n - \dim Y \geq 3$, and let $\varepsilon > 0$. Then Y has an open neighborhood $U \subset N(Y; \varepsilon)$ such that to any open neighborhood V of Y there corresponds another neighborhood Q of Y in V and for every finite k-complex K in U there exists a PL ε-push H_t of (M, Y) satisfying $H_0 = \mathrm{Id}_M$, $H_t | Q = \text{inclusion}$, and $H_1(V) \supset K$.*

Sketch of the proof.   Distinguishing this from the conclusions of most engulfing results is the claim that H_t can be obtained leaving the neighborhood Q of Y pointwise fixed. To do this, first one sets up according to the usual procedure, obtaining a neighborhood U such that, for every k-complex K' in U, every neighborhood V of Y, and every finite $(n - 3)$-complex L in V, there exists an $(\varepsilon/4)$-push θ_t of (M, Y) fixing a neighborhood of L, such that $\theta_1(V) \supset K'$. Using demension theory, one chooses a nice neighborhood Q of Y in V for which there exists a PL $(\varepsilon/4)$-map $c: M \to M$ fixed outside V and collapsing W to some finite complex L, where $\dim L = \text{dem } Y \leq n - 3$, such that c restricts to a homeomorphism of $M - Q$ onto $M - L$. [Here Q is a regular neighborhood of part of the (dem Y)-skeleton S of a small mesh triangulation T of M, chosen so the dual $(n - \text{dem } Y - 1)$-skeleton to S in T', the first barycentric subdivision of T, misses Y.] Then obtain θ_t engulfing $K' = c(K)$ from $V = c(V)$ while keeping a neighborhood of $L = c(Q)$ fixed, and define

$$H_t(x) = \begin{cases} x & \text{if } x \in Q \\ c^{-1}\theta_t c(x) & \text{if } x \in M - Q. \end{cases} \quad \blacksquare$$

Lemma A2. *Let Y be a compact ANR 1-LCC embedded in a (PL) n-manifold M, where $n - \dim Y \geq 4$ and $n \geq 5$, and let $\varepsilon > 0$. Then there exists an open neighborhood U_ε of Y in $N(Y; \varepsilon)$ with the property that for any neighborhood V of Y there exists an ε-push ϕ_t of (M, Y) fixed outside $N(Y; \varepsilon)$ and on some neighborhood of Y such that $\phi_1(U_\varepsilon) \subset V$.*

Sketch of the proof. Apply Lemma A1 with positive number $\varepsilon/5$ to obtain U^*. Restrict U^* to an $(\varepsilon/20)$-thickening U_ε^* of some (dem Y)-complex R, using the fact that Y is 1-LCC embedded, and delete a thin collar of U_ε^* to obtain a similar object U_ε, a closed PL regular neighborhood of Y with $Y \subset \text{Int } U_\varepsilon$. Given V, let Q be the neighborhood of Y, as promised in Lemma A1, corresponding to V and $\varepsilon/5$. Determine a triangulation T of U_ε having mesh less than $\varepsilon/5$ so small that any simplex of T touching Y is contained in Q; then $\text{St}(Y, T)$ contains another neighborhood Q' of Y in Q. Adjust T, using the hypothesis that Y is 1-LCC embedded, so that the dual 2-skeleton P to $T^{(n-3)}$ in T' misses Y, and invoke Lemma A1 to obtain an $(\varepsilon/5)$-push H_t of (M, Y) fixed on Q for which $H_1(V) \supset T^{(n-3)}$. Because of the extra restriction dem $Y \leq n - 4$, for each i-complex C $(i = 0, 1, 2)$ in $M - Y$ there exists an $(\varepsilon/20)$-homotopy pulling C into $M - U$ through $M - Y$, having support in $U_\varepsilon - Y$ and having $U_\varepsilon^* - Y$ as the range of its support. Consequently, by radial engulfing methods (Bing [11]) there exists an $(\varepsilon/5)$-push g_t of (M, Y), fixed outside U_ε^* and on Y, such that $g_1(M - U_\varepsilon) \supset P$. Stretch $g_1(M - U_\varepsilon)$ across the join structure of T via an $(\varepsilon/5)$-push p_t of (M, Y) fixed outside U_ε^* and on Q' such that

$$M = p_1 g_1(M - U_\varepsilon) \cup H_1(V).$$

Then $U_\varepsilon \subset (p_1 g_1)^{-1} H_1(V)$, and $\phi_t = (p_t g_t)^{-1} H_t$ is an ε-push of (M, Y) moving no point of Y, as required. $\quad \blacksquare$

Proof of Lemma 2.   This is now routine. Let $\varepsilon > 0$. For $j = 1, 2, \ldots$ let U_j be the open set containing Y given by Lemma A2 for $\varepsilon_j = \varepsilon/2^j$. Repeated applications of Lemma A2 give ε_j-pushes ϕ_t^j of (M, Y), fixing Y, such that $\phi_1^j(U_j) \subset U_{j+1}$. The limit of $\phi_t^k \cdots \phi_1^2 \phi_t^1$ (properly parameterized) shows Y to be constrictable.   ■

EXERCISES

1. If Y is a constrictable compact subset of an n-manifold M, show that $Y \times S^1$ has a manifold MCN in $M \times S^1$.
2. Construct the map f mentioned in the proof of Lemma 4.
3. Prove Theorem 6.

REFERENCES

ALEXANDER, J. W.

[1] An example of a simply connected surface bounding a region which is not simply connected. *Proc. Nat. Acad. Sci. U.S.A.* **10** (1924), 8-10.

ANCEL, F. D., AND CANNON, J. W.

[1] The locally flat approximation of cell-like embedding relations. *Ann. of Math. (2)* **109** (1979), 61-86.

ANDREWS, J. J., AND CURTIS, M. L.

[1] *n*-space modulo an arc. *Ann. of Math. (2)* **75** (1962), 1-7.

ANDREWS, J. J., AND RUBIN, L. R.

[1] Some spaces whose product with E^1 is E^4. *Bull. Amer. Math. Soc.* **71** (1965), 675-677.

ANTOINE, M. L.

[1] Sur l'homéomorphie de deux figures et de leurs voisinages. *J. Math. Pures Appl.* **86** (1921), 221-325.

ARMENTROUT, S.

[1] Monotone decompositions of E^3. *In* "Topology Seminar, Wisconsin 1965" (R. J. Bean and R. H. Bing, eds.), Ann. of Math. Studies No. 60, pp. 1-25. Princeton Univ. Press, Princeton, New Jersey, 1966.

[2] Homotopy properties of decomposition spaces. *Trans. Amer. Math. Soc.* **143** (1969), 499-507.

[3] UV Properties of compact sets. *Trans. Amer. Math. Soc.* **143** (1969), 487-498.

[4] A decomposition of E^3 into straight arcs and singletons. *Dissertationes Math. (Rozprawy Mat.)* **73** (1970).

[5] A three-dimensional spheroidal space which is not a sphere. *Fund. Math.* **68** (1970), 183-186.

[6] Cellular decompositions of 3-manifolds that yield 3-manifolds. *Mem. Amer. Math. Soc.*, No. 107 (1971).

[7] Saturated 2-sphere boundaries in Bing's straight-line segment example. *In* "Continua, Decompositions, Manifolds" (R. H. Bing, W. T. Eaton, and M. P. Starbird, eds.), pp. 96-110. Univ. of Texas Press, Austin, Texas, 1983.

ARMENTROUT, S., AND PRICE, T. M.

[1] Decompositions into compact sets with UV properties. *Trans. Amer. Math. Soc.* **141** (1969), 433-442.

BASS, C. D.

[1] Squeezing m-cells to $(m - 1)$-cells in E^n. *Fund. Math.* **110** (1980), 35–50.

[2] Some products of topological spaces which are manifolds, *Proc. Amer. Math. Soc.* **81** (1981), 641–646.

BEAN, R. J.

[1] Decompositions of E^3 with a null sequence of starlike equivalent nondegenerate elements are E^3. *Illinois J. Math.* **11** (1967), 21–23.

BEGLE, E. G.

[1] The Vietoris mapping theorem for bicompact spaces. *Ann. of Math.* *(2)* **51** (1950), 534–543.

BING, R. H.

[1] A homeomorphism between the 3-sphere and the sum of two solid horned spheres. *Ann. of Math.* *(2)* **56** (1952), 354–362.

[2] Upper semicontinuous decompositions of E^3. *Ann. of Math.* *(2)* **65** (1957), 363–374.

[3] A decomposition of E^3 into points and tame arcs such that the decomposition space is topologically different from E^3. *Ann. of Math.* *(2)* **65** (1957), 484–500.

[4] An alternative proof that 3-manifolds can be triangulated. *Ann. of Math.* *(2)* **69** (1959), 37–65.

[5] The Cartesian product of a certain nonmanifold and a line is E^4. *Ann. of Math.* *(2)* **70** (1959), 399–412.

[6] Tame Cantor sets in E^3. *Pacific J. Math.* **11** (1961), 435–446.

[7] A set is a 3-cell if its Cartesian product with an arc is a 4-cell. *Proc. Amer. Math. Soc.* **12** (1961), 13–19.

[8] Point-like decompositions of E^3. *Fund. Math.* **50** (1962), 431–453.

[9] Decompositions of E^3. *In* "Topology of 3-Manifolds and Related Topics" (M. K. Fort, Jr., ed.), pp. 5–21. Prentice-Hall, Englewood Cliffs, New Jersey. 1962.

[10] Each disk in E^3 contains a tame arc. *Amer. J. Math.* **84** (1962), 583–590.

[11] Radial engulfing. *In* "Conference on the Topology of Manifolds, 1967" (J. G. Hocking, ed.), pp. 1–18. Prindle, Weber & Schmidt, Boston, 1968.

BING, R. H., AND BORSUK, K.

[1] A 3-dimensional absolute retract which does not contain any disk. *Fund. Math.* **54** (1964), 159–175.

BING, R. H., AND KISTER, J. M.

[1] Taming complexes in hyperplanes. *Duke Math. J.* **31** (1964), 491–511.

BLANKINSHIP, W. A.

[1] Generalization of a construction of Antoine. *Ann. of Math.* *(2)* **53** (1951), 276–297.

BORSUK, K.

[1] "Theory of Retracts." Pol. Sci. Publ., Warsaw. 1967.

[2] Sur l'elimination de phénomènes paradoxaux en topologie général. *Proc. Int. Congr. Math., Amsterdam, 1954*, pp. 197–208. North-Holland Publ., Amsterdam, 1957.

BOTHE, H. G.

[1] Ein eindimensionales Kompaktum in E^3, das sich nicht lagetreu in die Mengershe Universalkurve einbetten lässt. *Fund. Math.* **54** (1964), 251–258.

BROWN, MORTON

[1] A proof of the generalized Schoenflies theorem. *Bull. Amer. Math. Soc.* **66** (1960), 74–76.

[2] Locally flat embeddings of topological manifolds. *Ann. of Math.* *(2)* **75** (1962), 331–341.

[3] Wild cells and spheres in high dimensions. *Michigan Math. J.* **14** (1967), 219–224.

BRYANT, J. L.

[1] Euclidean space modulo a cell. *Fund. Math.* **63** (1968), 43–51.

[2] On embeddings of compacta in Euclidean space. *Proc. Amer. Math. Soc.* **23** (1969), 46–51.

[3] Euclidean n-space modulo an $(n - 1)$-cell. *Trans. Amer. Math. Soc.* **179** (1973), 181–192.
BRYANT, J. L., AND HOLLINGSWORTH, J. G.
[1] Manifold factors that are manifold quotients. *Topology* **13** (1974), 19–24.
BRYANT, J. L., AND LACHER, R. C.
[1] Resolving 0-dimensional singularities in generalized manifolds. *Math. Proc. Cambridge Philos. Soc.* **83** (1978), 403–413.
BRYANT, J. L., AND SEEBECK, C. L., III
[1] Locally nice embeddings of polyhedra. *Quart. J. Math. Oxford Ser. (2)* **19** (1968), 257–274.
[2] Locally nice embeddings in codimension three. *Quart. J. Math. Oxford Ser. (2)* **21** (1970), 265–272.
CANNON, J. W.
[1] Taming cell-like embedding relations. *In* "Geometric Topology" (L. C. Glaser and T. B. Rushing, eds.), Lecture Notes in Math. No. 438, pp. 66–118. Springer-Verlag, Berlin and New York, 1975.
[2] Taming codimension-one generalized submanifolds of S^n. *Topology* **16** (1977), 323–334.
[3] The recognition problem: What is a topological manifold? *Bull. Amer. Math. Soc.* **84** (1978), 832–866.
[4] $E^3/X \times E^1 \approx E^4$ (X a cell-like set): an alternate proof. *Trans. Amer. Math. Soc.* **240** (1978), 277–285.
[5] $\Sigma^2 H^3 = S^5/G$. *Rocky Mountain J. Math.* **8** (1978), 527–532.
[6] Shrinking cell-like decompositions of manifolds. Codimension three. *Ann. of Math. (2)* **110** (1979), 83–112.
CANNON, J. W., BRYANT, J. L., AND LACHER, R. C.
[1] The structure of generalized manifolds having nonmanifold set of trivial dimension. *In* "Geometric Topology" (J. C. Cantrell, ed.), pp. 261–300. Academic Press, New York, 1979.
CANNON, J. W., AND DAVERMAN, R. J.
[1] A totally wild flow. *Indiana Univ. Math. J.* **30** (1981), 371–387.
[2] Cell-like decompositions arising from mismatched sewings: applications to 4-manifolds. *Fund. Math.* **111** (1981), 211–233.
CANNON, L. O.
[1] Sums of solid horned spheres. *Trans. Amer. Math. Soc.* **122** (1966), 203–228.
ČERNAVSKIĬ, A. V.
[1] Local contractibility of the homeomorphism group of a manifold. *Soviet Math. Dokl.* **9** (1968), 1171–1174.
[2] Local contractibility of the group of homeomorphisms of a manifold. *Math. Sb.* **8** (1969), 287–333.
[3] Coincidence of local flatness and local simple-connectedness for embeddings of $(n - 1)$-dimensional manifolds in n-dimensional manifolds when $n > 4$. *Mat. Sb. (N.S.)* **91** (133) (1973), 279–296 (*Math. USSR-Sb.* **20** (1973), 297–304).
CHAPMAN, T. A.
[1] Compact Hilbert cube manifolds and the invariance of Whitehead torsion. *Bull. Amer. Math. Soc.* **79** (1973), 52–56.
[2] Lectures on Hilbert cube manifolds. *CBMS Regional Conf. Ser. in Math.* No. 28, Amer. Math. Soc., Providence, Rhode Island, 1976.
[3] Controlled boundary and h-cobordism theorems. *Trans. Amer. Math. Soc.* **280** (1983), 73–95.
CHRISTENSON, C. O., AND OSBORNE, R. P.
[1] Pointlike subsets of a manifold. *Pacific J. Math.* **24** (1968), 431–435.
CONNELLY, R.
[1] A new proof of Brown's collaring theorem. *Proc. Amer. Math. Soc.* **27** (1971), 180–182.

CURTIS, M. L.

[1] Cartesian products with intervals. *Proc. Amer. Math. Soc.* **12** (1961), 819–820.

DAVERMAN, R. J.

[1] On the scarcity of tame disks in certain wild cells. *Fund. Math.* **79** (1973), 63–77.

[2] Locally nice codimension one manifolds are locally flat. *Bull. Amer. Math. Soc.* **79** (1973), 410–413.

[3] On cells in Euclidean space that cannot be squeezed. *Rocky Mountain J. Math.* **5** (1975), 87–94.

[4] Every crumpled n-cube is a closed n-cell-complement. *Michigan Math. J.* **24** (1977), 225–241.

[5] A nonshrinkable decomposition of S^n determined by a null sequence of cellular sets. *Proc. Amer. Math. Soc.* **75** (1979), 171–176.

[6] Products of cell-like decompositions. *Topology Appl.* **11** (1980), 121–139.

[7] Detecting the disjoint disks property. *Pacific J. Math.* **93** (1981), 277–298.

[8] A mismatch property in spherical decompositions spaces. *In* "Continua, Decompositions, Manifolds" (R. H. Bing, W. T. Eaton, and M. P. Starbird, eds.), pp. 119–127. Univ. of Texas Press, Austin, Texas, 1983.

DAVERMAN, R. J., AND EATON, W. T.

[1] An equivalence for the embeddings of cells in a 3-manifold. *Trans. Amer. Math. Soc.* **145** (1969), 369–381.

DAVERMAN, R. J., AND GARITY, D. J.

[1] Intrinsically $(n - 2)$-dimensional cellular decompositions of E^n. *Pacific J. Math.* **102** (1982), 275–283.

[2] Intrinsically $(n - 1)$-dimensional cellular decompositions of S^n. *Illinois J. Math.* **27** (1983), 670–690.

DAVERMAN, R. J., AND PRESTON, D. K.

[1] Shrinking certain sliced decompositions of E^{n+1}. *Proc. Amer. Math. Soc.* **79** (1980), 477–483.

[2] Cell-like 1-demensional decompositions of S^3 are 4-manifold factors. *Houston J. Math.* **6** (1980), 491–502.

DAVERMAN, R. J., AND ROW, W. H.

[1] Cell-like 0-dimensional decompositions of S^3 are 4-manifold factors. *Trans. Amer. Math. Soc.* **254** (1979), 217–236.

DAVERMAN, R. J., AND WALSH, J. J.

[1] A ghastly generalized n-manifold. *Illinois J. Math.* **25** (1981), 555–576.

[2] A nonshrinkable decomposition of S^n involving a null sequence of cellular arcs. *Trans. Amer. Math. Soc.* **272** (1982), 771–784.

DELYRA, C. B.

[1] On spaces of the same homotopy type as polyhedra. *Bol. Soc. Mat. São Paulo* **12** (1957), 43–62.

DENMAN, R.

[1] Countable starlike decompositions of S^3. *In* "Continua, Decompositions, Manifolds" (R. H. Bing, W. T. Eaton, and M. P. Starbird, eds.), pp. 128–131. Univ. of Texas Press, Austin, Texas, 1983.

DENMAN, R., AND STARBIRD, M.

[1] Shrinking countable decomposition of S^3. *Trans. Amer. Math. Soc.* **276** (1983), 743–756.

DUGUNDJI, J.

[1] "Topology." Allyn & Bacon, Boston, 1966.

DYER, E., AND HAMSTROM, M. E.

[1] Completely regular mappings. *Fund. Math.* **45** (1958), 103–118.

EATON, W. T.
[1] The sum of solid spheres. *Michigan Math. J.* **19** (1972), 193–207.
[2] A generalization of the dog bone space to E^n. *Proc. Amer. Math. Soc.* **39** (1973), 379–387.
[3] Applications of a mismatch theorem to decomposition spaces. *Fund Math.* **89** (1975), 199–224.

EATON, W. T., PIXLEY, C. P., AND VENEMA, G.
[1] A topological embedding which cannot be approximated by a piecewise linear embedding. *Notices Amer. Math. Soc.* **24** (1977), A-302.

EDWARDS, R. D.
[1] Demension theory. I. *In* "Geometric Topology" (L. C. Glaser and T. B. Rushing, eds.), Lecture Notes in Math. No. 438, pp. 195–211. Springer-Verlag, Berlin and New York, 1975.
[2] Approximating codimension ≥ 3 σ-compacta with locally homotopically unknotted embeddings (unpublished manuscript).
[3] Suspensions of homology spheres (unpublished manuscript).
[4] Approximating certain cell-like maps by homeomorphisms (unpublished manuscript).
[5] The topology of manifolds and cell-like maps. *In* "Proc. Internat. Congr. Mathematicians, Helsinki, 1978" (O. Lehto, ed.), pp. 111–127. Acad. Sci. Fenn., Helsinki, 1980.

EDWARDS, R. D., AND GLASER, L. C.
[1] A method for shrinking decompositions of certain manifolds. *Trans. Amer. Math. Soc.* **165** (1972), 45–56.

EDWARDS, R. D., AND KIRBY, R. C.
[1] Deformations of spaces of embeddings. *Ann. of Math.* (2) **93** (1971), 63–88.

EDWARDS, R. D., AND MILLER, R. T.
[1] Cell-like closed-0-dimensional decompositions of R^3 are R^4 factors. *Trans. Amer. Math. Soc.* **215** (1976), 191–203.

ENGELKING, R.
[1] "Dimension Theory." North-Holland Publ., Amsterdam, 1978.

EVERETT, D. L.
[1] Embedding theorems for decomposition spaces. *Houston J. Math.* **3** (1977), 351–368.
[2] Shrinking countable decompositions of E^3 into points and tame cells. *In* "Geometric Topology" (J. C. Cantrell, ed.), pp. 53–72. Academic Press, New York, 1979.

FERRY, S.
[1] Homotoping ε-maps to homeomorphisms. *Amer. J. Math.* **101** (1979), 567–582.

FREEDMAN, M. H.
[1] The topology of four-dimensional manifolds. *J. Differential Geom.* **17** (1982), 357–453.

FREUDENTHAL, H.
[1] Über dimensionserhohende stetige Abbildungen. *Sitaungsber. Preuss. Akad. Wiss.* **5** (1932), 34–38.

GARITY, D. J.
[1] A characterization of manifold decompositions satisfying the disjoint triples property. *Proc. Amer. Math. Soc.* **83** (1981), 833–838.
[2] A classification scheme for cellular decompositions of manifolds. *Topology Appl.* **14** (1982), 43–58.
[3] General position properties related to the disjoint disks property. *In* "Continua, Decompositions, Manifolds" (R. H. Bing, W. T. Eaton, and M. P. Starbird, eds.), pp. 132–140. Univ. of Texas Press, Austin, Texas, 1983.

GIFFEN, C. H.
[1] Disciplining dunce hats in 4-manifolds (unpublished manuscript).

GLASER, L. C.
[1] Contractible complexes in S^n. *Proc. Amer. Math. Soc.* **16** (1965), 1357–1364.

GLUCK, H.

[1] Embeddings in the trivial range. *Ann. of Math.* (2) **81** (1965). 195–210.

HANAI, S.

[1] On closed mappings. *Proc. Japan Acad.* **30** (1954), 285–288.

HARLEY, P. W.

[1] The product of an n-cell modulo an arc in its boundary and a 1-cell is an $(n + 1)$-cell. *Duke Math. J.* **35** (1968), 463–474.

HAVER, W. E.

[1] Mappings between ANR's that are fine homotopy equivalences. *Pacific J. Math.* **58** (1975), 457–461.

HIRSCH, M. W.

[1] On non-linear cell bundles. *Ann. of Math.* (2) **84** (1966), 373–385.

HOMMA, T.

[1] On tame embedding of 0-dimensional compact sets in E^3. *Yokohama Math. J.* **7** (1959), 191–195.

HUREWICZ, W.

[1] Über oberhalb-stetige Zerlegungen von Punktmengen in Kontinua. *Fund. Math.* **15** (1930), 57–60.

[2] Über dimensionserhöherende stetige Abbildungen, *J. Reine u. Angew. Math.* **169** (1933), 71–78.

HUREWICZ, W., AND WALLMAN, H.

[1] "Dimension Theory." Princeton Univ. Press, Princeton, New Jersey, 1941.

KERVAIRE, M. A.

[1] Smooth homology spheres and their fundamental groups. *Trans. Amer. Math. Soc.* **144** (1969), 67–72.

KIRBY, R. C.

[1] Stable homeomorphisms and the annulus conjecture. *Ann. of Math.* (2) **89** (1969), 575–582.

KIRBY, R. C., AND SIEBENMANN, L.C.

[1] "Foundational Essays on Topological Manifolds, Smoothings, and Triangulations," Ann. of Math. Studies No. 88. Princeton Univ. Press, Princeton, New Jersey, 1977.

KLEE, V. L.

[1] Some topological properties of convex sets. *Trans. Amer. Math. Soc.* **78** (1955), 30–45.

KOZLOWSKI, G.

[1] Images of ANRs (unpublished manuscript).

KOZLOWSKI, G. AND WALSH, J. J.

[1] Cell-like mappings on 3-manifolds. *Topology* **22** (1983), 147–151.

KWUN, K. W.

[1] Product of Euclidean spaces modulo an arc. *Ann. of Math.* (2) **79** (1964), 104–108.

KWUN, K. W., AND RAYMOND, F.

[1] Factors of cubes. *Amer. J. Math.* **84** (1962), 433–440.

LACHER, R. C.

[1] Cellularity criteria for maps. *Michigan Math. J.* **17** (1970), 385–396.

[2] Cell-like mappings and their generalizations. *Bull. Amer. Math. Soc.* **83** (1977), 495–552.

LASHOF, R. S.

[1] Problems in differential and algebraic topology (Seattle Conf. 1963). *Ann. of Math.* (2) **81** (1965), 565–591.

LAY, T. L.

[1] Shrinking decompositions of E^n with countably many 1-dimensional starlike equivalent nondegenerate elements. *Proc. Amer. Math. Soc.* **79** (1980), 308–310.

LICKORISH, W. B. R., AND SIEBENMANN, L. C.

[1] Regular neighborhoods and the stable range. *Trans. Amer. Math. Soc.* **139** (1969), 207–230.

LININGER, L. L.

[1] Actions on S^n. *Topology* **9** (1970), 301–308.

MCAULEY, L. F.

[1] Decompositions of continua into aposyndetic continua. Ph.D. Thesis, University of North Carolina, 1954.

[2] Upper semicontinuous decompositions of E^3 into E^3 and generalizations to metric spaces. *In* "Topology of 3-Manifolds and Related Topics" (M. K. Fort, Jr., ed.), pp. 21–26. Prentice-Hall, Englewood Cliffs, New Jersey, 1962.

MCMILLAN, D. R., JR.

[1] A criterion for cellularity in a manifold. *Ann. of Math (2)* **79** (1964), 327–337.

MCMILLAN, D. R., JR., AND ROW, H.

[1] Tangled embeddings of one-dimensional continua. *Proc. Amer. Math. Soc.* **22** (1969), 378–385.

MARIN, A., AND VISETTI, Y. M.

[1] A general proof of Bing's shrinkability criterion. *Proc. Amer. Math. Soc.* **53** (1975), 501–507.

MATSUMOTO, Y.

[1] A 4-manifold which admits no spine. *Bull. Amer. Math. Soc.* **81** (1975), 467–470.

MAZUR, B.

[1] A note on some contractible 4-manifolds. *Ann. of Math. (2)* **73** (1961), 221–228.

MEYER, D. V.

[1] More decompositions of E^n which are factors of E^{n+1}. *Fund. Math.* **67** (1970), 49–65.

MILLER, R. T.

[1] Mapping cylinder neighborhoods of some ANR's. *Ann. of Math. (2)* **103** (1976), 417–427.

MOISE, E. E.

[1] Affine structures in 3-manifolds. V. The triangulation theorem and Hauptvermutung. *Ann. of Math. (2)* **56** (1952), 96–114.

MOORE, R. L.

[1] Concerning upper semicontinuous collections of compacta. *Trans. Amer. Math. Soc.* **27** (1925), 416–428.

[2] "Foundations of Point Set Theory," *Amer. Math. Soc. Colloq. Publ.*, Vol. 13. Amer. Math. Soc., Providence, Rhode Island, 1970.

NEUZIL, J. P.

[1] Spheroidal decompositions of E^4. *Trans. Amer. Math. Soc.* **155** (1971), 35–64.

NEWMAN, M. H. A.

[1] The engulfing theorem for topological manifolds. *Ann. of Math. (2)* **84** (1966), 555–571.

PAPAKYRIAKOPOLOUS, C. D.

[1] On Dehn's Lemma and the asphericity of knots. *Ann. of Math. (2)* **66** (1957), 1–26.

PIXLEY, C. P., AND EATON W. T.

[1] S^1 cross a UV^∞ decomposition of S^3 yields $S^1 \times S^3$. *In* "Geometric Topology" (L. C. Glaser, and T. B. Rushing, eds.), Lecture Notes in Math. No. 438, pp. 166–194. Springer-Verlag, Berlin and New York, 1975.

POENARU, V.

[1] Les décompositions de l'hypercube en produit topologique. *Bull. Soc. Math. France* **88** (1960), 113–129.

PRESTON, D. K.

[1] A study of product decompositions of topological manifolds. Ph.D. dissertation, University of Tennessee, Knoxville, 1979.

PRICE, T. M.

[1] A necessary condition that a cellular upper semicontinuous decomposition of E^n yield E^n. *Trans. Amer. Math. Soc.* **122** (1966), 427–435.

[2] Decompositions of S^3 and pseudo-isotopies. *Trans. Amer. Math. Soc.* **140** (1969), 295–299.

PRICE, T. M., AND SEEBECK, C. L., III

[1] Somewhere locally flat codimension one manifolds with 1-ULC complements are locally flat. *Trans. Amer. Math. Soc.* **193** (1974), 111–122.

QUINN, F.

[1] Ends of maps. I. *Ann. of Math. (2)* **110** (1979), 275–331.

[2] Ends of maps. III. Dimensions 4 and 5. *J. Differential Geom.* **17** (1982), 503–521.

[3] Resolutions of homology manifolds, and the topological characterization of manifolds. *Invent. Math.* **72** (1983), 267–284.

[4] An obstruction to the resolution of homology manifolds (to appear).

ROBERTS, J. H.

[1] On a problem of C. Kuratowski concerning upper semi-continuous collections. *Fund. Math.* **14** (1929), 96–102.

ROURKE, C. P., AND SANDERSON, B. J.

[1] "Introduction to Piecewise-linear Topology," Ergebn. Math. No. 69. Springer-Verlag, Berlin and New York, 1972.

RUSHING, T. B.

[1] "Topological Embeddings." Academic Press, New York, 1973.

SEEBECK, C. L., III

[1] Tame arcs on wild cells. *Proc. Amer. Math. Soc.* **29** (1971), 197–201.

SHER, R. B.

[1] Tame polyhedra in wild cells and spheres. *Proc. Amer. Math. Soc.* **30** (1971), 169–174.

SIEBENMANN, L. C.

[1] Approximating cellular maps by homeomorphisms. *Topology* **11** (1972), 271–294.

SIEKLUCKI, K.

[1] A generalization of a theorem of K. Borsuk concerning the dimension of ANR-sets. *Bull. Acad. Polon. Sci. Ser. Sci. Math. Astronom. Phys.* **10** (1962), 433–436. See also *ibid.* **12** (1964), 695.

SINGH, S.

[1] 3-dimensional AR's which do not contain 2-dimensional ANR's. *Fund. Math.* **93** (1976), 23–36.

[2] Constructing exotic retracts, factors of manifolds, and generalized manifolds via decompositions. *Fund. Math.* **113** (1981), 81–89.

[3] Generalized manifolds (ANR's and AR's) and null decompositions of manifolds. *Fund. Math.* **115** (1983), 57–73.

SPANIER, E. H.

[1] "Algebraic Topology." McGraw-Hill, New York, 1966.

ŠTAN'KO, M. A.

[1] The embedding of compacta in Euclidean space. *Mat. Sb.* **83** (125) (1970), 234–255 (*Math. USSR-Sb.* **12** (1970), 234–254).

[2] Solution of Menger's problem in the class of compacta. *Dokl. Akad. Nauk SSSR* **201** (1971), 1299–1302 (*Soviet Math. Dokl.* **12** (1971), 1846–1849).

[3] Approximation of compacta in E^n in codimension greater than two. *Mat. Sb.* **90** (132) (1973), 625–636 (*Math. USSR-Sb.* **19** (1973), 615–626).

STARBIRD, M.

[1] Cell-like 0-dimensional decompositions of E^3. *Trans. Amer. Math. Soc.* **249** (1979), 203–216.

[2] Null sequence cellular decompositions of S^3. *Fund. Math.* **112** (1981), 81–87.

STARBIRD, M., AND WOODRUFF, E. P.

[1] Decompositions of E^3 with countably many nondegenerate elements. *In* "Geometric Topology" (J. C. Cantrell, ed.), pp. 239–252. Academic Press, New York, 1979.

STONE, A. H.

[1] Metrizability of decomposition spaces. *Proc. Amer. Math. Soc.* **7** (1956), 690–700.

TAYLOR, J. L.

[1] A counterexample in shape theory. *Bull. Amer. Math. Soc.* **81** (1975), 629–632.

TORUŃCZYK, H.

[1] On CE-images of the Hilbert cube and characterization of Q-manifolds. *Fund. Math.* **106** (1980), 31–40.

[2] Characterizing Hilbert space topology. *Fund. Math.* **111** (1981), 247–262.

WALSH, J. J.

[1] Dimension, cohomological dimension, and cell-like mappings. *In* "Shape Theory and Geometric Topology" (S. Mardešić, and J. Segal, eds.), Lecture Notes in Math. No. 870, pp. 105–118. Springer-Verlag, Berlin and New York, 1981.

[2] The finite dimensionality of integral homology 3-manifolds. *Proc. Amer. Math. Soc.* **88** (1983), 154–156.

WEST, J. E.

[1] Mapping Hilbert cube manifolds to ANR's: a solution to a conjecture of Borsuk. *Ann. of Math. (2)* **106** (1977), 1–18.

WHITEHEAD, J. H. C.

[1] A certain open manifold whose group is unity. *Quart. J. Math. Oxford Ser. (2)* **6** (1935), 268–279.

[2] Simplicial spaces, nuclei, and m-groups. *Proc. London Math. Soc.* **45** (1939), 243–327.

[3] A certain exact sequence. *Ann. of Math. (2)* **52** (1950), 51–110.

WILDER, R. L.

[1] "Topology of Manifolds," *Amer. Math. Soc. Colloq. Publ.*, Vol. 32, Amer. Math. Soc., Providence, Rhode Island, 1963.

WOODRUFF, E. P.

[1] Decomposition spaces having arbitrarily small neighborhoods with 2-sphere boundaries. *Trans. Amer. Math. Soc.* **232** (1977), 195–204.

[2] Decomposition spaces having arbitrarily small neighborhoods with 2-sphere boundaries. II. *Fund. Math.* **119** (1983), 185–204.

WRIGHT, D. G.

[1] AR's which contain only trivial ANR's. *Houston J. Math.* **4** (1978), 121–127.

[2] A decomposition of E^n ($n \geq 3$) into points and a null sequence of cellular sets. *General Topology Appl.* **10** (1979), 297–304.

[3] Countable decompositions of E^n. *Pacific J. Math.* **103** (1982), 603–609.

ZEEMAN, E. C.

[1] On the dunce hat. *Topology* **2** (1964), 341–358.

Index